Meta-Logics and Logic Programming

Logic Programming
Ehud Shapiro, editor
Koichi Furukawa, Jean-Louis Lassez, Fernando Pereira, and David H. D. Warren, associate editors

Parallel Logic Programming,
Evan Tick, 1991

Foundations of Disjunctive Logic Programming,
Jorge Lobo, Jack Minker, and Arcot Rajasekar, 1992

Types in Logic Programming,
edited by Frank Pfenning, 1992

Concurrent Constraint Programming,
Vijay A. Saraswat, 1993

Logic Programming Languages: Constraints, Functions, and Objects,
edited by K. R. Apt, J. W. de Bakker, and J. J. M. M. Rutten, 1993

Constraint Logic Programming: Selected Research,
edited by Frederic Benhamou and Alain Colmerauer, 1993

A Grammatical View of Logic Programming,
Pierre Deransart and Jan Maluszynski, 1993

The Gödel Programming Language,
Patricia Hill and John Lloyd, 1994

The Art of Prolog: Advanced Programming Techniques,
second edition, Leon Sterling and Ehud Shapiro, 1994

Inductive Logic Programming: From Machine Learning to Software Engineering,
Francesco Bergadano and Daniele Gunetti, 1995

Meta-Logics and Logic Programming,
edited by Krzysztof R. Apt and Franco Turini, 1995

Meta-Logics and Logic Programming

edited by
Krzysztof R. Apt and Franco Turini

The MIT Press
Cambridge, Massachusetts
London, England

© 1995 Massachusetts Institute of Technology

All rights reserved. No part of this book may be reproduced in any form by any electronic or mechanical means (including photocopying, recording, or information storage and retrieval) without permission in writing from the publisher.

This book was set by the contributors and was printed and bound in the United States of America.

Library of Congress Cataloging-in-Publication Data

Apt, Krzysztof R., 1949–
Meta-logics and logic programming / edited by Krzysztof R. Apt and Franco Turini.
   p.   cm.—(Logic programming)
  Includes bibliographical references.
  ISBN 0-262-01152-2 (hc: alk. paper)
  1. Logic programming. 2. Logic, Symbolic and mathematical. I. Turini, Franco, 1949– . II. Series.
QA76.63.A68   1995
005.13′1—dc20
                                                                                                                                              95-18954
                                                                                                                                                CIP

Contents

	Series Foreword	vii
	Preface	ix
	Acknowledgements	xv

I FOUNDATIONS

1 Correctness of the Vanilla Meta-Interpreter and Ambivalent Syntax ... 3
M. Kalsbeek

2 A Vademecum of Ambivalent Logic ... 27
M. Kalsbeek and Y. Jiang

3 Two Semantics for Definite Meta-Programs, Using the Non-Ground Representation ... 57
B. Martens and D. De Schreye

4 Meta-Logic for Program Composition: Semantics Issues ... 83
A. Brogi and F. Turini

5 Comparing Negation in Logic Programming and in Prolog ... 111
K. R. Apt and F. Teusink

II LANGUAGE SUPPORT FOR META-LOGICS

6 Towards Fast and Declarative Meta-Programming ... 137
A. F. Bowers and C.A. Gurr

7 Composing Logic Programs by Meta-Programming in Gödel ... 167
A. Brogi and S. Contiero

8 Meta-Programming with Theory Systems ... 195
J. Barklund, K. Boberg, P. Dell'Acqua and M. Veanes

III META-LOGICS FOR KNOWLEDGE MANAGEMENT

9   Using Meta-Logic to Reconcile Reactive with Rational Agents 227
R. A. Kowalski

10   Modal and Meta Languages: Consistency and Expressiveness 243
L. Carlucci Aiello, M. Cialdea, D. Nardi, M. Schaerf

11   Model-based Diagnosis Preferences and Strategies Representation with Logic Meta-Programming 267
C. V. Damásio, W. Nejdl, L. M. Pereira and M. Schroeder

12   The Generalized ChronoBase Temporal Data Model 309
S. M. Sripada and P. Möller

Contributors 337

Series Foreword

The logic programming approach to computing investigates the use of logic as a programming language and explores computational models based on controlled deduction.

The field of logic programming has seen a tremendous growth in the last several years, both in depth and in scope. This growth is reflected in the number of articles, journals, theses, books, workshops, and conferences devoted to the subject. The MIT Press series in logic programming was created to accommodate this development and to nurture it. It is dedicated to the publication of high-quality textbooks, monographs, collections, and proceedings in logic programming.

Ehud Shapiro
The Weizmann Institute of Science
Rehovot, Israel

Preface

Background

The term meta-programming refers to the ability of writing programs that have other programs as data. It is usual to refer to the programs that play the role of data as *object programs*, and to the manipulating programs as *meta-programs*. To be more precise, meta-programs work on a *representation* of object programs.

Meta-programming has played a fundamental role both in the foundations of computer science and in its practical developments. Its roots go back to mathematical logic, more specifically to Kleene's normal form theorem that states that for some primitive recursive functions T, U every partial recursive function equals $U(\mu y.T(e, x, y))$ (usually denoted by ϕ_e) for some natural number e. We can view ϕ_e as the function computed by the program e, and U and T as meta-programs that work on the object program e.

The choice of logic programming as a basis for meta-programming offers a number of practical and theoretical advantages. One of them is the possibility of tackling critical foundational problems of meta-programming within a framework with a strong theoretical basis. Another is the surprising ease of programming.

However, to formally deal with meta-programs, the usual framework of logic programming, or more generally, of first order logic has to be modified and appropriately extended. The reason is that various phenomenena relevant to meta-programming, like the representation of object programs and their syntax, the interplay between the object level and meta-level, the use of modules, the representation of proof strategies etc. require logics that are richer and more expressive.

We denote these extensions of logic programming and first order logic collectively by the term of *meta-logics*. Their definitions, formal properties and use form the main theme of this book. The other is meta-programming in logic programming.

Let us discuss now some of the issues mentioned above in more detail. This will provide us with a better insight into the problems that need to be solved on the meta-logical level.

From the viewpoint adopted above, it is certainly true that compilers can be considered as meta-programs in their full right. But, is it reasonable to consider a text editor as a meta-program, when it is applied to a file containing the representation of a program? In this case the most sensible answer seems to be "no", and, in order to cope with this contradiction, we propose to refine our informal definition into: *meta-programming refers to the ability of writing programs that have other programs as data and exploit their semantics*.

Among the examples of useful meta-programs we can list compilers, interpreters, program analyzers, and partial evaluators.

In the application field, things get even more interesting when the meta-program and the object program are written in the same language. The deep and far reaching phenomenon that arises in this case is that the very same piece of syntax can play, in principle, either the passive role of datum or the active role of program construct in different contexts and in different stages of the program execution.

In Prolog, the minimal example of this phenomenon is represented by the clause
```
eval(x) :- x
```
that allows a programmer to "lift" any ordinary term (a Prolog datum) to the status of being an atom (a Prolog programming construct).

Such a double role for the same piece of syntax has been typical for many knowledge representation systems that have been realized and experimented with.

The idea is simple. Write a program that takes as input the representation of a program written in a superset of the language, and execute it according to the following strategy: if a construct is a new one then call a special procedure to handle it, otherwise let the construct be executed as it is. Many books on the use of Lisp or Prolog for artificial intelligence applications contain extensive examples of this approach. What these books lack, however, is a justification of this approach from the logical point of view.

The most critical problem of meta-programming is certainly the *representation problem*, i.e. how object programs are represented within meta-program. The first part of the volume deals with this essential problem and discusses various meta-logical solutions to it. The second part of the volume is concerned with language extensions that make meta-programming easier and more elegant. Finally, the third part of the volume deals with the use of meta-logics for advanced knowledge representation purposes. Let us discuss now the individual contributions to this volume.

Part I: Foundations

A classical problem in the foundations of meta-logic programming is the justification of the formally incorrect (untyped) Vanilla meta-interpreter, which uses a non-ground representation of object variables. In particular, the unwell-typedness of Vanilla leads to the presence of unrelated atoms in the least Herbrand model of the Vanilla meta-interpreter.

The paper of **Kalsbeek** overviews and compares various approaches towards the problem of the occurrence of unrelated atoms in the semantics of the Vanilla meta-interpreter: the use of the (correct) typed version, restriction to language independent object programs, and the use of S-semantics. In particular, Kalsbeek argues that Hill and Lloyd's seminal procedural correctness result for the typed version Vanilla is also a proof for the procedural correctness for the untyped version of the Vanilla meta-interpreter. It is also shown that the various correctness proofs are insensitive to the precise representation of the object level language. In addition, she discusses the use of ambivalent syntax as the underlying syntax for the Vanilla meta-interpreter, in particular for amalgamated extensions and enhanced versions. She presents a separate proof for the declarative correctness of the Vanilla meta-interpreter with ambivalent syntax as the underlying language. This result is then used to prove the correctness of a simple amalgamation of the object program with the associated Vanilla meta-program.

While logic programming is formally based on first order predicate logic, many of its applications use non-standard syntaxes, which are characterised by syntactical ambivalence between formulas, terms, predicates, and functions. Examples are the meta-variable facility of Prolog, the overloading of predicate and function symbols allowed in Prolog, the identity naming of object level constructs used in Vanilla meta-programming, and the use of generic predicates in databases.

The paper of **Kalsbeek** and **Jiang** discusses Ambivalent Logic, which provides a general framework for first order predicate logic with various levels of syntactic ambivalence. A conservativity result shows that Ambivalent Logic is a conservative extension of first order predicate logic. They prove a series of results which justify the use of ambivalent syntax in logic programming. In particular, they prove termination and correctness of an appropriate version of the Martelli-Montanari unification algorithm, and show that resolution is a sound and complete inference method for Ambivalent Logic.

The two best known semantics for definite logic programs are least Herbrand semantics and S-semantics. It is however not a priori clear that these semantics lead to meaningful results for meta-programs in the Prolog-style non-typed tradition, using a non-ground representation for object level variables like the well-known vanilla meta-interpreter. Since this style of meta-programming seems to be of considerable practical importance, this situation must be judged unsatisfactory.

In their contribution, **Martens** and **De Schreye** study the relation between the semantics of definite object programs and the corresponding untyped vanilla meta-programs, both in the context of least Herbrand and S-semantics. They also investigate various enhanced meta-programs, some of which feature limited forms of amalgamation. The latter extension is enabled through the overloading of function and predicate symbols, a technique that essentially coincides with allowing a certain degree of syntactical ambivalence in the language. For these programs and semantics, they establish under which conditions there is a strong correspondence between object and meta-level semantics, thus shedding light on the question to what extent meta-programming of this kind can be judged meaningful.

Also the paper by **Brogi** and **Turini** addresses the representation problem. In their contribution they propose a semantic justification for a simple representation technique in the field of a generalised notion of meta-programming in logic. The representation technique is again based on the notion of ambivalent syntax, and the generalisation consists in specifying the meta-programs with respect to object programs defined via program expressions. The expressions are defined by means of a rich suite of operations on logic programs. The technique allows one to build straightforward and concise meta-programs via the representation of object level variables by meta-level variables.

One of the interesting features of Prolog is that it allows us to extend its syntax in a simple way, by means of meta-variables. This property is used to define negation in Prolog, using meta-variables, clause ordering and the cut operator. In logic programming negation is defined in quite a different way, by means of so-called SLDNF-resolution.

The paper by **Apt** and **Teusink** compares these two uses of negation – in Prolog and in logic programming. This requires a careful reexamination of the assumptions about the underlying syntax and a precise definition of the computational processes involved. After taking care of these matters, among others by adopting an ambivalent syntax, they prove an equivalence in appropriate sense between these two uses of negation. This result allows them to argue about correctness of Prolog programs that use negation.

Part II: Language Support for Meta-Logics

High-level languages such as Lisp and Prolog are often chosen because the syntactic similarity between programs and data makes it very convenient to write meta-programs in those languages. If we desire a truly declarative programming language however, it can be seen that we have been tempted down a blind alley in the approach these languages take to meta-programming, because they do not provide the means for a declarative treatment of object variables in the meta-program. Without a ground representation, any but the simplest of meta-programs can have no declarative semantics. Unfortunately, using a ground representation apparently incurs a significant overhead in program complexity and computation time.

Gödel is a new programming language aimed at narrowing the gap between theory and practice in logic programming, and with particular emphasis on declarative meta-programming. To achieve its aim, Gödel must make the ground representation attractive to programmers in both ease of use and execution time. The paper by **Bowers** and **Gurr** shows how this might be done, through the careful design of library modules, and the use of partial evaluation and low-level implementation techniques. Their experiments on simple Gödel meta-interpreters using the ground representation demonstrate some dramatic performance improvements from these methods.

The paper by **Brogi** and **Contiero** investigates the adequacy of Gödel as a meta-language for implementing various forms of logic program composition. Two alternative implementations of a set of program composition operations are presented, based on the non-ground and the ground representation of object programs respectively. The merits of Gödel as a meta-language are discussed by comparing the two implementations and by analysing the results of some experiments with the Gödel partial evaluator. Finally, some directions in which Gödel might be extended or improved are identified.

In the paper by **Barklund, Boberg, Dell'Acqua**, and **Veanes** *theory systems* are proposed as a device for writing software engineering applications and applications that involve reasoning and meta-reasoning. A theory is a set of sentences that is closed under inference and a theory system is then a collection of theories that are related through reflection principles.

The meta-logic programming language *Alloy* for defining theory systems is introduced with formal syntax, inference rules and a concept of models for Alloy programs. Several examples of Alloy programs that define theory systems are given.

Part III: Meta-Logics for Knowledge Management

Traditional logic is concerned with static theories, which do not change over the course of time. Deductive databases and knowledge bases extend this static form of logic to include the dynamics of database updates and knowledge assimilation. Such dynamic theories, however, are still essentially passive, in that, although they change their own internal state, they do not change the state of the environment.

Kowalski's paper proposes the use of meta-logic programming, within a concurrent logic programming framework, to extend such theories to active theories, which behave as intelligent agents. He presents a meta-logic program which defines the observation-thought-action cycle of such an agent, with the intention of giving the definition both a procedural (process) and declarative (logical) interpretation. Moreover, he argues that, by adjusting the amount of resources available for thought versus observation and action, it is possible to simulate both reactive agents (when the amount of resource is small) and rational agents (when the amount is sufficiently large).

In knowledge representation several formalisms for reasoning about knowledge in a multi agent scenario have been proposed. More specifically, we can identify a family of languages based on the use of a modal operator and another one based on the use of first-order logic enriched with meta-level capabilities.

The paper by **Carlucci Aiello, Cialdea, Nardi** and **Schaerf** considers these two approaches by addressing the issues of consistency that arise from selfreferentiality, their expressiveness and the methods for translating classical modal systems into meta-level first-order formalisms.

Preferences and strategies are fundamental to model-based diagnosis, for specifying preferred and fall-back approaches to the diagnosis task, both to capture general and domain specific criteria, but also to tackle the complexity issue by employing heuristics.

The paper by **Damásio, Nejdl, Pereira**, and **Schroeder** presents a formal framework based on extended logic programming and meta-programs for the representation of preferences and strategies required by model-based diagnosis. This framework is clearer and more expressive than other approaches that have addressed these problems. The authors show how the concepts of preferences and strategies are directly programmed and captured by logic meta-programming and meta-reasoning methods, and their implementation techniques. The paper is intended as proof-of-principle that all concepts needed by a model-based diagnosis system can represented declaratively and captured by a logic meta-program. Specialized more efficient algorithms can be substituted for the simpler proof-of-principle ones they include, and are the subject of ongoing work.

Meta-programming can also be used as a theoretical basis for defining more expressive data models. The paper by **Sripada** and **Möller** describes a rich temporal data model for advanced database applications. They illustrate the power of meta-programming for temporal knowledge representation and reasoning. They then describe how the relational model can be extended to provide support at the database level for the concepts derived from metalevel knowledge representation.

Acknowledgments

This volume presents an outcome of research carried out within the Esprit funded Basic Research Project "Compulog II". (For those interested in numbers - No. 6810.) The project has started August 1, 1992 and will finish July 31, 1995. The first editor is the project coordinator and the second one coordinator of the area "Meta- and non-monotonic reasoning" within the project.

The aim of "Compulog II" has been to study various extensions of logic programming which make it more amenable for knowledge representation and programming. One of the important elements has been the investigation of the meta-programming within the logic programming paradigm. This book is devoted to this aspect of Compulog II research. Many chapters were specially written for this book. They were all internally refereed. We would like to take this opportunity to thank all the authors for their contributions, refereeing work and assistance in preparing this preface. Bob Prior from the MIT Press provided us with the necessary help on the side of the publisher and, last but not least, Kees Doets and Bonnie Friedman helped us to win in our struggle with the MIT stylefiles.

We hope this volume will not satisfy but rather stimulate readers' interest in this exciting research area.

K.R.A.
F.T.

I FOUNDATIONS

1 Correctness of the Vanilla Meta-Interpreter and Ambivalent Syntax

Marianne Kalsbeek

Abstract

We discuss and compare various approaches to correctness of the Vanilla meta-interpreter (procedural, declarative, S-semantics). We compare the typed meta-interpreter with the untyped version, and argue that the procedural correctness proof for the typed interpreter has great generality. We present a detailed proof of declarative correctness in the context of the ambivalent syntax, which is the appropriate syntax underlying most amalgamated extensions of the Vanilla meta-interpreter.

1.1 Introduction

In this paper we study the simplest meta-interpreter for definite logic programs, usually known as the Vanilla meta-interpreter. In particular, we focus on correctness results for the Vanilla meta-interpreter. The Vanilla meta-interpreter is a definite logic program which consists of two parts: a general part $\mathbf{V}$, which consists of an intensional formalisation of derivability by SLD-resolution from definite object programs, and an object program specific part meta-P, which consists of a meta-level description of the clauses of an object program P.

Definition 1 The standard *Vanilla meta-interpreter* $\mathbf{V}_P$ for definite object programs P.

$$\begin{array}{lll}
\mathbf{V} & [\text{M1}] & demo(empty) \leftarrow \\
 & [\text{M2}] & demo(x) \leftarrow clause(x,y), demo(y) \\
 & [\text{M3}] & demo(x\&y) \leftarrow demo(x), demo(y) \\
\text{meta}-P & [\text{M4}] & clause(A, B_1 \& \ldots \& B_n) \leftarrow \\
 & & \quad \text{for every clause } A \leftarrow B_1, \ldots, B_n \text{ in } P \\
 & [\text{M5}] & clause(A, empty) \leftarrow \\
 & & \quad \text{for every clause } A \leftarrow \text{ in } P
\end{array}$$

$\square$

Additionally, a meta-interpreter for normal programs is obtained by extending $\mathbf{V}_P$ with the clause

$$[\text{M}\neg] \quad demo(not\ x) \leftarrow \neg demo(x).$$

In the context of the present paper however, we mainly concentrate on the interpreter for definite object programs.

A correctness result for the Vanilla-meta interpreter establishes that its intended behaviour is its observed behaviour. Basically, a correctness result for $\mathbf{V}_P$ is a relation between $\mathbf{V}_P$ and P of the form

$$\mathbf{V}_P \mathrel{\mid\!\sim} demo(A) \quad \text{iff} \quad P \mathrel{\mid\!\sim} A,$$

where $\mathrel{\mid\!\sim}$ is a semantical consequence relation w.r.t. a preferred semantics (declarative correctness), or a relevant derivational consequence relation such as refutability be SLD(NF)-resolution (procedural correctness). The two directions of the abstract correctness relation are usually distinguished as soundness (from left to right) and completeness (vice versa). Various correctness results and their proofs will be discussed and compared in Section 1.3. In addition, we will present one specific correctness proof in some detail in Section 1.5.

One of the problems involved in proving correctness results for the Vanilla meta-interpreter originates in the fact that $\mathbf{V}_P$ is not well-typed. As an example, consider the following object program Q:

$$\begin{aligned} p(c) &\leftarrow \\ r(x) &\leftarrow \\ q(x) &\leftarrow p(x) \end{aligned}$$

The associated Vanilla meta-interpreter $\mathbf{V}_Q$ is

$$\begin{aligned} demo(empty) &\leftarrow \\ demo(x) &\leftarrow clause(x,y), demo(y) \\ demo(x \& y) &\leftarrow demo(x), demo(y) \\ clause(p(c), empty) &\leftarrow \\ clause(r(x), empty) &\leftarrow \\ clause(q(x), p(x)) &\leftarrow \end{aligned}$$

Now consider the variables which occur in $\mathbf{V}_Q$ above. The variables occurring in the second and third clause are intended to range over object level atoms, whereas the variables which occur in the last three clauses are meant to range over the domain of the object program. Thus, intuitively, the Vanilla meta-interpreter is a typed program with two types: the variables occurring in the clauses [M2] and [M3] are intended as meta-variables ranging over object level atoms; the variables that occur in the part which represents the object level program, are meant to be object level variables ranging over object level terms.

Motivated by the observation that the intuitive interpretation of the Vanilla meta-interpreter is typed, Hill and Lloyd [HL89] advocate a typed version of the Vanilla meta-interpreter and prove its (declarative and procedural) correctness. We will discuss the correctness result for the typed version of $\mathbf{V}_P$ in some detail in Section 1.3. However, it is the *untyped* version of the Vanilla meta-interpreter, and extensions of it, that is used in general Prolog practice and in applications (see, for instance, programs discussed in Sterling and Shapiro [SS86] and Kowalski and Kim [KK91]). Therefore, a correctness result for the untyped interpreter is of interest.

Typically, the untyped Vanilla meta-interpreter does not distinguish between object- and meta-level variables and terms. As a result, the least Herbrand model of $\mathbf{V}_Q$ will contain 'unrelated' atoms, such as $demo(r(empty))$ and $demo(r(r(c)\&p(c)))$, while the least Herbrand model of the object program Q does not contain the atoms $r(empty)$ and $r(r(c)\&p(c))$. This illustrates that a declarative correctness result for the Vanilla meta-interpreter will, in general, not establish a complete correlation between the least Herbrand models of meta- and object programs.

Another issue is the representation of object level predicates and function symbols in the meta-program. In principle, two options are available. First, the function symbols f and relation symbols p that occur in the object program can be represented by function symbols f' and p' in the associated meta-program. Using this *functional representation*, object level clauses $A \leftarrow B_1, \ldots, B_n$ are represented in the meta-program as $clause(A', B_1', \ldots, B_n') \leftarrow$. For a more detailed discussion of the functional representation and a precise definition of the Vanilla meta-interpreter in this context we refer to Martens and De Schreye [MS95a].

Second, the naive *identity representation* can be used. In that case, the clauses of the object program are represented as in Definition 1 above. As long as the Vanilla meta-interpreter is not combined with (clauses from) the object program or extended with *amalgamated* clauses (in which both atoms from the object and the meta-level occur), the identity representation we use in Definition 1 is just a special case of the functional representation — every symbol from the alphabet of the object language being represented by itself. However, in the case of such amalgamated extensions, the underlying language of the meta-program is non-standard. As an example, consider an extension of $\mathbf{V}_Q$ above with the object program Q and the clause $demo(q(f(x)) \leftarrow q(x), demo(p(x))$. The non-standard syntax employed here is mainly characterised by the occurrence of atoms in term positions. We will refer to the syntax in which predicates double as function symbols as *ambivalent syntax*. The Vanilla meta-interpreter of Definition 1, which uses the identity representation of the object level language, is typically suitable for applications in which it is extended with such amalgamated clauses. (Again, we refer to Sterling and Shapiro [SS86] and Kowalski and Kim [KK91] for examples of such applications.) In contrast, the version using a non-identity functional representation is obviously not suitable for such extensions.

Prolog itself admits for several forms of syntactic ambivalence. The simple ambivalence we focus on in the present paper, corresponds to the overloading of predicate symbols admissible in Prolog. An example of overloading in Prolog is the definition of the predicate *system*, which is intended to be true for system predicates (cf. Sterling and Shapiro [SS86]). A further level of ambivalence admitted in Prolog implementations is the so-called meta-variable facility, which allows for the occurrence of variables as atoms in bodies of clauses and in conjunctive goals. Typical instances of the application of the meta-variable facility are the cut-fail definition of negation in Prolog, the definition of the Prolog disjunction ';', and the meta-interpreter consisting in the single clause $solve(X) \leftarrow X$. An intermediate level of ambivalence is needed for the Prolog built-in *assert*, which takes clauses as arguments. On this level of ambivalence, not only atoms, but also clauses are admitted as terms.

Given the interest of amalgamated extensions of the Vanilla meta-interpreter, and in view of the fact that Prolog itself employs ambivalent, rather than standard syntax, an account of meta-logic programming in the context of ambivalent syntax is of interest. This paper provides an exploration in this area.

Outline of this paper

First in Section 1.2 we give a formal definition of the ambivalent syntax suitable for (amalgamated extensions of) the Vanilla meta-interpreter. We define an appropriate (Herbrand style) semantics for first-order logic with ambivalent syntax, which also gives us a (least Herbrand) semantics for the Vanilla meta-interpreter.

Section 1.3 provides an overview of the various existing correctness results for the Vanilla meta-interpreter. In addition, this section contains a proof sketch of the procedural correctness of the untyped Vanilla meta-interpreter for normal object programs.

In Section 1.5 we prove a satisfying declarative correctness result for the untyped Vanilla meta-interpreter for definite object programs in the context of ambivalent syntax. Section 1.4 provides some technical preliminaries for this correctness proof.

In Section 1.6 we discuss a simple amalgamation. This paper is an extended version of Kalsbeek [Kal93].

1.2 Ambivalent Syntax

In the present section we discuss an ambivalent syntax suitable for the Vanilla meta-interpreter and various extensions, that is, an ambivalent syntax in which atoms are allowed in term positions. We develop some basic theory for first order predicate logic with ambivalent syntax. We restrict our discussion to issues which are of immediate relevance to the subject of the present paper. An extensive discussion of first order predicate logic with ambivalent syntax, focusing on more general levels of ambivalence, is given in Kalsbeek and Jiang [KJ95].

We consider an alphabet L, of which the non-logical constants are given as a triple (R_L, F_L, C_L), where R_L is a set of relation symbols, F_L is a set of function symbols, and C_L a set of constants. We assume that the sets R_L, F_L, and C_L have pairwise empty intersections. In addition, the alphabet has countably many variables. Using this alphabet, an ambivalent term-language $\mathcal{L}_{amb}$ consisting of atoms and terms, is defined as follows.

Definition 2

ATOM$_{\mathcal{L}_{amb}}$ (1) $p(t) \in$ ATOM$_{\mathcal{L}_{amb}}$ is an atom if $p \in R_L$ and $t \in$ TERM$_{\mathcal{L}_{amb}}$;
(2) there are no other atoms.

TERM$_{\mathcal{L}_{amb}}$ (1) $c \in$ TERM$_{\mathcal{L}_{amb}}$ if $c \in C_L$;
(2) $x \in$ TERM$_{\mathcal{L}_{amb}}$ if x is a variable;
(3) $t \in$ TERM$_{\mathcal{L}_{amb}}$ if $t \in$ ATOM$_{\mathcal{L}_{amb}}$;
(4) $f(t) \in$ TERM$_{\mathcal{L}_{amb}}$ if $t \in$ TERM$_{\mathcal{L}_{amb}}$ and $f \in F_L$;
(5) there are no other terms. □

Note that this definition differs from the corresponding definition for standard syntax in the following way. The definition of the set of terms refers to the set of atoms; also, the set of terms properly contains the set of atoms.

Example 3 Let the nonlogical constants of L be $(\{p\}, \{f\}, \{c\})$. Then $p(c)$, $f(p(c))$ and $f(x)$ are terms of $\mathcal{L}_{amb}$, whereas $f(x)$ is a term in the standard language $\mathcal{L}_{st}$ generated by L, but $p(c)$ and $f(p(c))$ are not; $p(c)$ is both an atom of $\mathcal{L}_{st}$ and of $\mathcal{L}_{amb}$; $p(p(c))$ is both an atom and a term of $\mathcal{L}_{amb}$, but neither of those in $\mathcal{L}_{st}$; $f(p(x))$ is a term—but not an atom—in $\mathcal{L}_{amb}$.

This set-up for an ambivalent syntax for the Vanilla meta-interpreter results in a language that is essentially a term-language. By extending Definition 2 with the usual inductive definition for the set of formulas, a complete (ambivalent) syntax for first order predicate logic is obtained. This syntax is characterised by the occurrence of atoms in term positions. It can alternatively be thought of a a syntax in which all the predicate symbols are overloaded.

One part of the Unique Readability Property holds for ambivalent syntax: formulas and terms can only be written in one unique way. But by the very definition of ambivalent syntax, the part of the property that states that expressions of the language are either formulas or terms or none of these, does not hold, of course: all atoms are terms.

Consider first-order predicate logic for languages with an alphabet $L = (R_L, F_L, C_L)$, and with the atoms-as-terms ambivalent syntax. Take a standard system of natural deduction. Observe that $\forall x\, p(x)/p(p(x))$ is now an instance of the standard rule "$\forall x\, \phi/\phi(t)$, for all terms t such that t is free for x in ϕ". We will provide this logic with a closed term semantics with respect to which it is sound and strongly complete.

We need a (standard) definition of free variables and ground terms and atoms.

Definition 4 The set FV(t) of free variables of a term t is inductively defined as follows:
1. FV(x) = x, for all variables x;
2. FV(c) = $\emptyset$, for all $c \in C_L$;
3. FV($f(t_1, \ldots, t_n)$) = FV(t_1) $\cup \ldots \cup$ FV(t_n),
   for $f \in F_L \cup P_L$ and $\{t_1, \ldots, t_n\} \subseteq \text{TERM}_{\mathcal{L}_{amb}}$. □

Definition 5 The sets GRATOM$_{\mathcal{L}_{amb}}$ and CLTERM$_{\mathcal{L}_{amb}}$ of ground atoms and closed terms of $\mathcal{L}_{amb}$ are defined as follows:

$$\text{GRATOM}_{\mathcal{L}_{amb}} = \{\phi \in \text{ATOM}_{\mathcal{L}_{amb}} : \text{FV}(\phi) = \emptyset\}$$
$$\text{CLTERM}_{\mathcal{L}_{amb}} = \{t \in \text{TERM}_{\mathcal{L}_{amb}} : \text{FV}(t) = \emptyset\}.$$
□

These definitions can be extended with the usual clauses defining the set of ground formulas. Note that, according to Definition 5, x occurs as a free variable in the atom $p(p(x))$.

We are now in position to define a semantics for first order logic with ambivalent syntax.

Definition 6 A *pre-interpretation* J for an ambivalent language $\mathcal{L}_{amb}$ consists of
1. a *domain* D such that
   (a) $D = \text{CLTERM}_{\mathcal{L}'}$ for some ambivalent language $\mathcal{L}'$;
   (b) $\text{CLTERM}_{\mathcal{L}_{amb}} \subseteq D$.
2. for each term $t \in \text{CLTERM}_{\mathcal{L}_{amb}}$, the assignment of $t \in D$.

An *interpretation* M of $\mathcal{L}_{amb}$ consists of a pre-interpretation J with domain D and, for each n-ary predicate symbol in P_L, the assignment of a subset p^M of D^n. A *Herbrand pre-interpretation* J of $\mathcal{L}_{amb}$ is a pre-interpretation such that $D = \text{CLTERM}_{\mathcal{L}_{amb}}$. A *Herbrand interpretation* is an interpretation based on a Herbrand pre-interpretation. □

Observe that the above interpretations are in fact closed term interpretations. In fact, a more general semantics can be defined, with non-identity interpretation functions. In the context of the present paper, however, we will not pursue this issue.

Definition 7 Let M be an interpretation of a language $\mathcal{L}_{amb}$. Satisfaction of closed formulas of $\mathcal{L}_{amb}$ in M is inductively defined as follows:

$M \models p(t)$ for $p(t) \in \text{ATOM}_{\mathcal{L}_{amb}}$ iff $t \in p^M$
$M \models \forall x\, \phi$ iff $M \models \phi[x/a]$, for all $a \in D_M$
$M \models \exists x\, \phi$ iff there is an $a \in D_M$ such that $M \models \phi[x/a]$
and the interpretation of the logical constants $\vee\ \wedge\ \neg\ \rightarrow$ is as usual. □

Definition 8 Let T be a theory in a language $\mathcal{L}_{amb}$. Let M be an interpretation of $\mathcal{L}_{amb}$. M is a *model* for T if all the axioms of T are true in M. A *Herbrand model* is a model based on a Herbrand interpretation. □

Theorem 9 *First order predicate logic with ambivalent syntax is sound and (strongly) complete with respect to the semantics defined in Definitions 6–8.*

The soundness and completeness proof uses a standard Henkin construction. For a full proof we refer to Kalsbeek and Jiang [KJ95]. For infinitely extendible theories a model can be constructed by using those closed terms of the language that do not occur as terms in the theory as Henkin constants. The resulting model is a Herbrand model for the theory. Observe that logic programs, being finite, are infinitely extendible.

A language with ambivalent syntax was first defined by Richards [Ric74] in the context of intensional predicate logics. Kowalski and Kim [KK91] advanced ambivalent language as appropriate for the Vanilla meta-interpreter, especially for extensions of it in the field of meta-logic programming for muti-agent knowledge, and consequently, Richards [Ric74] has become a standard reference in the literature on the Vanilla meta-interpreter. It should be observed however, that Richards' syntax was devised for a different purpose and has a calculus adapted to this goal. As a result, standard first order logic with ambivalent syntax is unsound with respect to the semantics proposed in Richards [Ric74][1]. A broader perspective on ambivalent syntax and its semantics is given by Gabbay [Gab92], where several meta-languages are proposed which allow for great freedom of expression. Among them is HFP, which is essentially first-order predicate language enriched with meta-level predicate and function symbols. Ambivalent syntax can be considered as a version of HFP. A more general semantics for ambivalent syntax than the one proposed in the present paper can be obtained by considering models for HFP. Another logic with ambivalent syntactic phenomena is Hilog (Chen et al. [CKW93]). The relation between first order predicate logic with ambivalent syntax and Hilog is explored in Kalsbeek and Jiang [KJ95].

Conventions. We will adhere to the following conventions regarding syntax and language. The underlying language of an object program P is identified with $\mathcal{L}_P$, the language generated by $(\mathcal{R}_P, \mathcal{F}_P, \mathcal{C}_P)$ and standard syntax, where $\mathcal{R}_P$, $\mathcal{F}_P$, and $\mathcal{C}_P$ are,

---

[1] Let $\mathcal{L}$ be an ambivalent language with one binary relation symbol R, one unary function symbol f, and one constant symbol c. Then $\forall x\, R(x, Q(x)) \rightarrow \forall x\, R(f(x), Q(f(x)))$ should be valid in every model. However, one can construct a model M according to Richards' definitions which validates the antecedent, and in which the consequent is false. Take, according to the definitions in Richards [Ric74], as the domain of M, $D_M = \{c\} \cup \{\phi : \phi \in \text{CLFORM}_{\mathcal{L}}\}$. The extension of Q can be taken arbitrary. $V(f)$ is the identity function on D_M. The extension of R is taken as follows: $V(R) = \{\langle d, Q(d)\rangle | d \in D_M\}$. By Richards' definition, the interpretation function $*$ is the identity function on constants and closed formulas. Thus, M validates $\forall x\, R(x, Q(x))$. However, $(f(c))^* = V(f)(c^*) = c$, while $(Q(f(c)))^* = Q(f(c))$. So $\langle (f(c))^*, (Q(f(c)))^* \rangle = \langle c, Q(f(c)) \rangle$, which does not belong to $V(R)$.

respectively, the set of relation symbols, function symbols and constants occurring in P. We assume that none of the predicates $demo(\cdot)$ and $clause(\cdot,\cdot)$ is in $\mathcal{R}_P \cup \mathcal{F}_P$. For the underlying language $\mathcal{L}_{V_P}$ of the associated meta-program $\mathbf{V}_P$ we take the language generated by $(\mathcal{R}_{V_P}, \mathcal{F}_{V_P}, \mathcal{C}_{V_P})$ and ambivalent syntax as in Definition 2, where $\mathcal{R}_{V_P} = \mathcal{R}_P \cup \{demo(\cdot), clause(\cdot,\cdot)\}$, $\mathcal{F}_{V_P} = \mathcal{F}_P \cup \{(\cdot)\&(\cdot)\}$, and $\mathcal{C}_{V_P} = \mathcal{C}_P \cup \{empty\}$.

Other conventions regarding the language underlying the meta-program are possible. The logical connective &, which we choose to represent as a function symbol in the language of the meta-programs, could also be represented as a relation symbol. Alternatively, instead of restricting ambivalence to atoms occurring as tems, we could more liberally allow conjunctions of atoms to occur as terms in the ambivalent language. This would result in a more general level of ambivalence, for which however, the theory developed above applies without any significant modification. It should be stressed that none of these particular choices affects the results reported below.

We use some of the basic concepts and results of the theory of logic programs. We adhere to the terminology used in Lloyd [Llo87]. As observed in Martens and De Schreye [MS95a], the basic theory for definite programs as developed in Lloyd [Llo87] holds without any restriction for programs with ambivalent language, and thus for the Vanilla meta-interpreter, as long as all the relevant concepts are defined with respect to the underlying ambivalent language.

1.3 Correctness of the Vanilla Meta-Interpreter

In the present section and the two subsequent sections, we set out to prove the correctness of the untyped Vanilla meta-interpreter for definite object programs. In particular, we will consider the meta-program with underlying ambivalent syntax allowing for atoms to occur as terms. Two different correctness results will be proved and discussed: procedural correctness, which relates computed answer substitutions (via SLD(NF)-resolution) of the object program and the associated meta-program, and declarative correctness, which relates the intended interpretations (i.e., the least Herbrand models, as we will mainly concern definite object programs) of object and meta-program. The procedural correctness will be shown to be a corollary of the procedural correctness result for the *typed* Vanilla meta-interpreter, as proven by Hill and Lloyd [HL89]. For the declarative correctness we will present a direct proof in Section 1.5.

1.3.1 Procedural correctness

Hill and Lloyd [HL89] prove procedural correctness of the typed meta-interpreter for normal programs. In their approach, a functional representation of the object program is used. The procedural correctness of the untyped Vanilla meta-interpreter (Theorem 10 below) can be obtained as a corollary of the proof given in Hill and Lloyd [HL89]. To see this, the following observations suffice. The crucial step in the proof is to consider, instead of the original typed meta-program V_P, a partial evaluation V_P^* with respect to the atom $demo(x)$. It can be easily shown that, for any object level goal $\leftarrow Q$, and any derivation d for $V_P^* \cup \{\leftarrow Q\}$, any variable occurring in d is of object type. As a consequence, although both V_P and V_P^* are typed, the typing plays a negligible role for the queries we are interested in ($demo(Q)$, where Q is an object level query). Thus, the procedural correctness result of Hill and Lloyd immediately translates to the untyped Vanilla meta-interpreter for normal object programs (and thus, by inclusion, for definite object programs). Similarly, the proof is insensitive to their particular choice of representation of the object level language (functional representation), and goes through, without modification, for the above approach with ambivalent syntax.

As a consequence of the above observations, we have the following result:

Theorem 10 (Procedural correctness) *Let P be a normal program and $\leftarrow Q$ a normal goal. Let $\mathbf{V}_P$ be as in Definition 1, extended with the clause [M¬]. Then the following hold:*

1. *θ is a computed answer substitution for $P \cup \{\leftarrow Q\}$ iff*
   θ is a computed answer substitution for $\mathbf{V}_P \cup \{\leftarrow demo(Q)\}$.
2. *$P \cup \{\leftarrow Q\}$ has a finitely failed SLDNF-tree iff $\mathbf{V}_P \cup \{\leftarrow demo(Q)\}$ has a finitely failed SLDNF-tree.*

Mutatis mutandis, this result goes through, via the same arguments, for the Vanilla meta-interpreter with a functional representation of the object program.

1.3.2 Declarative correctness for normal object programs

In addition to their result on procedural correctness, Hill and Lloyd [HL89] prove declarative correctness of the typed Vanilla meta-interpreter for normal object programs. As the intended meaning of a normal program, the completion is taken. More precisely, it is shown in Hill and Lloyd [HL89] that correct answers for $comp(P) \cup \{\leftarrow Q\}$ coincide with correct answers for $comp(V_P) \cup \{\leftarrow demo(Q)\}$ for object level queries Q. Unlike the proof of procedural correctness, Hill and Lloyd's proof of declarative correctness does not generalise to the case of the *untyped* Vanilla meta-interpreter. The reason for this failure is that the sortedness of domains of models for the completion of the meta-program plays an essential role in this proof.

The difficulty in getting a satisfactory result on declarative completeness for the untyped case is illustrated by the restrictions on the results in Martens and De Schreye [MS95]. There, the class of normal programs for which the associated Vanilla meta-interpreter is shown to be complete, is the class of stratifiable, language independent programs. The correctness result for this class of programs relates the perfect Herbrand model of an object program with (a suitable subset of) the weakly perfect Herbrand model of the associated meta-program. A functional representation of the object programs is used. The results do not depend on this particular choice of language, and also hold if ambivalent syntax is used as the underlying language for the Vanilla meta-interpreter. It appears that the restriction to stratifiable object programs in the results of Martens and De Schreye can be liberated somewhat to local stratifiability and weak stratifiability. In contrast, the restriction of language independence, which is closely linked to typing, seems to be crucial. At the same time, the restriction to language independent programs seriously limits the applicability of the result. In practice, as language independence is undecidable, the result applies to a syntactically defined subclass—the class of range restricted programs. Range restriction, however, is the exception rather than the rule among the basic logic programs; e.g., the programs *set*, *list*, and *member* are not range restricted.

Concluding, the declarative correctness result proven in Martens and De Schreye [MS95] applies only to a seriously limited class of programs, but it seems that it defines the boundaries of what can be obtained with respect to the untyped Vanilla meta-interpreter for normal programs.

1.3.3 Declarative correctness for definite object programs

For *definite* object programs, a satisfactory declarative correctness result can be obtained in several ways. First, declarative correctness of the untyped version is a corollary of the procedural correctness Theorem 10, by the soundness and (strong) completeness of SLD-resolution. Again, this result holds both for the Vanilla meta-interpreter using ambivalent syntax and for the version using a functional representation of the object program. Second, a direct proof can be given, in which, by comparing different stages of the relevant fixpoint operators, the following relation between the least Herbrand models $\mathbf{M}_P$ en $\mathbf{M}_{V_P}$ of respectively the object program and the associated meta-program is established: For all object level atoms A, $A \in \mathbf{M}_P$ iff $demo(A) \in \mathbf{M}_{V_P}$. We will present a proof of this correctness result (Theorem 32) in Section 1.5.

Typically, as observed in the introduction, this declarative correctness result for definite object programs relates the least Herbrand model $\mathbf{M}_P$ of an object program P to the subset of the least Herbrand model $\mathbf{M}_{V_P}$ of the associated meta-program, not containing 'unrelated atoms'. In a corollary to the declarative correctness result Theorem 32, we obtain a satisfactory interpretation of unrelated atoms: the metalevel terms occurring in these atoms can be interpreted as variables ranging over the object level terms (Corollary 29).

A stronger correlation between the least Herbrand models $\mathbf{M}_P$ and $\mathbf{M}_{V_P}$ is established in Martens and De Schreye [MS95a] for the class of *language independent* definite object programs. A program is language independent if its least Herbrand model is not affected by extensions of the underlying language. More practically, the class of language independent programs can be characterised as the class of programs for which every computed answer substitution is ground. For language independent programs P, not only the above correctness result holds, but in addition the following property. If $demo(p(t)) \in \mathbf{M}_{V_P}$, and p is a predicate occurring in P then t must be a closed term in the language $\mathcal{L}_P$ underlying P. In other words, unrelated atoms do not occur in $\mathbf{M}_{V_P}$, provided the object program P is language independent.

The real interest of the Vanilla meta-interpreter lies in extensions. Enhanced meta-interpreters are obtained from the Vanilla meta-interpreter by allowing extra arguments in the *demo*-predicate and extra body atoms in the clauses of the general part **V**. The resulting program is then extended with clauses defining the new predicates. In general, enhanced meta-interpreters can not be expected to be complete w.r.t. object programs. In many applications, soundness rather than completeness is desired. Martens and De Schreye [MS95a] prove declarative soundness for enhanced meta-programs w.r.t. language independent programs. In addition, they obtain several correctness results for amalgamations of language independent object programs and the Vanilla meta-interpreter.

1.3.4 S-semantics

A different approach to proving correctness of the Vanilla meta-interpreter for definite programs is the use of S-semantics. The S-semantics was introduced by Falaschi et al. [FLMP93] to close the gap between the procedural and the declarative interpretations of definite programs. Essentially, the (unique) least S-herbrand model $\mathbf{M}_P^S$ of a program P consists, modulo renaming of variables, of those (not necessarily closed) atoms $Q(t)$ such that $t = x\theta$, where θ is the computed answer substitution for $P \cup \{\leftarrow Q(x)\}$. Analogous to the usual least Herbrand models, the least S-Herbrand models are the least fixed points of a proper version of the T-operator on subsets of the S-Herbrand base. Two properties are of interest in the present context. First, as observed in Levi and Ramundo [LR93], the S-semantics is independent of the language. Second, while the S-semantics is a declarative way to express the procedural interpretation of a program, the least Herbrand model of a program can be computed from its S-semantics (by taking all the ground instances of its elements).

Independently, Levi and Ramundo [LR93] and Martens and De Schreye [MS95a] proved the following strong correspondence between the S-semantics of a definite object program and the S-semantics of its related Vanilla meta-interpreter: $demo(p(t)) \in \mathbf{M}_{V_P}^S$ iff t is an object level term and $p(t) \in \mathbf{M}_P^S$. As observed by Levi and Ramundo [LR93], both the procedural and the declarative correctness of the Vanilla meta-interpreter for definite programs are a consequence of this correctness theorem w.r.t. the S-semantics.

The approach via S-semantics has various interesting applications. Brogi and Turini [BT95] obtain in this context elegant equivalence proofs for the procedural and the meta-logical definition of basic composition operators for the construction of composite object programs. A binary demo-predicate is defined, the extra argument of which explicitly denotes the interpreted object program as a term built up from program constants and the composition operators. This predicate also differs from the Vanilla meta-interpreter in that it represents the object clauses, so that an extra predicate clause is not used. The above cited correctness result for the Vanilla meta-interpreter is a special case of the more general results in Brogi and Turini [BT95]. Levi and Ramundo [LR93] obtain a correctness result for enhanced meta-interpreters defining inheritance mechanisms on structured object programs. In addition, they show that the linear overhead of meta-computation via the Vanilla meta-interpreter can be eliminated by specialising the meta-interpreter. In contrast, Martens and De Schreye give a counterexample, showing that in the context of S-semantics even soundness cannot be expected for the general class of enhanced meta-interpreters. They prove that language independence of the object program is a sufficient condition for soundness.

1.4 Preliminaries on Substitution

As we observed in the Introduction, one of the problems involved in proving correctness of the Vanilla meta-interpreter $\mathbf{V}_P$ results from in the fact that the $\mathcal{L}_{V_P}$, the language underlying $\mathbf{V}_P$, is in several ways an extension of the language $\mathcal{L}_P$ underlying the object program P. First, the ambivalent syntax of $\mathcal{L}_{V_P}$ can be considerd as an extension of the standard syntax of $\mathcal{L}_P$. Second, the sets of predicates $\mathcal{R}_{V_P}$ occurring in $\mathbf{V}_P$ is a proper extension of the set of predicates $\mathcal{R}_P$ occurring in p. Similarly, $\mathcal{F}_P \subset \mathcal{F}_{V_P}$, and $\mathcal{C}_P \subset \mathcal{C}_{V_P}$. We will first formalise this notion of extension of a language.

For this purpose, we identify languages $\mathcal{L}$ with pairs (L, S), where L is a triple (R_L, F_L, C_L), indicating the non-logical constants of $\mathcal{L}$, and where S indicates the particular syntax of $\mathcal{L}$ (standard or ambivalent). We will only consider languages, which, like $\mathcal{L}_P$ and $\mathcal{L}_{M_P}$, have no logical constants—that is, languages of which the set of formulas consists of atoms only. This is not a necessary restriction, and the theory developed below can be generalised to the case of languages with connectives and quantifiers. In addition, we confine the discussion to atoms-as-terms ambivalence. Again, this is not a necessary restriction. Also, the theory below easily generalises to suit comparison of two languages with a different level of ambivalence. In the remainder of this section, if we mention ambivalent syntax, we always intend atoms-as-terms ambivalence.

Definition 11 For two languages $\mathcal{L} = (L, S)$ and $\mathcal{L}' = (L', S')$, we say that $\mathcal{L}$ is *part of* $\mathcal{L}'$ ($\mathcal{L} \subseteq \mathcal{L}'$) if

1. $R_L \subseteq R_{L'}$, $F_L \subseteq F_{L'}$, and $C_L \subseteq C_{L'}$;
2. S is standard syntax and S' is ambivalent syntax. □

Typically, the underlying language of an object program P is part of the language of the associated meta-interpreter $\mathbf{V}_P$. An easy consequence of the above definition is the following:

Lemma 12 Let $\mathcal{L}$ and $\mathcal{L}'$ be two languages such that $\mathcal{L} \subseteq \mathcal{L}'$. Then $\text{TERM}_\mathcal{L} \subseteq \text{TERM}_{\mathcal{L}'}$ and $\text{ATOM}_\mathcal{L} \subseteq \text{ATOM}_{\mathcal{L}'}$.

In the remainder of the present section, we will always suppose $\mathcal{L} \subseteq \mathcal{L}'$.

A crucial step in the correctness proof for the Vanilla meta-interpreter which we present in Section 1.5, relies on a shift from (sub)terms of the language underlying $\mathbf{V}_P$ to terms of the language underlying an object program P. In the remainder of the present section we define and discuss the necessary transformation (Definition 13).

If one takes a closer look at the (tree-)structure of a term t of $\mathcal{L}'$, one sees that certain subterms of t are terms of $\mathcal{L}$, while others are proper $\mathcal{L}'$-terms. By substituting in an $\mathcal{L}'$-term t all of its subterms which are proper $\mathcal{L}'$-terms by variables, one can transform t into an $\mathcal{L}$-term, as follows: One can descend in the term-tree of t until one encounters a symbol s that is either a predicate in $\mathcal{L}'$, or a function symbol or constant, but not a constant or function symbol of $\mathcal{L}$; replace the subtree starting with s with a fresh variable which does not occur in t. This procedure yields a term of $\mathcal{L}$. Moreover, this can be done in a systematic way, by replacing equal terms by equal variables. The $\mathcal{L}'/\mathcal{L}$-*abstraction*, which formalises this idea, thus transforms terms from $\mathcal{L}'$ into terms from $\mathcal{L}$ (Definition 13). Moreover, the transformation is reversible (Lemma 16).

Definition 13 Let $\mathcal{L} \subseteq \mathcal{L}'$, and let t be a term of $\mathcal{L}'$. The $\mathcal{L}'/\mathcal{L}$ *-abstraction of* t, $\text{abs}_{\mathcal{L}'/\mathcal{L}}(t)$, is inductively defined as follows:

$$\begin{aligned}
&\text{abs}_{\mathcal{L}'/\mathcal{L}}(x) = x && \text{if } x \text{ is a variable;} \\
&\text{abs}_{\mathcal{L}'/\mathcal{L}}(c) = c && \text{if } c \in C_L; \\
&\text{abs}_{\mathcal{L}'/\mathcal{L}}(c) = x_c && \text{if } c \in C_{L'} \setminus C_L, \\
&\text{abs}_{\mathcal{L}'/\mathcal{L}}(f(t_1, \ldots, t_n)) = f(\text{abs}_{\mathcal{L}'/\mathcal{L}}(t_1), \ldots, \text{abs}_{\mathcal{L}'/\mathcal{L}}(t_n)) && \text{if } f \in F_L; \\
&\text{abs}_{\mathcal{L}'/\mathcal{L}}(f(t_1, \ldots, t_n)) = x_{f(t_1, \ldots, t_n)} \\
&&& \text{if } f \in R_{L'} \cup (F_{L'} \setminus F_L).
\end{aligned}$$

Here, x_c and $x_{f(t_1, \ldots, t_n)}$ are variables. □

Example 14 Let $\mathcal{L}$ be the underlying (standard) language of an object program P such that $\mathcal{R}_P = \{p\}$, $\mathcal{F}_P = \{f(\cdot), g(\cdot, \cdot)\}$ and $\mathcal{C}_P = \{c\}$. Let $\mathcal{L}'$ be the underlying (ambivalent) language of the associate meta-interpreter $\mathbf{V}_P$. Then
$\text{abs}_{\mathcal{L}'/\mathcal{L}}(demo(f(c))) = x_{demo(f(c))}$
$\text{abs}_{\mathcal{L}'/\mathcal{L}}(f(c)) = f(c)$
$\text{abs}_{\mathcal{L}'/\mathcal{L}}(p(f(c))) = x_{p(f(c))}$
$\text{abs}_{\mathcal{L}'/\mathcal{L}}(f(empty)) = f(x_{empty})$
$\text{abs}_{\mathcal{L}'/\mathcal{L}}(g(demo(x), c)) = g(\text{abs}_{\mathcal{L}'/\mathcal{L}}(demo(x)), \text{abs}_{\mathcal{L}'/\mathcal{L}}(c)) = g(x_{demo(x)}, c)$. □

Implicitly we have assumed that we have extended the set of variables of $\mathcal{L}$ with a set of fresh variables which are indexed by $\mathcal{L}'$-terms. This means that, while our purpose was to get a terms of $\mathcal{L}$ as the result of an abstraction operation, we get a term in a language which is in fact an extension of $\mathcal{L}$. However, by renaming the fresh variables in $\text{abs}_{\mathcal{L}'/\mathcal{L}}(t)$, we can get a term of $\mathcal{L}$ proper. The following lemma immediately follows from the above definition:

Lemma 15 *Let t be a term of $\mathcal{L}$. Then the following hold:*

1. $\text{abs}_{\mathcal{L}'/\mathcal{L}}(t)$ *is unique;*
2. $\text{abs}_{\mathcal{L}'/\mathcal{L}}(t)$ *is, modulo renaming of variables, a term of $\mathcal{L}$.*

The abstraction operation is reversible, as the following lemma shows. We will sometimes write the application of a substitution σ to a term t as $t \cdot \sigma$, to increase readability.

Lemma 16 *For every term t in $\mathcal{L}'$, there is an $\mathcal{L}$-substitution σ_t, such that $t = \text{abs}_{\mathcal{L}'/\mathcal{L}}(t) \cdot \sigma_t$.*

Proof: Define σ_t by induction on the term structure of t, as follows:

$\sigma_c = \emptyset$ if $c \in C_L$;
$\sigma_x = \emptyset$ if x is a variable without an index from $\text{TERM}_{\mathcal{L}'}$;
$\sigma_{x_c} = \{x_c/c\}$ if $c \in C_{L'} \setminus C_L$;
$\sigma_{f(t_1,\ldots,t_n)} = \sigma_{t_1} \cup \ldots \cup \sigma_{t_n}$ if $f \in F_L$;
$\sigma_{x_{f(t_1,\ldots,t_n)}} = \{x_{f(t_1,\ldots,t_n)}/f(t_1,\ldots,t_n)\}$ if $f \in R_{L'} \cup (F_{L'} \setminus F_L)$. □

Example 17 Let $\mathcal{L}$ and $\mathcal{L}'$ be as in Example 14. Then
$f(demo(c)) = f(x_{demo(c)})\{x_{demo(c)}/demo(c)\}$
$g(f(demo(c)), empty) = g(f(x_{demo(c)}), x_{empty})\{(x_{demo(c)}/demo(c), x_{empty}/empty\}$ □

In the correctness proof, we will use not only the concept of abstraction of terms, but also the more general concept of abstraction of a substitution. The abstraction of a substitution is obtained by applying the abstraction function to its substituents.

Definition 18 Let θ be an $\mathcal{L}$-substitution. The $\mathcal{L}'/\mathcal{L}$ -abstraction of θ, written as $\text{abs}_{\mathcal{L}'/\mathcal{L}}(\theta)$, is defined as follows:

$$\text{abs}_{\mathcal{L}'/\mathcal{L}}(\theta) = \{x/\text{abs}_{\mathcal{L}'/\mathcal{L}}(t) : x/t \in \theta\}.$$

The following lemma characterises the connection between the $\mathcal{L}/\mathcal{L}'$ -abstraction of an $\mathcal{L}'$-ground term t and $\mathcal{L}$-terms s of which t is an instantiation.

Lemma 19 Let s be a term of $\mathcal{L}$, t a term of $\mathcal{L}'$, and θ an $\mathcal{L}'$-substitution such that $s \cdot \theta = t$. Then $s \cdot \text{abs}_{\mathcal{L}'/\mathcal{L}}(\theta) = \text{abs}_{\mathcal{L}'/\mathcal{L}}(t)$.

Lemma 19 can be read as follows, writing θ^* for $\text{abs}_{\mathcal{L}'/\mathcal{L}}(\theta)$ and t^* for $\text{abs}_{\mathcal{L}'/\mathcal{L}}(t)$: $s \cdot \theta^* = (s \cdot \theta)^*$.

Example 20 Let again $\mathcal{L}$ and $\mathcal{L}'$ be as in Example 14. Then
$g(f(empty), demo(c)) = g(f(x), y)\{x/empty, y/demo(c)\}$ and
$g(f(x_{empty}), x_{demo(c)}) = g(f(x), y)\{x/x_{empty}, y/x_{demo(c)}\}$ □

We extend the usual definition of ground substitution in the following way:

Definition 21 A substitution σ is $\mathcal{L}$-ground if $\sigma = \{x_i/t_i : i \in [1,n]\}$, where t_i is a ground term of $\mathcal{L}$.

The following lemma is an application of a well-known fact about ground substitutions.

Lemma 22 Let t be a ground term of $\mathcal{L}'$. Let σ be an $\mathcal{L}$-ground substitution such that $dom(\sigma) \supseteq \text{FV}(\text{abs}_{\mathcal{L}'/\mathcal{L}}(t))$. Then $\text{abs}_{\mathcal{L}'/\mathcal{L}}(t) \cdot \sigma$ is a ground term of $\mathcal{L}$.

1.5 Declarative Correctness

The present section is devoted to a proof of the declarative correctness of the untyped Vanilla meta-interpreter using ambivalent syntax. This is basically a result about the relation between the least Herbrand model of a definite program and the least Herbrand model of its associated meta-program, and the proof we present proceeds by comparing stages of the respective fixpoint-operators. Hence we will be mainly concerned with ground versions of programs. As we remarked in Section 1.1, variables occurring in the object-program specific part of the Vanilla interpreter, the clauses [M4] and [M5], can be instantiated with terms from the meta-language. Therefore, we need the following extension of the usual concept of the ground version of a program.

Definition 23 Let P be a definite program. Let $\mathcal{L}$ be a language containing $\mathcal{L}_P$. With $\mathcal{L}$-$ground(P)$ we mean the set of all clauses that are $\mathcal{L}$-ground instantiations of clauses of P. The natural ground version of P, $\mathcal{L}_P$-$ground(P)$, will be indicated simply as $ground(P)$. In the remainder of this section, P will always be a definite program.   □

The next lemma is an immediate application of the Lemmas 19 and 22 to clauses of definite programs. It establishes some simple relations between a definite program P, its natural ground version, and $\mathcal{L}_{V_P}$-$ground(P)$. We will speak about P-ground and $\mathbf{V}_P$-ground substitutions to refer to $\mathcal{L}_P$-ground and $\mathcal{L}_{V_P}$-ground substitutions, respectively. We will also leave out indices of the abstraction operator, that is, we will write $\mathbf{abs}(\cdot)$ instead of $\mathbf{abs}_{\mathcal{L}_{V_P}/\mathcal{L}_P}$.

Lemma 24 *Let $C = p(s) \leftarrow p_1(s_1), \ldots, p_n(s_n)$ be a clause of P. Let $C' = p(t) \leftarrow p_1(t_1), \ldots, p_n(t_n)$ be in $\mathbf{V}_P$-$ground(P)$, and let θ be an $\mathbf{V}_P$-ground substitution with $dom(\theta) \supseteq \mathrm{FV}(C)$, such that $C \cdot \theta = C'$. Then the following hold:*

1. *$C \cdot \mathbf{abs}(\theta) = p(\mathbf{abs}(t)) \leftarrow p_1(\mathbf{abs}(t_1)), \ldots, p_n(\mathbf{abs}(t_n))$;*
2. *for all $\mathbf{V}_P$-ground substitutions σ with $dom(\sigma) \supseteq ran(\mathbf{abs}(\theta))$, $C \cdot \mathbf{abs}(\theta) \cdot \sigma$ is in $\mathbf{V}_P$-$ground(P)$;*
3. *for all P-ground substitutions σ with $dom(\sigma) \supseteq ran(\mathbf{abs}(\theta))$, $C \cdot \mathbf{abs}(\theta) \cdot \sigma$ is in $ground(P)$.*

Proof: (1) immediately follows from Lemma 19. (2) follows from the definitions, and (3) follows from (1) and Lemma 22. □

We will first establish some easy facts about $\mathbf{M}_{V_P}$, the least Herbrand model of $\mathbf{V}_P$.

Lemma 25 *If $demo(p(t_1, \ldots, t_n)) \in \mathbf{M}_{V_P}$, and p is not $\&$, then $p \in P_{\mathcal{L}_P}$.*

Proof: Suppose p is not $\&$, and $demo(p(t_1, \ldots, t_n)) \in \mathbf{M}_{V_P}$. Then there is a d such that $demo(p(t_1, \ldots, t_n)) \in \mathrm{T}_{V_P} \uparrow d \setminus \mathrm{T}_{V_P} \uparrow (d-1)$. Also, as p is not $\&$, $demo(p(t_1, \ldots, t_n))$ must have entered by an application of the meta-clause [M2]. Therefore, for some E, $\{clause(p(t_1, \ldots, t_n), E), demo(E)\} \subseteq \mathrm{T}_{V_P} \uparrow (d-1)$. Thus $p(t_1, \ldots, t_n) \leftarrow E$ (or, if E is the constant $empty$, $p(t_1, \ldots, t_n) \leftarrow$) belongs to $\mathbf{V}_P$-$ground(P)$. So p must be a predicate in the language of P. □

An immediate consequence of this lemma is that atoms of the form $demo(demo(t))$ and $demo(clause(s,t))$ do not occur in $\mathbf{M}_{V_P}$. By essentially the same argument one sees that:

Lemma 26 *Let $demo(A\&B) \in \mathrm{T}_{V_P} \uparrow n$, for some n. Then there are $m, k < n$ such that $demo(A) \in \mathrm{T}_{V_P} \uparrow m$ and $demo(B) \in \mathrm{T}_{V_P} \uparrow k$.*

Corollary 27 *Let $demo(A_1 \& \ldots \& A_m) \in \mathrm{T}_{V_P} \uparrow n$ for some n. Then, for $i \in [1, k]$, $demo(A_i) \in \mathrm{T}_{V_P} \uparrow (n-i)$, and, for $i \in [1, m-1]$, $demo(A_{i+1} \& \ldots \& A_m) \in \mathrm{T}_{V_P} \uparrow (n-i)$ for some n.*

In the remainder of this section, we will abstract from arities of predicates. That is, we will only consider unary (object level) predicates. This simplifies notation, while it does not affect the generality of proofs and results.

The next lemma is crucial for the proof of the soundness part of the correctness theorem. It expresses the main idea behind the soundness proof.

Lemma 28 *Let p be a predicate in $\mathcal{L}_P$, and let t be a closed term in $\mathcal{L}_{V_P}$. Suppose $demo(p(t)) \in T_{V_P} \uparrow n$, for some n. Then, for all $\mathbf{V}_P$-ground substitutions σ such that $dom(\sigma) \supseteq \mathrm{FV}(\mathbf{abs}(t))$, $demo(p(\mathbf{abs}(t) \cdot \sigma)) \in T_{V_P} \uparrow n$.*

Proof: By induction on the stages of the T_{V_P}-operator. Let P, T, p, and σ be as in the formulation of the lemma. The lemma trivially holds for $n < 2$. Let, for some n, $demo(p(t)) \in T_{V_P} \uparrow n$, and suppose that Then, by definition of the T_{V_P}-operator, there are the lemma is true for all $m < n$. Then, by definition of the T_{V_P}-operator, there are $p_1, \ldots, p_k \in P_{\mathcal{L}_P}$, $\mathcal{L}_P$-terms $s, s_1, \ldots, s_k$, and an $\mathbf{V}_P$-ground substitution θ such that
(1) $p(s) \leftarrow p_1(s_1), \ldots, p_n(s_n) \in P$;
(2) $(p(s) \leftarrow p_1(s_1), \ldots, p_n(s_n)) \cdot \theta = p(t) \leftarrow p_1(t_1), \ldots, p_n(t_n)$;
(3) $clause(p(t), p_1(t_1)\& \ldots \& p_n(t_n)) \in T_{V_P} \uparrow (n-1)$;
(4) $demo(p_1(t_1)\& \ldots \& p_n(t_n)) \in T_{V_P} \uparrow (n-1)$.
(For $n = 2$, the situation is slightly, but not essentially, different: the P-clause in (1) has an empty body.)
Now let τ be an $\mathbf{V}_P$-ground substitution with $dom(\tau) \supseteq \mathrm{FV}(\mathbf{abs}(\theta))$ and $\tau_{|\mathrm{FV}(\mathbf{abs}(t))} = \sigma$. By (1), (2), Lemma 24, and the definition of $\mathbf{V}_P$, we see that
(a) $clause(p(\mathbf{abs}(t)\tau), p_1(\mathbf{abs}(t_1)\tau)\& \ldots \& p_n(\mathbf{abs}(t_n)\tau)) \in T_{V_P} \uparrow (n-1)$.
From (4) and Corollary 27, it follows that $demo(p_i(t_i)) \in T_{V_P} \uparrow (n-1-i)$ for $i \in [1, k]$. Thus, by inductive hypotheses, we have that $demo(p_i(\mathbf{abs}(t_i)\tau)) \in T_{V_P} \uparrow (n-1-i)$ for $i \in [1, k]$. By $k-1$ applications of the meta-clause for conjunction [M3], it follows that
(b) $demo(p_1(\mathbf{abs}(t_1)\tau)\& \ldots \& p_k(\mathbf{abs}(t_k)\tau)) \in T_{V_P} \uparrow (n-1)$.
By definition of the T_{V_P}-operator (using [M2], (a) and (b)), we conclude that $demo(p(\mathbf{abs}(t)\sigma)) = demo(p(\mathbf{abs}(t)\tau)) \in T_{V_P} \uparrow n$. □

This lemma has the following obvious generalisation:

Corollary 29 *Let $p_1, \ldots, p_k \in \mathcal{L}_P$, and let $t_1, \ldots, t_k$ be terms of $\mathcal{L}_{V_P}$. Suppose that $demo(p_1(t_1)\& \ldots \& p_k(t_k)) \in T_{V_P} \uparrow n$, for some n. Let σ be an $\mathbf{V}_P$-ground substitution such that $dom(\sigma) \supseteq \mathrm{FV}(\mathbf{abs}(t_1)) \cup \ldots \cup \mathrm{FV}(\mathbf{abs}(t_k))$. Then $demo(p_1(\mathbf{abs}(t_1)\sigma)\& \ldots \& p_k(\mathbf{abs}(t_k)\sigma)) \in T_{V_P} \uparrow n$.*

We are now in position to prove the completeness theorem. The proof of the soundness part heavily depends on Lemma 28 and its corollary. The proof of the completeness lemma is inspired by the corresponding proof in Martens and De Schreye [MS95a]. We indicate the Herbrand base of a definite program P as $\mathbf{B}_P$.

Lemma 30 (Soundness) *Let P be a definite program and let $\mathbf{V}_P$ be the associated Vanilla meta-interpreter Then for all $p(t) \in \mathbf{B}_P$, $\forall n \in \mathbf{N}\,[demo(p(t)) \in T_{V_P} \uparrow n \Longrightarrow \exists m \in \mathbf{N}.\ p(t) \in T_P \uparrow m]$.*

Proof: By induction on n. Let $p(t) \in \mathbf{B}_P$. The lemma trivially holds for $n < 2$.

Suppose $demo(p(t)) \in T_{V_P} \uparrow 2$. There must be a P-term s and an $\mathbf{V}_P$-ground instantiation θ such that $p(s) \leftarrow \in P$ and $s\theta = t$. However, by the assumption that $p(t) \in \mathbf{B}_P$, (the relevant part of) θ must be a P-ground substitution. So $p(t) \leftarrow \in ground(P)$ and $p(t) \in T_P \uparrow 1$.

Suppose that, for some $n > 2$, $demo(p(t)) \in T_{V_P} \uparrow n \setminus T_{V_P} \uparrow (n-1)$. Assume that the lemma holds for all $n' < n$. Then the $demo(p(t))$ must have entered the Herbrand model by an application of [M3]. So there must be ground atoms $p_1(t_1), \ldots, p_k(t_k) \in \mathbf{B}_{M_P}$ such that
(1) $clause(p(t), p_1(t_1)\& \ldots \& p_n(t_n)) \in T_{V_P} \uparrow 1$;
(2) $demo(p_1(t_1)\& \ldots \& p_n(t_n)) \in T_{V_P} \uparrow (n-1)$.
From the definition of $\mathbf{V}_P$ it follows that $p_1, \ldots, p_k \in P_{\mathcal{L}_P}$, and that there are P-terms $s_1, \ldots, s_k$, and an $\mathbf{V}_P$-ground substitution θ such that
(3) $(p(t) \leftarrow p_1(s_1), \ldots, p_n(s_n)) \cdot \theta = p(t) \leftarrow p_1(t_1), \ldots, p_n(t_n) \in \mathbf{V}_P\text{-}ground(P)$. Now take a P-ground substitution σ such that $dom(\sigma) = ran(\mathbf{abs}(\theta))$. By Lemma 24,
$$p(t) \leftarrow p_1(s_1\mathbf{abs}(\theta)\sigma), \ldots, p_n(s_n\mathbf{abs}(\theta)\sigma) \in ground(P). \quad [a]$$
From (2), Corollary 29 and the inductive hypothesis, we can conclude that
$$\exists m_i \in \mathbf{N}.\ p_i(\mathbf{abs}(t_i)\sigma) \in T_P \uparrow m_i,\ \text{for } i \in [1, k]. \quad [b]$$
Now from [a] and [b], we can conclude that there is an m such that $p(t) \in T_P \uparrow m$. □

Lemma 31 (Completeness) *Let P be a definite program and let $\mathbf{V}_P$ be the associated Vanilla meta-interpreter. Then for all $p(t) \in \mathbf{B}_P$,*
$$\forall n \in \mathbf{N}\ [p(t) \in T_P \uparrow n \Longrightarrow \exists m \in \mathbf{N}.\ demo(p(t)) \in T_{V_P} \uparrow m].$$

Proof: By course of values induction on n. The lemma trivially holds for $n = 0$.

Suppose $p(t) \in T_P \uparrow 1$. Then $clause(p(t), empty) \leftarrow \in ground(\mathbf{V}_P)$ and therefore, $clause(p(t), empty) \in T_{V_P} \uparrow 1$. Also, $demo(empty) \in T_{V_P} \uparrow 1$, so, by [M3], $demo(p(t)) \in T_{V_P} \uparrow 2$. Suppose that $p(t) \in T_P \uparrow n \setminus T_P \uparrow (n-1)$, for some $n > 1$. And suppose that the lemma holds for all $n' < n$. Then there are $p_1(s_1), \ldots p_k(s_k) \in \mathbf{B}_P$, such that $p(t) \leftarrow p_1(s_1), \ldots p_k(s_k) \in ground(P)$, and $p_i(s_i) \in T_P \uparrow (n-1)$ for $i \in [1, k]$. Thus, by the inductive hypothesis, there is an m such that $demo(p_i(s_i)) \in T_{V_P} \uparrow m$, for $i \in [1, k]$. Thus, by repeated use of [M3], there is an m' such that $demo(p_1(s_1)\& \ldots \& p_k(s_k)) \in T_{V_P} \uparrow m$. Also, $clause(p(t), p_1(s_1)\& \ldots \& p_k(s_k)) \leftarrow \in ground(\mathbf{V}_P)$. As a consequence, $clause(p(t), p_1(s_1)\& \ldots \& p_k(s_k)) \in T_{V_P} \uparrow m'$.
We conclude that $demo(p(t)) \in T_{V_P} \uparrow (m' + 1)$. □

Theorem 32 (Correctness) *Let P be a definite program and $\mathbf{V}_P$ the Vanilla meta-interpreter associated with P. Then for all $p(t) \in \mathbf{B}_P$,*
$$p(t) \in \mathbf{M}_P \text{ iff } demo(p(t)) \in \mathbf{M}_{V_P}.$$

As an immediate consequence of the correctness theorem and Lemma 28 is the following corollary:

Corollary 33 *Let $demo(p(t)) \in \mathbf{M}_{V_P}$, where p is a predicate in $\mathcal{L}_P$. Let σ be an $\mathcal{L}_P$-ground substitution with $dom(\sigma) = var(\mathbf{abs}(t))$. Then $p(\mathbf{abs}(t) \cdot \sigma) \in \mathbf{M}_P$.*

The above corollary shows that the occurrence of unrelated atoms in the least Herbrand model of the Vanilla meta-program is less of a problem than usually thought: the meta-level terms occurring in unrelated atoms can be interpreted as free variables, ranging over the object level terms.

It should be observed that, by Lemma 25, the proof of the correctness theorem above is also a proof of the correctness theorem for the Vanilla meta-interpreter with a functional representation of the object program. The above corollary likewise translates into the functional case.

The above correctness theorem should be compared with the corresponding result for the class of language independent programs (Theorem 13 of Martens and De Schreye [MS95a].) The latter is stronger, in the sense that it additionally shows that for language independent programs, no unrelated atoms can occur in the least Herbrand model of the meta-program.

In the above, we have silently assumed that the language underlying the object program has a non-empty set of constants $C_{\mathcal{L}_P}$. Adopting the usual convention that, in case $C_{\mathcal{L}_P}$ is empty, the Herbrand universe is built up using a generic constant $*$, the following special case of the correctness theorem holds: for all $p \in R_{\mathcal{L}_P}$,

$$p(*) \in \mathbf{M}_P \text{ iff, for all } t \in \text{CLTERM}_{\mathcal{L}_P}, demo(p(t)) \mathbf{M}_{V_P}.$$

This follows easily from adaptations of the proofs of the lemma's 30 and 31.

The above completeness proof provides evidence of the linear overhead of meta-level computation. In particular, an explicit measure for this overhead (on the declarative level) can be extracted from the proof. We need the following definitions.

Definition 34 *Let P be a program consisting of the clauses $C_1, \ldots, C_k$.*
The body depth of $d(C)$ of a clause C is the number of its body atoms, i.e.,
$d(C) = n$ if C is a clause $A \leftarrow B_1, \ldots, B_n$, and
$d(C) = 0$ if C is a fact clause $A \leftarrow$.
The body depth d_P of P is the maximum of the depths of the clauses of P, i.e.,
$d_P = max\{d(C_i) : i \in [1, k]\}$. □

Proposition 35 *Let P be a program.*
For $n > 0$, if $p(t) \in T_P \uparrow n$, then $demo(p(t)) \in T_{V_P} \uparrow (2 + d_P(n-1))$.

Proof: The proof proceeds by induction, and follows the proof of Lemma 31. Clearly, if $p(t) \in T_P \uparrow 1$, the $demo(p(t)) \in T_{V_P} \uparrow 2$, so the proposition holds for $n = 1$. Let $n > 1$, and suppose the proposition holds for $n - 1$. Let $p(t) \in T_P \uparrow n \setminus T_P \uparrow (n-1)$.

Then there are $p_1(s_1), \ldots p_k(s_k) \in \mathbf{B}_P$, such that $p(t) \leftarrow p_1(s_1), \ldots p_k(s_k) \in ground(P)$, and $p_i(s_i) \in \mathrm{T}_P\uparrow(n-1)$ for $i \in [1,k]$. By the inductive hypothesis, $demo(p_i(s_i)) \in \mathrm{T}_{V_P}\uparrow(2+d_P(n-2))$, for $i \in [1,k]$. This implies, by $k-1$ applications of [M3], that $demo(p_1(s_1)\&\ldots\&p_k(s_k)) \in \mathrm{T}_{V_P}\uparrow(2+d_P(n-2)+k-1)$. Then, by an application of [M2], $demo(p(t)) \in \mathrm{T}_{V_P}\uparrow(2+d_P(n-2)+k)$. Observe that $k \leq d_P$. The proposition now follows from the monotonicity of the fixpoint operator. □

A similar linear relation between the length of object level and meta level SLD-derivations can be established. We refer for this result to Levi and Ramundo [LR93].

1.6 An Amalgamation

The correctness of the Vanilla meta-interpreter in fact shows that it proves no more nor less than the object program. It is extensions of the Vanilla meta-interpreter where the real interest lies. The theory developed in the present paper can be used to obtain correctness results for those extensions. We mention the following preliminary result on a simple amalgamation, the textual combination of P with $\mathbf{V}_P$.

Theorem 36
For all (ground) A in $\mathcal{L}_{V_P}$, $demo(A) \in \mathbf{M}_{V_P}$ iff $demo(A) \in \mathbf{M}_{V_P \cup P}$;
For all (ground) A in $\mathcal{L}_P$, $A \in \mathbf{M}_P$ iff $A \in \mathbf{M}_{V_P \cup P}$.
Proof: Along the same lines as the proof of Theorem 32, while using the following observations: The underlying language of $\mathbf{V}_P \cup P$ is $\mathcal{L}_{V_P}$; Clauses of P can now be instantiated with ground terms from $\mathcal{L}_{V_P}$; Lemma 28 also holds for atoms $p(t) \in \mathbf{M}_{V_P}$ for which $p \in P_{\mathcal{L}_P}$; Atoms from the bodies of clauses of $\mathbf{V}_P$ do not unify with heads of clauses from P—and the converse. □

Alternatively, another proof of the above theorem is obtained by examining properties of the fixpoint operator $T_{V_P \cup P}$ of the amalgamated program. As observed in Brogi and Turini [BT95], it is in general not true that the least Herbrand model of the textual combination $P \cup Q$ of two programs P and Q is the union of their respective least Herbrand models. However, for certain pairs of programs, the least Herbrand model of their union does equal the union of the respective least Herbrand models w.r.t. the union of the languages. We will use the following notion:

Definition 37 Two programs P and Q are *non-connected* if
(i) none of the atoms occurring in the bodies of clauses of $\mathcal{L}_{P \cup Q} - ground(P)$ occurs as the head of a clause in $\mathcal{L}_{P \cup Q} - ground(Q)$, and
(ii) none of the atoms occurring in the bodies of clauses of $\mathcal{L}_{P \cup Q} - ground(Q)$ occurs as the head of a clause in $\mathcal{L}_{P \cup Q} - ground(P)$. □

We need the following generalisation of the notion of T_P-operator and its fixed point.

Definition 38 Let P be a program, and let $\mathcal{L} \supseteq \mathcal{L}_P$.
$T_{P,\mathcal{L}}(I) := \{A : A \leftarrow B_1, \ldots, B_n \in \mathcal{L} - ground(P) \text{ and } \{B_1, \ldots, B_n\} \subseteq I\}$
$T_{P,\mathcal{L}}^\omega(\emptyset) := \bigcup T_{P,\mathcal{L}}^n(\emptyset)$. □

In addition, we can, as usual, identify the least $\mathcal{L}$-Herbrand model $\mathbf{M}_P^\mathcal{L}$ with $T_{P,\mathcal{L}}^\omega(\emptyset)$.

Lemma 39 Let P and Q be two non-connected programs, and let $\mathcal{L}$ be the union of their respective underlying languages. Then $\mathbf{M}_{P \cup Q} = \mathbf{M}_P^\mathcal{L} \cup \mathbf{M}_Q^\mathcal{L}$.

Proof: By induction on the stages of the $T_{P \cup Q}$-operator.
$T_{P \cup Q} \uparrow 0 = T_{P,\mathcal{L}} \uparrow 0 \cup T_{Q,\mathcal{L}} \uparrow 0 = \emptyset$, by definition.
Suppose that for some n, $T_{P \cup Q} \uparrow n = T_{P,\mathcal{L}} \uparrow n \cup T_{Q,\mathcal{L}} \uparrow n$. Then

$T_{P \cup Q} \uparrow n+1 =$
{ by definition}
$= T_{P \cup Q}(T_{P \cup Q} \uparrow n)$
{by the inductive hypothesis}
$= T_{P \cup Q}(T_{P,\mathcal{L}} \uparrow n \cup T_{Q,\mathcal{L}} \uparrow n)$
{ by definition of $T_{P \cup Q}$}
$= \{A : A \leftarrow B_1, \ldots, B_k \in ground(P \cup Q) \& \{B_1, \ldots, B_k\} \subseteq T_{P,\mathcal{L}} \uparrow n \cup T_{Q,\mathcal{L}} \uparrow n\}$
{because $ground(P \cup Q) = \mathcal{L} - ground(P) \cup \mathcal{L} - ground(Q)$}
$= \{A : A \leftarrow B_1, \ldots, B_k \in ground(P) \& \{B_1, \ldots, B_k\} \subseteq T_{P,\mathcal{L}} \uparrow n \cup T_{Q,\mathcal{L}} \uparrow n\}$
$\bigcup \{A : A \leftarrow B_1, \ldots, B_k \in ground(Q) \& \{B_1, \ldots, B_k\} \subseteq T_{P,\mathcal{L}} \uparrow n \cup T_{Q,\mathcal{L}} \uparrow n\}$
{because P and Q are non-connected }
$= \{A : A \leftarrow B_1, \ldots, B_k \in ground(P) \& \{B_1, \ldots, B_k\} \subseteq T_{P,\mathcal{L}} \uparrow n\}$
$\bigcup \{A : A \leftarrow B_1, \ldots, B_k \in ground(Q) \& \{B_1, \ldots, B_k\} \subseteq T_{Q,\mathcal{L}} \uparrow n\}$
{by definition of $T_{P,\mathcal{L}} \uparrow n+1$ and $T_{Q,\mathcal{L}} \uparrow n+1$}
$= T_{P,\mathcal{L}} \uparrow (n+1) \cup T_{Q,\mathcal{L}} \uparrow (n+1)$

Thus,
$\mathbf{M}_{P \cup Q} = T_{P \cup Q}^\omega(\emptyset) = \bigcup T_{P \cup Q} \uparrow n =$
$= \bigcup (T_{P,\mathcal{L}} \uparrow n \cup T_{Q,\mathcal{L}} \uparrow n) = \bigcup T_{P,\mathcal{L}} \uparrow n \cup \bigcup T_{Q,\mathcal{L}} \uparrow n = T_{P,\mathcal{L}}^\omega \cup T_{Q,\mathcal{L}}^\omega = \mathbf{M}_P^\mathcal{L} \cup \mathbf{M}_Q^\mathcal{L}$. □

The above general lemma in particular applies to the amalgamation of object and meta-program.

Theorem 40 $\mathbf{M}_{V_P \cup P} = \mathbf{M}_{V_P} \cup \mathbf{M}_P^{\mathcal{L}_{V_P}}$.

Proof: We leave if to the reader to check that $\mathbf{V}_P$ and P are indeed non-connected. (Essential here is the assumption we have made throughout the paper on the language of the object paper, which implies that $\{demo(\cdot), clause(\cdot, \cdot)\} \cap \mathcal{R}_P = \emptyset$.) By the fact that $\mathcal{L}_{V_P} \supseteq \mathcal{L}_P$, the theorem is now immediately follows from the above lemma. □

As an immediate consequence of the above theorem and the completeness theorem 32, we have the following corollary.

Corollary 41 *Let A be a closed term of $\mathcal{L}_P$. The following are equivalent:*
- (i) $demo(A) \in \mathbf{M}_{V_P \cup P}$
- (ii) $demo(A) \in \mathbf{M}_{V_P}$
- (iii) $A \in \mathbf{M}_P$
- (iv) $A \in \mathbf{M}_{V_P \cup P}$.

Other examples involving amalgamated and non-amalgamated extensions of the Vanilla meta-interpreter can be found in Martens and De Schreye [MS95a]. A theorem similar to the above (Theorem 15) is proven there for language independent object programs P.

1.7 Conclusions

In the present paper we have studied correctness of the untyped Vanilla meta-interpreter in the context of ambivalent syntax. Ambivalent syntax, characterised by the occurrence of atoms as terms, is the proper underlying syntax of amalgamated extensions of the Vanilla meta-interpreter. Also, Prolog uses ambivalent syntax rather than the standard syntax for first order logic. We have defined ambivalent syntax with several levels of ambivalence, and we showed that first order logic with ambivalent syntax is sound and strongly complete with respect to a suitable semantics.

We have shown that the proof of procedural correctness for the typed version of the Vanilla meta-interpreter for normal programs has great generality, and implies correctness for the (standard) untyped Vanilla meta-interpreter for normal programs. In contrast, the declarative correctness result for the typed Vanilla meta-interpreter does not translate into declarative correctness for the untyped version(s), due to the essential role of types. Declarative correctness for the untyped Vanilla meta-interpreter has been proven for a limited class of normal programs, which excludes most of the basic Prolog programs. For definite object programs however, a satisfying declarative correctness result can be obtained in two ways. By the usual soundness and completeness of SLD-resolution, declarative correctness for definite object programs is a consequence of the procedural correctness. We have presented a direct proof of declarative correctness for the Vanilla meta-interpreter for definite object programs with ambivalent syntax as the underlying language. The proof generalises to a corresponding proof of the correctness of the variant of the Vanilla meta-interpreter that uses a functional representation of the object program. Finally, we have presented two different proofs of the correctness of a simple amalgamation of object and meta-program.

The interest of the reported results is two-fold. First, we have shown that various correctness results (procedural correctness of the typed interpreter and declarative correctness of the untyped interpreter using ambivalent syntax) have a great generality. Second, we have shown that ambivalent syntax is useful and appropriate in the context of meta-programming. The strong similarities between the various correctness proofs suggest further research, investigating necessary and sufficient conditions for soundness and completeness of Vanilla style demo-predicates. In addition, a further investigation of sufficient conditions for soundness and completeness of extended meta-interpreters may be of some practical importance.

Acknowledgements

I want to thank Frank van Harmelen for many pleasant discussions and his careful reading of various drafts of this paper. The critical opposition of Pat Hill and John Lloyd has been very stimulating. In addition, comments from and discussions with Krzysztof Apt, Johan van Benthem, Dov Gabbay, Bern Martens, Danny De Schreye, and two anonymous referees were very helpful.
The author was supported by the Dutch Organisation for Scientific Research (N.W.O.).

References

[BT95]   A. Brogi and F. Turini. Meta-logic for program composition: semantics issues. This volume.

[CKW93]   W. Chen, M. Kifer, and D.S. Warren. Hilog: A foundation for higher-order logic programming. *Journal of Logic Programming*, 15(3):187–230, 1993.

[FLMP93]   M. Falaschi, G. Levi, M. Martelli, and C. Palamidessi. A model-theoretic reconstruction of the operational semantics of logic programs. *Information and Computation*, 102(1):86–113, 1993.

[Gab92]   D. Gabbay. Metalevel features in the object level: modal and temporal logic programming III. In L.Farinas del Cerro and M. Penttonen, editors, *Intensional logics for programming*, pages 85–124. Clarendon Press, 1992.

[HL88]   P. M. Hill and J. W. Lloyd. Analysis of meta-programs. In H.D. Abramson and M.H. Rogers, editors, *Proceedings of the Meta88 Workshop*, pages 23–52. MIT Press, 1988.

[Kal93]   M. Kalsbeek. The Vanilla meta-interpreter for definite logic programs and ambivalent syntax. Technical Report CT-93-01, Department of Mathematics and Computer Science, University of Amsterdam, The Netherlands, 1993.

[KJ95]   M. Kalsbeek and Y. Jiang. A vademecum of ambivalent logic. This volume.

[KK91]   R.A. Kowalski and J. Kim. A metalogic programming approach to multi-agent knowledge and belief. In V. Lifschitz, editor, *Artificial Intelligence and Mathematical Theory of Computation*, pages 231–246. Academic Press, 1991.

[Llo87]   J. W. Lloyd. *Foundations of Logic Programming*. Springer-Verlag, Berlin, second edition, 1987.

[LR93]   G. Levi and D. Ramundo. A formalization of metaprogramming for real. In D. S. Warren, editor, *Proceedings ICLP '93*, pages 354–373. MIT Press, 1993.

[MS95a] B. Martens and D. De Schreye. Two semantics for definite meta-programs, using the non-ground representation. This volume.

[MS95]   B. Martens and D. De Schreye. Why untyped non-ground meta-programming is not (much of) a problem. *Journal of Logic Programming*, 22(1):47–99, 1995.

[Ric74] B. Richards. A point of reference. *Synthese*, 28:431–445, 1974.

[SS86]   L. Sterling and E. Shapiro. *The Art of Prolog*. MIT Press, 1986.

2 A Vademecum of Ambivalent Logic

Marianne Kalsbeek and Yuejun Jiang

Abstract

Ambivalent Logic AL, first introduced in its full sense in Jiang [Jia94] is obtained from first order predicate logic FOL by relaxing several restrictions on its usual syntax. In particular, the usual distinctions between predicates, functions, formulas and terms are not made in AL. We show that Ambivalent Logic provides a general and flexible framework for various ambivalent syntactic phenomena that occur in Prolog, in meta-logic programming and in formalisations of knowledge and belief. A series of formal results justifies the use of syntaxes with ambivalent phenomena in these areas. We discuss a closed term semantics for AL, and show that the standard derivational calculus for first order predicate logic is sound and complete w.r.t. this semantics. A conservativity result shows that AL should be considered as a conservative extension of FOL. We define a version of the Martelli Montanari unification algorithm for AL, and show it has the usual properties. In combination with various other basic proof theoretic results, this shows that resolution is a complete and sound inference method for AL. We also discuss the relation with Hilog.

2.1 Introduction

A (mostly tacit) assumption in the syntax for first order predicate logic (FOL) is that the sets of predicates, functions, and constants have mutually empty intersections. As a consequence, the syntactic categories of formulas and terms are mutually exclusive, as the Unique Reading Lemma for FOL witnesses. While Logic Programming is in principle based on FOL, in various of its application areas, syntaxes are used which do not satisfy this property of the usual syntax for FOL.

A principal example, analysed in Apt and Teusink [AT95], is the syntax underlying Prolog. Prolog's meta-variable facility allows for variables to occur both in terms and in atoms positions. Two well-known examples are the cut-fail definition of negation: $neg(X)\text{:-}X,!,fail$, $neg(X)\text{:-}$ and the definition of the solve-predicate as $solve(X)\text{:-}X$. Clearly, this use of variables is not allowed in the standard syntax of FOL. Additionally, queries which involve instantiations of the heads of the above clauses with terms $p(t)$, presuppose a syntax which allows for function symbols to be accepted as predicate symbols. Conversely, an inductive definition of a predicate involving its negation presupposes that Prolog's syntax allows for predicate symbols to be accepted as function symbols. Also, in Prolog's syntax, predicates and functions do not have fixed arities.

Another example of the use of a deviant syntax in (meta-)Logic Programming practice is the untyped, non-ground Vanilla meta-interpreter which uses identity naming for atoms and terms of the object level language (cf. Kalsbeek [Kal95]). While in this simple, unamalgamated case the (partial) correctness of Vanilla could still be guaranteed by the possibility of renaming the object level predicates to meta-level function symbols, this solution does not generalise to various interesting extensions of Vanilla. For example, extending Vanilla with reflective clauses like $demo(X) :- X$ or instances of these clauses, eliminates the possibility of renaming and introduces ambivalence between function and predicate symbols and variables and terms, similar to the ambivalences observed above in the case of Prolog's syntax. A related example is the formalisation of the Three Wise Man problem as an extension of Vanilla in Kowalski and Kim [KK91]. This formalisation involves clauses of the form $demo(wise0, not\ demo(wise1, white1)) :-$, in which the predicates $demo$ and not occur in function positions.

The above examples require only limited forms of ambivalence: atoms-as-terms ambivalence (atoms occurring in term positions) and terms-as-atoms ambivalence (terms occurring in atom positions). As observed in Chen et al. [CKW93] and Jiang [Jia94], more advanced ambivalent features are required to obtain a syntax which allows for efficient formalisations of generic predicates. An example is the generic closure predicate $(cl(Z))(X, Y)$, which, given any binary predicate R, returns its transitive closure $cl(R)$. Its definition, $((cl(Z))(X, Y) \leftrightarrow Z(X, Y) \vee (Z(X, V) \wedge (cl(Z))(V, Y)))$, presupposes a syntax which allows for predicates to occur in term positions, and for variables and atoms to occur in predicate positions. Similar ambivalent phenomena occur in languages which allow for data retrieval and schema browsing. In databases, such forms of syntactic ambivalence are a desirable option, allowing caching of data.

Other areas where ambivalent phenomena occur, are formalisations of knowledge and belief. While the use of a real naming function, avoiding syntactic ambivalence between terms and formulas, is often possible, it is not always desirable. Also, the above atoms-terms ambivalence is not always sufficient. In the predicate case, it is desirable that a reflective axiom like $x \to K(x)$ can be instantiated with all formulas, not just atomic formulas. This then supposes a syntax in which all formulas, and not just atomic formulas, are allowed to occur as terms. While the quantifiers could, in principle, be represented by functions, the effect of a functional representation does not in all cases give the desired effect. The obvious reason is that quantifiers, in contrast to functions, bind variables. For example, consider the formula $\forall x.bel(John, (friend(John, x) \to exists(y, loves(x, y))))$, which expresses the proposition 'John believes about all his friends that they are loved by somebody'. Instantiation of both x and y with the same constant is possible, yielding unintended statements like

$$bel(John, (friend(John, Mary) \to exists(Mary, loves(Mary, Mary)))).$$

In contrast, the use of a quantifier representation for the same proposition yields the formula $\forall x.bel(John, (friend(John, x) \to \exists y.loves(x, y)))$. Here y is bound by the existential quantifier, and cannot be instantiated. Thus, unintended instantiations like the above are not possible.

In the present paper, we discuss Ambivalent Logic AL as a general framework for first order logic with a syntax in which all of the above mentioned ambivalent phenomena occur. AL has a fully ambivalent syntax, in which the usual distinctions between the syntactic categories of terms, functions, predicates, and formulas, cannot be made. While the part of Unique Reading that says that a well-formed string in the language is either a formula or a term, but not both, does not hold for AL, these syntactic distinctions do retain their usual contextual meaning.

AL has a standard (first order predicate logic) derivational calculus, and a (first order) closed term semantics. We develop some basic proof theory for AL, including soundness and completeness of the derivational calculus w.r.t. the semantics, an s-equivalence theorem, and Herbrand's theorem. We discuss unification for AL. In particular, we show how the Martelli-Montanari unification algorithm can be adapted for AL. We prove that the appropriate AL version has the usual properties. In combination with various other results discussed in this paper, this shows that resolution is a complete and sound inference method for AL. In addition, we show that AL is a conservative extension of FOL. We also show how various known ambivalent syntaxes (such as Prolog's syntax and the syntax discussed in Kalsbeek [Kal95]) can be obtained as special instances of the full ambivalent syntax of AL. The proof theoretic results we obtain for AL also hold for the various specialisations of AL. These results justify the use of AL and various of its subsystems as a basis for (meta-) logic programming. In addition, we argue that AL provides an interesting framework for formalisations of knowledge and belief.

AL was first introduced in Jiang [Jia94] and [Jia94b]. Some of the results we present are improvements of results announced in Jiang [Jia94]. A proper subsystem of AL, appropriate for amalgamated extensions of the Vanilaa meta-interpreter, was introduced in Kalsbeek [Kal95].

There are various examples of logics which, like AL, have a syntax incorporating ambivalent phenomena. Recently, Hilog was proposed by Chen et al. [CKW93] as a basis for higher order logic programming. Both AL and Hilog combine a second order syntax with a first order semantics. While the syntax of AL extends Hilog syntax, Hilog is, in contrast to AL, not an extension of FOL. Another example is the logic proposed by

Richards [Ric74], which is intended for formalisations of intensional logics. For a comparison between AL and Richards logic, we refer to Jiang [Jia94]. The reader is also referred to Gabbay [Gab92], where other flexible meta-languages are proposed.

The outline of this paper is as follows. In Section 2.2, we introduce the fully ambivalent syntax of AL and we show how specialisations of it may be obtained. In Section 2.3 we develop a closed term semantics for AL. In Section 2.4 we discuss how equality can be incorporated in AL. Section 2.5 is devoted to various basic proof theoretic results for AL. In Section 2.6 we discuss the appropriate version of the Martelli-Montanari unification algorithm for AL. In Section 2.7 we discuss the relations between Ambivalent Logic and Hilog.

2.2 Syntax for Ambivalent Logic

We define in Section 2.2.1, a syntax which is fully ambivalent, that is, in which every well formed expression can act as a formula, as a term, as a function, and as a predicate. Whether an expression is evaluated as a term, a formula, a function, or a predicate will be determined by the context. We allow for free arity of predicates and functions. This syntax generates a multi-purpose language.

The full ambivalent syntax extends the syntax for the Vanilla meta-interpreter (cf. Kalsbeek [Kal95]), in which can occur in term positions but not vice versa, and in which predicates and functions are always symbols with a fixed arity. It also extends Prolog's syntax (cf. Apt and Teusink [AT95]), which shares with full ambivalent syntax the full atoms-terms ambivalence, and the free arity of functions and predicates, but in which predicates and functions are always parameters.

To obtain versions of ambivalent syntax which generate languages that are adapted to a particular purposes such as the above, the definitions for full ambivalent syntax can be adapted and specialised. In Section 2.2.2 we discuss several of these refinements.

2.2.1 Full ambivalent syntax

A fully ambivalent language $L_\mathcal{G}$ is generated by a set of non-logical constants (parameters) $\mathcal{G}$, an infinite set of variables $x, y, z, \ldots$, and the usual logical connectives and quantifiers of first-order logic.

Definition 1 (Expressions)
 1. The variables $x, y, z, \ldots$ and individual constants $a, b, c, \ldots \in \mathcal{G}$ are expressions.
 2. If $t, t_1, \ldots, t_n$ are expressions, then $(t)(t_1, \ldots, t_n)$ is an expression.

3. If A and B are expressions, then so are $\neg A$, $A \wedge B$, $A \vee B$, $A \leftarrow B$, $A \rightarrow B$ and $A \leftrightarrow B$.
4. If A is an expression and x is a variable, then $\forall x(A)$ and $\exists x(A)$ are expressions. □

The above ambivalent syntax thus extends standard syntax for first order predicate logic in several ways. The usual requirement is dropped which states that the sets of predicate symbols and function symbols are disjoint. In addition, the usual requirement is dropped that predicates and functions have a fixed arity. As an example, $p(p(p,p))$ is a well-formed ambivalent expression. In addition, variables may occur in formula positions. As an example, the Prolog clause $solve(x) \leftarrow x$ is a well-formed expression in full ambivalent syntax. Also, 'second order' quantification is allowed. That is, expressions like $\exists x.x(c)$ are well-formed in full ambivalent syntax. (We will see in Section 2.3 that, semantically, quantification over predicates will not be interpreted as second order quantification.) Moreover, not only parameters, but also more complex expressions are allowed as predicates and functions. An example is the generic closure predicate discussed in the introduction. This feature allows for the formation of new predicates, which can be useful in databases. Other examples are: $p(x) \wedge q(x) \rightarrow (p \wedge q)(x)$ and $\forall x \forall z(p(x,z) \rightarrow (\exists y.p(y))(x))$. The latter example shows that quantified expressions are allowed in predicate places.

In many cases, we will omit brackets if this does not lead to confusion. For example, we will write $c(x)$ instead of $(c)(x)$, and $\forall x p(x)$ (or also $\forall x.p(x)$) instead of $\forall x(p(x))$. In various cases, however, brackets cannot be omitted without altering an expression. For instance, $(p \vee q)(t,s)$ is an atomic expression, while $p \vee q(t,s)$ is a disjuctive expression. Similarly, we distinguish between the expressions $p((x)(a))$ and $(p(x))(a)$. In the former, the symbol p predicates over the expression (x)(a), while in the latter, $p(x)$ predicates over a.

Definition 2 (Atomic expressions)
An *atomic expression* is either a constant, or a variable, or an expression of the form $(t)(t_1, \ldots, t_n)$.
Atomic expressions of the form $(t)(t_1, \ldots, t_n)$ are *functional atoms*. □

The set $FV(t)$ of free variables of an expression t is defined analogous to the definition for standard syntax, except that it is additionally defined for expressions in predicate and function places.

Definition 3 (Free Variables) Let x be a variable, $c \in \mathcal{G}$, and let A, B, and $(t)(t_1, \ldots, t_n)$ be expressions. The set of free variables occurring in an expression is defined as follows:

$$FV(x) = x$$
$$FV(c) = \emptyset$$

$$FV((t)(t_1,\ldots,t_n)) = FV(t) \cup FV(t_1) \cup .. \cup FV(t_n)$$
$$FV(A \wedge B) = FV(A) \cup FV(B)$$
$$FV(\neg A) = FV(A)$$
$$FV(\exists x A) = FV(A) \setminus \{x\}$$

Similarly for the other binary connectives and the universal quantifier.

A variable x occurring in an expression t is *bound* in t if $x \notin FV(t)$; otherwise we say x occurs free in t. We write $A\{x/t\}$ to denote the result of replacing every free occurrence of x in A by t. □

Definition 4 An expression t is *closed* (or a *sentence*) if $FV(t) = \emptyset$

By Definition 3, x occurs free in $(p(x))(a)$. Also, x is not free in $p(\exists x p(x))$ because x is not free in the argument $\exists x p(x)$. In this sense, quantifiers in term-positions behave like ordinary quantifiers.

The above definition can be refined in a standard way to distinguish between various occurrences. For example, in $p(x) \vee \exists x f(x)$, the first occurrence of x is free, while the second occurrence is bound by the existential quantifier. The scope of quantifiers can be defined in the usual way. For example, in $\exists x q(p(x) \wedge \exists x f(x))$, the first occurrence of x is bound by the outermost existential quantifier, while the second occurrence is bound by the rightmost quantifier.

In Section 2.3 we will need the following standard notion.

Definition 5 An expression t is *free for* a variable x in an expression A iff t does not contain any free variable that is bound by some quantifier in A when every free occurrence of x in A is replaced by t. □

Due to the nature of full ambivalent syntax, the role of an expression can be determined only by the context of the expression. For example, consider the expression $(q(d))(a, q(a)) \wedge q(c)$. It can be evaluated both as a term and as formula, depending on the context. In the latter case, q(d) serves as a predicate, q(a) as a term, and q(c) as a formula.

Clearly, the Unique Reading Lemma does not hold for ambivalent syntax, as every AL expression can be both a formula and a term. We can, however, define in some cases which role a subexpression assumes in the context of an expression in which it appears. In some cases, it is trivial to determine the contextual role of subexpressions. For example, $a \vee b$ occurs as a term in $a(a \vee b)$, independent of whether the latter is evaluated as a term or as a formula. But we have the choice whether or not to consider a as a term in $c(a \vee b)$. The choice we make is inspired by the semantics we define in the next section. In this semantics, there will be no connection between the interpretation of the 'term' a and the interpretation of the 'term' $a \vee b$. Therefore, we will not consider a as a subterm in $c(a \vee b)$. Similar considerations lead to the following definitions of the notions of subformula, term, function, and predicate. The definition of the notion of subformula is standard.

Definition 6 (Subformula)
1. Every expression is a subformula of itself.
2. Let A be an expression. Then every subformula of A is a subformula of $\neg A$, $\exists x A$, and $\forall x A$.
3. Let A and B be expressions. If E is a subformula of A or B, then E is a subformula of $A \vee B$, $A \wedge B$, $A \to B$, $A \leftarrow B$, and $A \leftrightarrow B$. □

By this definition, a does not occur as a subformula in $c(a \vee b)$.

Definition 7 (Occurrence as a term)
1. In an expression of the form $(t)(t_1, \ldots, t_n)$, $t_1, \ldots, t_n$ occur as terms.
2. If t occurs as a term in A and A is a subformula of F, then t occurs as a term in F.
3. If t occurs as a term in s and s occurs as a term in F, then t occurs as a term in F. □

By the above definition, a does not occur as a term in $c(a \vee b)$, while $a \vee b$ does. Also, c does not occur as a term in $c(a \vee b)$.

Definition 8 (Occurrence as a function)
1. t occurs as a function in an expression $(t)(t_1, \ldots, t_n)$.
2. If t occurs as a function in s and s occurs as a term in F, then t occurs as a function in F. □

Definition 9 (Occurrence as a predicate)
1. t occurs as a predicate in an expression $(t)(t_1, \ldots, t_n)$.
2. If t occurs as a predicate in A and A is a subformula of F, then t occurs as a predicate in F. □

By the above definitions, t occurs both as a predicate and as a function in $(t)(t_1, \ldots, t_n)$. In $\forall x(t(x))$, t occurs as a predicate, but not as a function. In $p(\forall x(t(x)))$, t occurs neither as a predicate nor as a function.

All of the above definitions can be refined to distinguish between the various occurrences. For example, in $a(a) \vee a$, the first occurrence of a is as a predicate (but not as a function), the second occurrence is as a term, while the third occurrence is as a subformula.

It is useful to make the following distinction between two kinds of occurrences of quantifiers. The first kind of quantifier, which we will call 'outside quantifier', occurs in places where they are also allowed in FOL formulas. The second kind occurs only in 'ambivalent' places.

Definition 10 Let $Qx.s$, where Q is a quantifier $\forall$ or $\exists$, be an occurrence of a subexpression of an AL expression t. Q is an *outside quantifier* if it is an occurrence as a subformula of t. Otherwise, Q is an *inside quantifier*.
Inside and outside connectives are defined similarly. □

For example, in $\forall x.(x \vee \exists y.f(y))$, the first quantifier is an outside quantifier, while the second is an inside quantifier. Observe that both inside and outside quantifiers do bind variables in their scope. In Section 2.3, where we discuss semantics for AL, we will see that the outside quantifiers will be interpreted as real quantifiers, ranging over elements of the domain; this in contrast to inside quantifiers.

2.2.2 Refinements

For many purposes, the full ambivalent syntax defined in the previous subsection admits too many expressions. Refined versions of ambivalent syntax, well-tuned to particular domains of application, can be obtained by specialising one or more of the clauses in Definition 1, and by the use of sets of special constants in addition to the set of generating constants. In defining special versions of full ambivalent syntax, it is sometimes useful to introduce new syntactic categories.

We will give some examples of specialisation and the use of special constant sets.

- A version of ambivalent syntax which does not admit all its expressions to occur as predicates and functions, but only its constants, can be obtained by modifying clause (2) of Definition 1 as follows: If $t_1, \ldots, t_n$ are expressions and $c \in \mathcal{G}$, then $c(t_1, \ldots, t_n)$ is an expression. In this particular version, variables and more complex expressions do not occur as predicates and functions, and as a result there is no 'second order' quantification.

- It may be useful to select special roles for some of the parameters. As an example, in some domains, it may be useful to have one or more symbols that occur only as predicates. The negation predicate in Prolog is an example of such a special predicate (see below). In that case, a set of special symbols $\mathcal{S}$ is added to the signature of the language, and an extra syntactic category is introduced: special expressions. The following modification of Definition 1 is used:

    1 Variables x and parameters $c \in \mathcal{G}$ are expressions.

    2 If $t, t_1, \ldots, t_n$ are expressions, then $(t)(t_1, \ldots, t_n)$ is an expression.

    2' If $p \in \mathcal{S}$, and $t_1, \ldots, t_n$ are expressions, then $p(t_1, \ldots, t_n)$ is a special expression.

    3' If A is an expression, then $\neg A$ is an expression; If A is a special expression, then $\neg A$ is a special expression.

    $A \vee B$ is a special expression if both A and B are special expressions or if one

among them is a special expression and the other is an expression. $A \vee B$ is an expression if both A and B are expressions.
Similarly for the other binary connectives.

4' $\forall x(A)$ is an expression if A is an expression; it is a special expression if A is a special expression.

In particular, special expressions do not occur as terms, predicates, or functions, in expressions and special expressions. Special parameters only occur as predicates in special expressions.

Special definitional clauses can also introduce parameter symbols which only occur with fixed arities.

- The syntax of Prolog is a specialisation of full ambivalent syntax, and shares some of the properties of the above specialisations. In Prolog's syntax, only parameters are allowed in predicate and function positions. In addition, Prolog has a special predicate *not*, which is only allowed to occur in predicate positions. A representative part of Prolog's syntax can be described as follows:

1 variables x and parameters $c \in G$ are expressions.

2 $c(t_1, \ldots, t_n)$ is an expression if $c \in G$ and $t_1, \ldots, t_n$ are expressions.

3 $not(t)$ is a special expression if t is an expression.

4 $fail$ and ! are special expressions.

5 every expression is a special expression.

6 $t \leftarrow t_1, \ldots, t_n$ is a Prolog clause if t is an expression and $t_1, \ldots, t_n$ are expressions.

7 every expression that is not a variable is a Prolog clause.

Other features of Prolog's syntax can be incorporated in this framework, such as the Prolog built in predicates *assert* and *retract*, which take Prolog clauses as arguments.

- The ambivalent syntax described in Kalsbeek [Kal95] is a specialised version of full ambivalent syntax. It differs from full ambivalent syntax in the following ways:
  1. Only atomic expressions are allowed to occur in term positions.
  2. Only parameters are allowed to occur as functions and predicates.

In particular, variables are not allowed to occur in formula positions. A full description can be found in Kalsbeek [Kal95].

2.3 Semantics

We have chosen to develop a closed term semantics for AL for the purpose of this paper. All of the definitions we give can be adapted to special versions of AL such as Prolog's syntax.

Formally, a structure $\mathcal{M}$ for an ambivalent language $L_\mathcal{G}$ (the *underlying language* of $\mathcal{M}$) is a tuple (D, T), where

1. D, the domain of $\mathcal{M}$, is the set of *closed* expressions of $L_\mathcal{G}$;
2. T, the truth set of $\mathcal{M}$, is a subset of the set of *closed atomic* expressions of $L_\mathcal{G}$.

The satisfiability relation for closed expressions in a structure $\mathcal{M} = (D, T)$ is inductively defined as follows:

Definition 11

$\mathcal{M} \models_{AL} A$ iff $A \in T$ where A is a closed atomic expression
$\mathcal{M} \models_{AL} A \wedge B$ iff $\mathcal{M} \models_{AL} A$ and $\mathcal{M} \models_{AL} B$
$\mathcal{M} \models_{AL} \neg A$ iff $\mathcal{M} \not\models_{AL} A$
$\mathcal{M} \models_{AL} \forall x A$ iff for all $d \in D$, $\mathcal{M} \models_{AL} A\{x/d\}$

Disjunction, implications, equivalence and existential quantification are defined in the standard way. □

Several things are worth noting at this stage.

- Closed expressions can be evaluated as sentences in a model. At the same time, they constitute elements of the domain.

- A closed expression, evaluated as a sentence, in a model, has a unique truth value. Thus, while there is syntactic ambivalence, there is no semantic ambiguity.

- The truth values of atoms are, in principle, not related to the structure of expressions which occur in them as subterms. That is, for example, the truth value of an atom $p(s \wedge v)$ is in principle not related to the respective truth values of $p(s)$ and $p(t)$. A similar distinction holds for the relation between outside and inside quantifiers: The truth value of $\forall x.p(f(x))$ is independent of the truth value of $p(\forall x.f(x))$. The latter is decided by a checking whether the atom $p(\forall x.f(x))$ belongs to the truth set, while the former is decided by checking whether $p(f(d))$ belongs to the truth set, for each element d in the domain. In certain applications, such as formalisations of knowledge and belief, it might be desirable to have explicit relations between inside and outside quantifiers, or between outside and inside connectives such as the above. The soundness and strong completeness theorem 20 that we will prove in Section 2.5 shows that such relations may be implemented by axioms.

- It should be noted that 'similar' expressions like, for example, $\forall x.f(x)$ and $\forall y.f(y)$ constitute different, and unrelated, objects in the domains of models. That is, the truth values of the closed expressions $t(\forall x.f(x))$ and $t(\forall y.f(y))$ need not be the same. This may be counterintuitive, and even undesirable in some applications. Extra assumptions on the truth sets T may impose that expressions which are similar under appropriate renaming of the bound variables, behave similarly as elements of the domain. This in its turn should then be matched with an appropriate rule in the derivational calculus. In contrast, for outside quantifiers the following relation, familiar from FOL, holds: $\mathcal{M} \models \forall x A$ iff $\mathcal{M} \models \forall y A\{x/y\}$, if y is free for x in A.
- It should be observed that in the above interpretation of the quantifiers (the substitution interpretation), there is no real semantic second order quantification: the quantifiers range over object in domains, and not over subsets of domains. Thus, while syntactically AL allows for second order features such as quantification over functions and predicates, its closed term semantics is first order.

We will use the following standard notions.

Definition 12 Let A be a closed expression and let S be a set of AL sentences. S is *satisfiable* in AL iff there exists an interpretation $\mathcal{M}$ such that $\mathcal{M} \models_{AL} \phi$ for all $\phi \in S$. $\mathcal{M}$ is called a *model* of S in this case.
A is *valid* in AL, denoted by $\models_{AL} A$, iff $\neg A$ is not satisfiable in AL.
We say that A is a *logical consequence* of S, denoted by $S \models_{AL} A$, iff A is true in every model of S. □

Definition 13 A structure $\mathcal{M}$ is a *Herbrand model* for a theory S ($\mathcal{M} \models^h_{AL} S$) if $\mathcal{M}$ is a structure for S and the underlying language of $\mathcal{M}$ is the underlying language of S (that is, if D coincides with the set of closed expressions of the full ambivalent syntax generated by S). □

In addition, we can define semantics for all expressions, using assignments (Definition 15). We need the following standard definitions.

Definition 14 Let σ be a substitution $\{x_1/t_1, \ldots, x_n/t_n\}$. Then its *domain* $dom(\sigma)$ and *range* $ran(\sigma)$ are defined as follows:
$dom(\sigma) = \{x_1, \ldots, x_n\}$,
$ran(\sigma) = \{t_1, \ldots, t_n\}$
The σ-image of a variable x, denoted as $x\sigma$, is given as follows:
$x_i\sigma = t_i$, for $i \in [1, n]$, and
$x\sigma = x$, for $x \notin dom(\sigma)$.
For expressions t and substitutions σ, $t\sigma$ is the result of replacing each free occurrence of x in t by the σ-image of x. □

Definition 15 A substitution σ is an *assignment* for an expression s and an interpretation $\mathcal{M} = \langle D, T \rangle$, if $dom(\sigma) \supseteq FV(s)$ and $ran(\sigma) \subseteq D$. □

Definition 16 Let σ be an assignment for t and $\mathcal{M}$. We say that $\mathcal{M}$ *satisfies t with σ* ($\mathcal{M}, \sigma \models t$) if $\mathcal{M} \models t\sigma$. □

As a *derivational calculus* for Ambivalent Logic we take some standard system of natural deduction. The resulting derivability relation will, as usual, be indicated by $\vdash$. The deduction rules are defined in a standard way. Consider the usual elimination rule for the universal quantifier, $\forall x.t / t\{x/s\}$. It has a side condition, which prevents unintended bindings resulting from the substitution of s for x. In the case of AL, unintended bindings can result, not only from quantifiers in standard places, but also from inside quantifiers. As an example, $\forall x p(\exists y f(x, y)) / p(\exists y f(y, y))$ is not a correct instance of the rule for elimination of the universal quantifier: y is not free for x in $p(\exists y f(x, y))$.

Example 17 The ambivalent expression $p(a)$ can, among others, be considered as either $p(x)\{x/a\}$ or $x(a)\{x/p\}$. The former corresponds to the following (standard) instance of the introduction rule for the existential quantifier: $p(a)/\exists x(p(x))$. The latter in its turn corresponds to $p(a)/\exists x(x(a))$.

We remind the reader of the following two notions of completeness.

Definition 18 Let L be a logic with semantic consequence relation $\models_L$ and derivational consequence relation $\vdash_L$.
L is *weakly complete* if, for any sentence B, $\models_L B$ implies $\vdash_L B$.
L *strongly complete* if, for any sentence B and any set of sentences S, $S \models_L B$ implies $S \vdash_L B$. □

In the case of first order logic with standard syntax, a restriction to Herbrand semantics results in the weakening of some of the standard results. A well-known example is the loss of strong completeness in its full sense. While strong completeness w.r.t. Herbrand models holds for finite and infinitely extendible theories, it does not hold for theories in which every closed term of the underlying language occurs in the theory. The same phenomenon occurs in the context of Ambivalent Logic, as witnessed by the following proposition.

Proposition 19 *AL is not strongly complete for infinite theories with respect to the Herbrand semantics.*

Proof: Let $t_1, t_2, t_3, \ldots$ be an enumeration of all the closed expressions of an ambivalent language L. Then, by the definition of validity in Herbrand models, $t_1, t_2, t_3, \ldots \models_{AL}^h \forall x.x$. However, $t_1, t_2, t_3, \ldots \not\vdash \forall x.x$. □

While a restriction to Herbrand semantics is sensible in the case of logic programming (logic programs being finite theories), in a broader context it is, in view of the above proposition, sensible to consider the more general closed term semantics as the appropriate semantics for Ambivalent Logic.

A more general semantics for AL, with arbitrary domains and, consequently, non-identity interpretation functions, will not be considered in the context of the present paper. The reason is that, (as remarked in Chen et al. [CKW93]) interpretation of quantified terms using a non-identity interpretation function requires something like lambda-abstraction, which considerably complicates the construction of models. In the next section we argue that the restriction to closed term semantics is not a severe limitation.

2.4 Equality and Identity

The usual semantics for FOL with standard syntax is different from the closed term semantics we described above. In particular, in the semantics for FOL, any set can in principle serve as the domain of a model, and the interpretation functions, mapping closed terms of the language on elements of the domain, are not necessarily identity functions. In this semantics, equality is usually interpreted as identity on the domains, although the equality axioms in themselves do not force this interpretation. In contrast, closed term models are not appropriate for the identity interpretation of equality.

Closed term models, however, can be used to represent other models. Hence they play an important, albeit hidden, role in FOL. Consider the usual proof of the completeness theorem for FOL, using the Henkin method. The proof in fact consists of two separate stages, each the proof of an independent lemma. In the first stage, a consistent theory T is extended to a maximally consistent Henkin theory T^*, for which a Herbrand model H is constructed. H is a closed term model for the theory T, and, in presence of equality in the language, equality is interpreted in H as an equivalence relation which is also a congruence with respect to the predicates and functions. The second stage of the proof is merely motivated by the convention to interpret equality as identity. From H, a model M for T^* is constructed, the domain of which consists of the equivalence classes in H under equality. The valuation function of M is 'inherited' from H. This second stage can be interpreted as the proof of a representation lemma, which expresses that for every closed term model H there exists an elementary equivalent model M in which equality is interpreted as identity. This representation lemma can be reversed. Let $M = \langle D_M, V_M \rangle$ be a model. A closed term model H is now constructed as follows. Define a function f from the closed terms of the language to be interpreted to the domain of M, which is defined by $f(t) := t^M$. Now define the valuation function V_H as follows:

$V_H(P(\bar{t})) := V_M(P(\overline{f(t)}))$. Now H and M are elementary equivalent, and equality is interpreted as just another binary predicate in H, satisfying properties dictated by the equality axioms.

The two sides of this representation result together in fact show that there is a bijection between the class of closed term models and the class of general models (modulo isomorphism). It also shows that there is no inherent need to interpret identity as equality. The interpretation of equality as identity forces the choice of the usual semantics of FOL — if this restriction is abandoned, closed term models are a sound and complete semantics for first order logic with equality, provided the truth sets satisfy the properties dictated by the equality axioms. What we have argued here are several things. First, the usual axioms for equality do in themselves not force the identity interpretation. Second, if the usual identity interpretation of equality is abandoned, we can restrict ourselves to consideration of closed term models without loss of generality.

In particular, we can incorporate equality in Ambivalent Logic without changing the style of the semantics for Ambivalent Logic. In the closed term semantics, equality will not correspond to real identity on the domains.

Let us consider, in some more detail, the usual equality theory ET for FOL. It can be axiomatised (abstracting from arities) as follows:

(I) $\forall x. x = x$
$\forall x \forall y (x = y \to y = x)$
$\forall x \forall y \forall z (x = y \to (y = z \to x = z))$
(II) $\forall x \forall y \forall t (x = y \to t = t\{x/y\})$
(III) $\forall x \forall y \forall t (x = y \to (t \to t\{x/y\}))$

We obtain an appropriate closed term semantics for AL extended with these equality axioms, by additionally restricting the truth sets to satisfy the corresponding properties. In particular, ET(I) requires truth sets on which $=$ is an equivalence relation. ET(I,II) requires truth sets T (say, with underlying language L_G) which additionally satisfy the following: if $d = c \in T$, and t is an L_G-expression with only x free, then $t\{x/d\} = t\{x/c\} \in T$. ET as a whole requires truth sets T which in addition also satisfies the following property: for all closed L_G-expressions d and c, and for all L_G-expressions t with only x free, if $d = c \in T$, then $t\{x/d\} \in T$ iff $t\{x/c\} \in T$.

Observe that, in the context of the closed term semantics style, we are not committed to ET as a whole. In particular, there are interesting differences between AL + ET(I,II) and AL + ET, which we will more closely consider in the next section.

In the sequel, when we consider AL, we explicitly mean AL without equality. If we consider either of AL + ET(I), AL + ET(II), and AL + ET, we will implicitly assume that the semantics satisfies the corresponding conditions mentioned above.

2.5 Formal Results

In the present section, we set out to develop some basic theory for Ambivalent Logic. First, we prove soundness and completeness of AL w.r.t. the closed term semantics.

Theorem 20
AL is sound and strongly complete with respect to the closed term semantics;
AL is strongly complete for infinitely extendible theories with respect to Herbrand semantics.

Proof: Soundness is left to the reader. The completeness proof follows the standard completeness proof for FOL.

Let S be a consistent AL theory with underlying language $L = L_\mathcal{G}$. First extend S to a Henkin theory S^*, as follows.

Extend the set of generating constants $\mathcal{G}$ with a new constant d, and let $L' = L_{\mathcal{G} \cup \{d\}}$.

Let $\phi_1, \phi_2, \phi_3, \ldots$ be an enumeration of all the expressions of L' with only x free. Let $d_1, d_2, d_3, \ldots$ be an enumeration of all the closed expressions of L' that are not expressions of L. A carefully chosen subset of these expressions will serve as Henkin constants. Define C_n as the set of closed expressions that occur as terms in $\phi_1 \wedge \ldots \wedge \phi_n$.

Define
$$\begin{aligned}
S_0 &:= S, \\
m_1 &:= min\{k | d_k \notin C_1\}, \\
S_1 &:= S_0 \cup \{\exists x.\phi_1(x) \to \phi_1(d_{m_1}), \\
m_{n+1} &:= min\{k > m_n | d_k \notin C_{n+1}\}, \\
S_{n+1} &:= S_n \cup \{\exists x.\phi_{n+1}(x) \to \phi_{n+1}(d_{m_{n+1}}), \\
S^* &:= \cup S_n.
\end{aligned}$$

S^* is a Henkin theory. Conservativity of S_{n+1} over S_n can be proven in the usual way. Therefore, S^* is conservative over S and thus consistent.

Next, extend S^* to a maximally consistent theory S^m, by the Lindenbaum lemma. The language of S^m is L', and S^m is a Henkin theory. A model $\mathcal{M} = \langle D, T \rangle$ for S^m is obtained as follows. Let D consist of all the closed expressions of L'. Let T consist of the closed atomic expressions that belong to S^m. By induction on the complexity of sentences, $\mathcal{M}$ is a Herbrand model for S^m. By conservativity of S^m over S, $\mathcal{M}$ is a closed term model for S.

For infinitely extendible theories a model can be constructed in a slightly more elegant way. The closed expressions of L that do not occur as terms in the theory can be used as the Henkin constants. The resulting model is a Herbrand model for the theory. □

We leave it to the reader to check that the above result also goes through for AL + ET(I), AL + ET(I,II), and AL + ET.

Proposition 21 *Let S be a finite set of AL expressions, in which at least one parameter occurs. Then S is infinitely extendible.*

Proof: Let g be a parameter occurring in S. Then $g, g(g), g(g(g)), \ldots$ is an infinite set of closed expressions in the language underlying S. In contrast, the number of closed expressions of this language that occur as terms in expressions of S, is finite, as S is a finite set. □

An immediate consequence of the above theorem and proposition is the following.

Corollary 22 *Every satisfiable AL sentence has a Herbrand model.*

The following notion of normal form is the appropriate version for AL.

Definition 23 *An expression F is in normal form if $F \equiv Q_1 x_1, \ldots, Q_n x_n.t$, where*

1. The Q_i are quantifiers $\forall$ or $\exists$.
2. t is built up from atomic expressions and the connectives $\neg$, $\vee$, and $\wedge$; in particular, in any subexpression of t of the form $Qy.s$, where Q is a quantifier, Q is an inside quantifier.
3. The variables x_i are mutually distinct.
4. If $Qy.s$ is a subexpression of t, where Q is an (inside) quantifier, then y is distinct from any of the x_i. □

Lemma 24 *For every formula F there is an equivalent formula F', such that F' is in normal form.*

Proof: By the usual methods. □

The *Skolem form* F^s of a normal form expression F can be obtained by the usual methods (cf., for instance, Schöning [Sch89] for an algorithm to obtain Skolem forms). Observe that Skolemisation does leave the inside quantifiers and connectives intact.

Next we prove an AL version of the usual s-equivalence theorem. The proof differs from the usual proof in a crucial aspect. In the usual proof, essential use is made of the option to use non-trivial interpretation functions on domains. In particular, a model $\mathcal{M}$ for a normal form formula F is extended to a model for its Skolem form by extending the signature of $\mathcal{M}$ with functions on its domain that interpret witnesses of the existential quantifiers. In the case of Ambivalent Logic, this construction cannot be used, as we do not allow for non-trivial interpretation functions. For clarity of this exposition, assume that $\mathcal{M}$ is a Herbrand model for F. The introduction of the Skolem functions then results in a proper extension of the domain of $\mathcal{M}$. As a result, instead of obtaining a model

$\mathcal{M}'$ for the Skolem form by a proper extension the interpretation valuation functions of $\mathcal{M}$, a new model has to be constructed. The upshot is that this new model $\mathcal{M}'$ is a Herbrand model for the Skolem form F^s, in this sense the result below is stronger than the corresponding usual result for FOL. However, the construction of $\mathcal{M}'$ is a bit more involved in the ambivalent case. The truth set T' of $\mathcal{M}'$ is obtained by projecting the expressions of the extended language on the expressions of the original language, and by the taking T' as the pre-image of T under this projection. The projection (translation) is the identity function on the original language.

The proof we give here is given for the case of AL, but can be generalised to FOL and intermediate cases.

Theorem 25 (s-equivalence) *For every closed expression F in normal form, F is satisfiable iff its Skolem form is satisfiable.*

Proof: Let $F = \forall x_1 \exists y_1 \ldots \forall x_n \exists y_n.\phi$, where all outside quantifiers are indicated. Let $L_\mathcal{G}$ be the underlying language of F.
Let $f_1, \ldots, f_n$ be new parameters not occurring in ϕ.
Let $L' = L_{\mathcal{G} \cup \{f_1, \ldots, f_n\}}$.
Let $F^s = \forall x_1 \ldots \forall x_n.\phi\{y_1/f_1(x_1), \ldots, y_n/f_n(x_1, \ldots, x_n)\}$.
Let $\phi' = \phi\{y_1/f_1(x_1), \ldots, y_n/f_n(x_1, \ldots, x_n)\}$, that is, $F^s = \forall x_1 \ldots \forall x_n.\phi'$.

The proof of the left-to-right direction is as usual. For the converse direction, let, (by Corollary 22), $\mathcal{M} = \langle D, T \rangle$ be a Herbrand model for F; here, D consists of the closed expressions of $L_\mathcal{G}$. We can now, as usual, by the axiom of choice, define (external) functions $v_1, \ldots, v_n$ on D, such that for all $d_1, \ldots, d_n \in D$

$$\mathcal{M}, \{x_1/d_1, y_1/v_1(d_1), \ldots, x_n/d_n, y_n/v_n(d_1, \ldots, d_n)\} \models \phi. \tag{I}$$

We construct, using $\mathcal{M}$ and the functions $v_1, \ldots, v_n$, a Herbrand model $\mathcal{M}' = \langle D', T' \rangle$, where D' consists of all the closed expressions in L', such that $\mathcal{M}' \models F^s$. Observe that D is a strict subset of D'.

We will define the truth-set T' using the following translation (or projection) $(\cdot)^*$ of expressions of L' into expressions of $L_\mathcal{G}$.

- For all $c \in \mathcal{G}$, $c^* = c$
- For all variables x, let $x^* = x$
- For f_i, choose a $c_i \in \mathcal{G}$, and let $f_i^* = c_i$
- If $t = f_i$ and $n = i$, let $(t)(t_1, \ldots, t_n)^* = v_i(t_1^*, \ldots, t_i^*)$;
- Otherwise, $(t)(t_1, \ldots, t_n)^* = (t^*)(t_1^*, \ldots, t_i^*)$.
- $(A \vee B)^* = A^* \vee B^*$ and similarly for other binary connectives

- $(\neg A)^* = \neg A^*$
- $(\forall x(A))^* = \forall x(A^*)$
- $(\exists x(A))^* = \exists x(A^*)$.

Now for all expressions t in $L_{\mathcal{G}}$, $t^* = t$. In particular, $(\phi)^* = \phi$. Also, let t be an expression in $L_{\mathcal{G}}$, and let $\sigma = \{z_1/s_1, \ldots, z_k/s_k\}$ be a substitution such that the s_i are L'-expressions. We leave it to the reader to check that $(t\sigma)^* = t^*\{z_1/s_1^*, \ldots, z_k/s_k^*\} = t\{z_1/s_1^*, \ldots, z_k/s_k^*\}$. Therefore, the following equation (II) holds:

$$\begin{aligned}(\phi'\{x_1/d_1\}, \ldots, x_n/d_n\})^* &= \\ &= (\phi\{x_1/d_1, y_1/f_1(d_1), \ldots, x_n/d_n, y_n/f_n(d_1, \ldots, d_n)\})^* \\ &= \phi^*\{x_1/d_1^*, y_1/(f_1(d_1))^*, \ldots, x_n/d_n^*, y_n/(f_n(d_1, \ldots, d_n))^*\} \quad (II) \\ &= \phi\{x_1/d_1^*, y_1/v_1(d_1^*), \ldots, x_n/d_n^*, y_n/v_n(d_1^*, \ldots, d_n^*)\}.\end{aligned}$$

Now define the truth set T' as follows:

$$T' = \{t \in L' : t^* \in T\}$$

By induction it follows that for all sentences A in $L_{\mathcal{G}}$,

$$\mathcal{M}' \models A \quad \text{iff} \quad \mathcal{M} \models A^*.$$

In particular, by (I) and (II), $\mathcal{M}' \models F^s$. □

Another folklore result is Herbrand's theorem:

Theorem 26 *A closed formula in Skolem form with matrix F is unsatisfiable iff there is a finite subset of the Skolem expansion of F which is unsatisfiable.*

The usual proof of the above theorem goes through, without modification, for Ambivalent Logic.

All of the above results (21 to 26) also hold for extensions of AL with any of the above-mentioned equality theories.

We have proven some of the essential ingredients to prove soundness and completeness of resolution for Ambivalent Logic: the construction of Skolem forms, the s-equivalence theorem and the Herbrand theorem. The missing ingredients for the completeness of resolution are the lifting lemma (which allows a transformation of resolution refutation on clauses in propositional logic to a resolution refutation on clauses in predicate logic), and decidability of unification for Ambivalent Logic. The traditional proof of the lifting lemma goes through, without modification, for Ambivalent Logic. Unification theory for Ambivalent Logic will be discussed in Section 2.6. The traditional proof of soundness of resolution for FOL (see for instance Schöning [Sch89]) goes through, without modification, for Ambivalent Logic. We conclude that, modulo the results on unification in the next section, we have the following result:

Theorem 27 *Resolution is a sound and complete inference method for AL.*

We conclude this section by proving one more result. Given the fact that AL is a syntactic extension of FOL, and has the same derivational calculus, the question of conservativity of AL over FOL arises, that is: if a formula ϕ in standard syntax is derivable in AL, is it then also derivable in FOL — and vice versa? The answer, as the following theorem shows, is affirmative. This result shows that AL is an extension of FOL. The proof uses both directions of the above s-equivalence theorem 25. The technique used in the proof is similar to that used in the proof of Theorem 25: from a model, a new, ambivalent model is defined, the truth-set of which is a pre-image of a syntactic projection to the truth-set of the original model. In this case, a Herbrand model for a language with standard syntax, is extended to an elementary equivalent Herbrand model for the associated ambivalent language.

Theorem 28 *Let ϕ be a formula in standard syntax. Then $\models_{FOL} \phi$ iff $\models_{AL} \phi$.*

Proof: The left-to-right direction follows from the fact that every derivation in FOL is also a derivation in AL, combined with completeness of FOL, and soundness of AL. For the converse direction (conservativity of AL over FOL), it suffices, by contraposition, to show that if ϕ is satisfiable in FOL, then it is also satisfiable as an AL formula.

So let ϕ be satisfiable in FOL. Without loss of generality we can assume that ϕ is in normal form. By the s-equivalence theorem for FOL, the Skolem form ϕ^s is also satisfiable. In particular, there is a Herbrand model $\mathcal{M} = \langle D, T \rangle$ satisfying ϕ^s. Using $\mathcal{M}$, we will construct an ambivalent Herbrand model $\mathcal{M}' = \langle D', T' \rangle$ satisfying ϕ^s.

Let C, F, and P be the set of constants, respectively functions, respectively relation symbols occurring in ϕ^s. Then the universe D of $\mathcal{M}$ is generated by $\langle C, F \rangle$. Let D' be the ambivalent universe generated by the parameter set $C \cup F \cup P$. In order to define the truth set T', we will make use of a (total and surjective) function $(\cdot)^*$ from D' to D. (The definition of $*$ is inspired by the abstraction function defined in [Kal95]; here, we use, instead of fresh constants, some element of D.)

Choose an arbitrary $d \in D$. Define $* : D' \to D$ as follows:

1. For $c \in C \cup F \cup P$,
    $c^* := c$ if $c \in C$
    $c^* := d$ otherwise.

2. for closed terms $(t)(t_1, \ldots, t_n)$,
    $((t)(t_1, \ldots, t_n))^* := t(t_1^*, \ldots, t_n^*)$ if t is an n-ary predicate in P
    $(t)(t_1, \ldots, t_n)^* := d$ otherwise.

3. For all other $t \in D'$, $t^* := d$.

Observe that, for every $t \in D$, $t^* = t$.

Using the function *, the truth set T' of $\mathcal{M}'$ can now be defined as follows:

$$T' = \{t \in D' : t^* \in T\}.$$

Now let $\phi^s = \forall x_1 \ldots \forall x_n.A$, where A is a conjunctive normal form.

$\mathcal{M} \models_{FOL} \forall x_1 \ldots \forall x_n.A$
$\implies$ {by definition of the satisfaction relation}
for all $d_1, \ldots, d_n \in D$ $\mathcal{M} \models_{FOL} A\{x_1/d_1, \ldots, x_n/d_n\}$
$\implies$ {by definition of * and by the form of A}
for all $d_1, \ldots, d_n \in D'$, $\mathcal{M} \models_{FOL} A\{x_1/d_1{}^*, \ldots, x_n/d_n{}^*\}$
$\implies$ {by definition of T'}
for all $d_1, \ldots, d_n \in D'$, $\mathcal{M}' \models_{AL} A\{x_1/d_1, \ldots, x_n/d_n\}$
$\implies$ {by definition of the satisfaction relation}
$\mathcal{M}' \models_{AL} \forall x_1 \ldots \forall x_n.A$.

From the above implication and the easy side of the s-equivalence theorem 25, it immediately follows that $\mathcal{M}'$ is a closed term model satisfying ϕ. □

The soundness direction of the above theorem generalises to AL + ET(I), AL + ET(I,II), and AL + ET. The completeness side however, trivially does not hold for either of AL + ET(I) and AL + ET(I,II), as FOL incorporates all of ET.

Although AL + EQ(I,II) is not conservative over FOL for formulas with equality, this logic is of some interest in the context of intensional logic. By the soundness and completeness w.r.t. the appropriate semantics, it does not satisfy ET(III) — therefore, it is opaque w.r.t. substitutions of equal terms. In addition, by the ambivalent syntax of AL, which allows for all expressions to occur in term positions, AL + EQ(I,II) has identity naming, that is, expressions can be represented by themselves. In the domain of knowledge and belief this is a useful property. In contrast, AL + ET is, by the above theorem, a conservative extension of FOL.

2.6 Unification

In the present section we show how the Martelli-Montanari unification algorithm can be adapted to the case of ambivalent syntax. We show that the adapted algorithm (which is defined on page 49) has all the desired properties, in particular, termination and, in case of successful termination, generation of unifiers which are most general within an appropriate class of unifiers. We follow the outlines of the theory for unification as given in Doets [Doe94], to which we also refer the reader for the usual definitions of unifier and most general unifier.

There are several differences between unification for standard syntax and unification for full ambivalent syntax.

1. In ambivalent syntax, functions and predicates are arityless. Unification of two atoms with different arities has to be excluded. An extra action (9) is sufficient: halt with failure on an equation $(t)(t_1, \ldots, t_n) = s(s_1, \ldots, s_m)$, if $n \neq m$.

2. In the Martelli-Montanari algorithm for standard syntax, there are two actions for functional atoms:
halting with failure on $f(t_1, \ldots, t_n) = g(s_1, \ldots, s_n)$, if f is unequal to g; replacement with $t_1 = s_1, \ldots, t_n = s_n$, otherwise.
In ambivalent syntax, all expressions can occur in function positions. The above two actions are replaced by one single action (8), accounting for the unification of the expressions in the function positions as well as the arguments. An extra action (10) is needed to prevent unification of functional atoms with expressions that are not functional atoms or variables. In addition, two extra actions (6) and (7) deal with unification of parameters.

3. In ambivalent syntax, conjunctive, disjunctive, implicational, and negated expressions occur as terms and as subexpressions, thus they are candidates for unification. For example, $x \vee y$ and $a \vee c(z)$ are unified by $\{x/a, y/c(z)\}$. This is accounted for in the unification algorithm by the actions (11) and (13). Appropriate failure conditions are reflected by the actions (12) and (14).

4. Likewise, quantified expressions have to be dealt with. For closed quantified expressions, semantical identity coincides with syntactical identity. However, quantified expressions in general are still liable for unification. First, they can unify with variables. For example, $\exists x(x)$ and y are unified by $\{y/\exists x(x)\}$. Further, quantified expressions can contain free variables, which are candidates for unification. As an example, $\exists x(x(y))$ and $\exists x(x(c))$ are unified by $\{y/c\}$. In contrast, $\exists x(x)$ and $\exists y(y)$ cannot be unified, as both are closed expressions.

Unification of quantified expressions is partly engineered by action (15), which eliminates identical quantifiers: $\exists x.t = \exists x.s$ is repaced by $t = s$, and similarly for the universal quantifier. In contrast, the semantic distinction between $\exists x.t$ and $\exists y.t\{x/y\}$ is reflected in action (16), which halts with failure on equations $\exists x.t = \exists y.s$, if x and y are different (and likewise for the universal quantifier).

However, the (necessary) action (15) might lead to incorrect results, as it releases bound variables, which subsequently become, incorrectly, candidates for non-trivial unification. As an example, action (15) replaces the unsolvable equation $\exists x.x = \exists x.c$ with the solved equation $x = c$. (Below we will formally define the notion of solved equation.) This problem is solved in two ways.

First, before the algorithm is run on a set of equations, all the bound variables should be renamed to marked variants. An appropriate renaming function will be given in Definition 29 below. This enables us to keep track of the origin of variables during execution of the algorithm.

Second, the algorithm should treat marked and unmarked variables differently. In particular, marked variables, which should be thought of as bound variables, only unify with themselves and with unmarked variables. Trivial unification of marked and unmarked variables is dealt with in action (1). Non-trivial unification of marked variables with anything but unmarked variables is excluded by action (2). However, unification of unmarked variables with expressions in which marked variables occur free, should be possible. For example, consider the expressions $\exists x(p(x) \vee q(x))$ and $\exists x(y \vee q(x))$, where y is an unmarked variable. These expressions are unified by $\{y/p(x)\}$. That is, unification of unmarked variables with expressions in which unmarked variables occur free, should be allowed. This is taken care of by action (4). The usual actions (3) and (4) only apply if the left-hand variables are unmarked.

The appropriate marking of bound variables is obtained by applying a marking function $(\cdot)^m$, which replaces all occurrences of bound variables with marked copies of these variables. The effect of the marking function is a renaming of the bound variables in an expression, with the effect that no variable occurs both bound and free after marking.

Before we define the marking function, we introduce some notation. Recall that $(t)(t_1, \ldots, t_n)$ is a functional atom with n-ary predicate (or function) t, and arguments t_i. We will use the notation $t(x_1, \ldots, x_n)$ to indicate an expression t for which $FV(t) \subseteq \{x_1, \ldots, x_n\}$.

Definition 29 (Marking bound variables)

$$c^m = c \qquad \text{for parameters } c,$$
$$x^m = x \qquad \text{for unmarked variables } x,$$
$$x'^m = x' \qquad \text{for marked variables } x',$$
$$((t)(t_1, \ldots, t_n))^m = (t^m)(t_1^m, \ldots, t_n^m)$$

Martelli-Montanari unification algorithm for ambivalent syntax

1. $x = x$ where x is a (marked or unmarked) variable
         remove
2. $x' = t$ where x' is a marked variable, t is not an unmarked variable, and t is different from x'
         halt with failure
3. $x = t$ where x is unmarked, t is different from x, and x occurs in t
         halt with failure
4. $x = t$ where x is unmarked, t is different from x, and x does not occur in t
         replace x by t in all other equations
5. $t = x$ where t is not an unmarked variable
         replace by $x = t$
6. $c = c$ where c is a parameter
         remove
7. $c = t$ where c is a parameter, t is not a variable and c and t are different
         halt with failure
8. $(t)(t_1, \ldots, t_n) = (s)(s_1, \ldots, s_n)$
         replace by $t = s, t_1 = s_1, \ldots, t_n = s_n$
9. $(t)(t_1, \ldots, t_n) = (s)(s_1, \ldots, s_m)$ where $n \neq m$
         halt with failure
10. $(t)(t_1, \ldots, t_n) = s$ where s is not a variable or a functional atom
          halt with failure
11. $\neg t = \neg s$
          replace by $t = s$
12. $\neg t = s$ where s is not a variable or a negated expression
          halt with failure
13. $t_1 \diamond t_2 = s_1 \diamond s_2$ where $\diamond$ is one of the binary connectives
          replace by $t_1 = s_1, t_2 = s_2$
14. $t_1 \diamond t_2 = s$ where s is not a variable or of the form $s_1 \diamond s_2$
          halt with failure
15. $Qx'(t) = Qx'(s)$ where Q is one of the quantifiers $\exists$ and $\forall$
          replace by $t = s$
16. $Qx'(t) = s$ where Q is one of the quantifiers $\exists$ and $\forall$, s is not a variable and s is not of the form $Qx'(v)$
          halt with failure

$(\exists x(t))^m = \exists x'(t\{x/x'\})^m$ where x is an unmarked variable,
$(\exists x'(t))^m = \exists x'(t^m)$ where x' is a marked variable,
and similarly for universally quantified expressions.
In addition, $(\cdot)^m$ commutes with the logical connectives. □

The unification algorithm for AL will only yield the desired results if applied to terms in which all the bound variables are marked.

We will use the following terminology:

Definition 30 An expression t is *clean* if $t^m = t$. A set of equations $\{t_1 = s_1, \ldots, t_n = s_n\}$ is clean if all of the t_i and s_i are clean. A pair of sequents $\langle t_1, \ldots, t_n \rangle$, $\langle s_1, \ldots, s_n \rangle$ is clean if the associated set of equations $\{t_1 = s_1, \ldots, t_n = s_n\}$ is clean. □

In particular, marked variables can occur free in clean expressions, but in contrast, the latter do not contain bound occurrences of unmarked variables.

In the correctness proof of the algorithm we need the following notions of ambivalent substitution and ambivalent unifier:

Definition 31 σ is an *ambivalent substitution* if no marked variables occur in the domain of σ. □

Definition 32 Let t and s be expressions. σ is an *ambivalent unifier* for t and s, if σ is an ambivalent substitution and $t\sigma = s\sigma$. □

Observe that not every unifier is also an ambivalent unifier. For example, $\{x'/f(y)\}$ unifies $g(x')$ with $g(f(y))$, while no ambivalent unifiers exist for this tuple. In contrast, every ambivalent unifier is a unifier.

Proposition 33 *Let $L = \langle t_1, \ldots, t_n \rangle$ and $R = \langle s_1, \ldots, s_n \rangle$ be two sequents of clean expressions such that no marked variables occur free in any of the t_i and s_i. Then L and R are unifiable iff there is an ambivalent unifier for L and R.*

Proof: Suppose σ unifies L and R. Now let τ be the restriction of σ to $FV(L) \cup FV(R)$. Then, by assumption, τ is an ambivalent unifier of L and R. For the other direction, notice that every ambivalent unifier is a unifier. □

We relativise the notion of most general unifier to the class of ambivalent unifiers.

Definition 34 An ambivalent unifier θ for a set of (clean) expressions E is a *most general ambivalent unifier* for E (or, in short, an m.g.a.u.) if for every ambivalent unifier σ for E there is an ambivalent substitution τ such that $\sigma = \theta\tau$. An m.g.a.u. θ for E is *strong* if for every ambivalent unifier σ of E, $\sigma = \theta\sigma$. □

An m.g.a.u. is not necessarily an m.g.u., as the following counterexample witnesses.

Counterexample 35 Let $\theta = \{y/f(x')\}$. It is easy to check that θ is an m.g.a.u. for $E = \{p(\exists x'.f(x')) = p(\exists x'.y)\}$. Also, $\sigma = \{y/f(x'), x'/y\}$ is a unifier of E. However, suppose there is a substitution τ such that $\sigma = \theta\tau$. Then $f(x')\tau = f(x')$, and thus $x' \notin dom(\tau)$. But this contradicts $\{x'/y\} \in \{y/f(x')\}\tau$.

In addition, we need to adapt the notions of solved equation and equivalence between sets of equations.

Definition 36 A set of equations $\{t_1 = s_1, \ldots, t_n = s_n\}$ is *solved* if
1. the t_i are pairwise different, unmarked variables, and
2. no t_i occurs in any of the s_j. □

A solved set of equations $E = \{x_1 = s_1, \ldots, x_n = s_n\}$ determines a most general ambivalent unifier $\{x_1/s_1, \ldots, x_n/s_n\}$ for E (or, more precisely, for the associated pair of sequences $\langle x_1, \ldots, x_n \rangle$ and $\langle s_1, \ldots, s_n \rangle$). We need one more definition.

Definition 37 Two sets of equations of ambivalent terms are *equivalent* if they have the same ambivalent unifiers. □

Now we are in position to prove correctness of the unification algorithm.

Theorem 38 *The Martelli-Montanari unification algorithm for ambivalent syntax, when applied to a finite set of clean equations, results in a solved set of equations, determining a strong most general ambivalent unifier for the associated sequences in case an ambivalent unifier exists, and terminates with failure otherwise.*

Proof: The theorem follows from the following claims.

Claim 1 *Every non-hulling action applied to a set of clean equations produces an equivalent set of clean equations.*

Proof: None of the actions 1, 4, 5, 6, 8, 11, 13, and 15 introduces a bound, unmarked variable. Therefore any of these actions transforms a set of clean equations into a new set of clean equations.

Preservation of equivalence is trivial for the actions 1, 5, 6, 8, 11, and 13. For action 4, preservation of equivalence is proven as usual.
For action 15, let σ be an ambivalent substitution. Then the following holds:

$(\exists x'(t))\sigma = (\exists x'(s))\sigma \iff \quad \{x' \notin dom(\sigma)\}$
$\exists x'(t\sigma) = \exists x'(s\sigma) \iff t\sigma = s\sigma.$ □

Claim 2 *The algorithm terminates.*

Proof: Consider the lexicographic order $<_3$ on $\mathbf{N}^3$. That is,

$$(n_1, n_2, n_3) <_3 (m_1, m_2, m_3)$$

iff
$$n_1 < m_1$$
$$\text{or}\quad n_1 = m_1 \ \& \ n_2 < m_2$$
$$\text{or}\quad n_1 = m_1 \ \& \ n_2 = m_2 \ \& \ n_3 < m_3.$$

Given a set of equations E, we call an unmarked variable x *solved in E* if, for some expression t, $x = t \in E$, and this is the only free occurrence of x in E. We call a variable x *unsolved in E* if x is unmarked and x is not solved in E.

With each set of clean equations E we now associate the following three functions:

$uns(E) :=$ the number of unsolved variables in E,

$lfun(E) :=$ the total number of occurrences of parameter symbols on the left hand side of the equations in E,

$lsym(E) :=$ the total number of symbols (including brackets) occurring on the left hand side of equations of E.

We claim that each of the non-halting actions of the algorithm strictly reduces the triple $(uns(E), lfun(E), lsym(E))$.

Indeed, action 4 decreases $uns(E)$ by 1, while none of the successful actions increase $uns(E)$. Also, none of the other successful actions increase $lfun(E)$. Action 5 decreases $lfun(E)$ by 1 if t is a parameter, and decreases $lsym(E)$ by at least 1 otherwise. The actions 1, 6, 8, 11, 13, and 15 all decrease $lsym(E)$. Termination of the algorithm now follows from the well-foundedness of $<_3$.
□

Claim 3 *If the algorithm terminates successfully on a set of clean equations, then the final set of equations is solved.*

Proof: Suppose the algorithm terminates successfully, resulting in a set of clean equations E. Then none of the actions 5 – 16 applies to E, so the left hand expressions are all variables. Action 2 does not apply to E, so these are all unmarked variables. Finally, actions 1, 3, and 4 do not apply to E, so none of these variables occurs on the right hand side of any of the final equations.
□

Claim 4 *If the algorithm halts with failure on a clean set of equations, then this set does not have an ambivalent unifier.*

Proof: Suppose the algorithm halts on the clean set E. If failure is the result of action 2, then the equation $x' = t$, where t is different from x' and t is not a variable, is a member of E. Clearly there are no ambivalent unifiers for $x' = t$. If failure is the result of action 3, then $x = t$ is an element of E, and there is no unifier σ for x and t, as $x\sigma$ is a proper subterm of $t\sigma$. If failure is the result of action 7, then $c = t$ is an element of E, and there is no unifier σ for c and t, because $c\sigma = c$, and $t\sigma$ is either a parameter

different from c or an expression that is neither a variable nor a parameter. In the other cases, similar arguments apply. □
This completes the proof of Theorem 38. □

Clearly, this result is less general than the corresponding result for the original version of the unification algorithm. First of all, it applies only to those equations in which the sets of free and bound variables are distinct. It is an open question whether this restriction can be dropped. Second, the unifiers generated are most general only within the class of ambivalent unifiers, that is, those unifiers for which the domain does not contain any variable that occurs bound in the original set of equations. Counterexample 35 above suggests that this restriction can not be dropped.

As a corollary of the above theorem we now have a decidable unification algorithm for Prolog syntax. Clearly, in the Prolog case, where quantified expressions do not occur in term positions, we do not have to deal with the renaming of bound variables. The relevant actions in the Prolog version of the unification algorithm are the *Prolog actions* 1, 3, 4, 5, 6, 7, 8, 9, and 10. We leave it to the reader to check that the following holds.

Theorem 39 *The Martelli Montanari unification algorithm for Prolog's syntax, consisting of the Prolog actions 1, 3, 4, 5, 6, 7, 8, 9, and 10 results in a solved set of equations, determining a strong most general unifier in case a unifier exists, and terminates with failure otherwise.*

2.7 Comparison with Hilog

Hilog (Chen et al. [CKW93]) was developed as a language for higher order Logic Programming. In the discussion below, we assume that the reader is familiar with Hilog. We will here mainly point out some of the differences and similarities between Hilog and AL.

AL syntax is an extension of the syntax of Hilog. While the syntax of AL is fully ambivalent, Hilog syntax is characterised by the occurrence of terms in formula-, function- and predicate positions. (For example, $(c(a))(c)$, $x \to p(x)$, and $\forall x \exists y.(x(p))(y)$ are Hilog formulas.) Other than AL, Hilog does not admit quantified expressions in any unusual positions. As an example, $p(\forall x p(x))$ is an AL expression, but not a well-formed Hilog expression. In either syntax, the parameters are arityless.

While syntactically AL can be considered an extension of Hilog, the two logics differ considerably in semantics. The distinguishing feature of the (first-order) Hilog semantics is that each parameter of the language has a unique intension—that is, the interpretation function associates with each individual parameter (regardless of its contextual role), exactly one object in the domain of a model. With each intension then, several extensions are associated that capture the different contextual roles of the parameter. This extends to general terms. As a consequence, the schema $\forall x \forall y.(x = y \to \phi(x) \leftrightarrow \phi(y))$, and also $\forall x \forall y \forall z.(x = y \to x(z) \leftrightarrow y(z))$. In contrast, in the context of AL we have a *choice* between validating the above schema or not, by either taking all of ET as the equality theory, or restricting the equality to ET(I,II). Thus, unlike AL, Hilog is not appropriate for intensional logics, where opaqueness is usually desirable. (In contrast, observe that equivalence of terms does not imply their equality in neither Hilog nor AL.)

Another difference between AL and Hilog is found in comparing their respective relations to FOL. As we have seen, AL is a conservative extension of FOL without equality, and AL with the usual equality theory ET is a conservative extension of FOL with equality (Theorem 28). In Hilog, conservativity over FOL is restricted to the classes of equality free formulas and definite clauses in which equality only occurs in body-atoms. A standard derivational calculus for FOL is sound and complete w.r.t. AL semantics (Theorem 20). In contrast, in accordance to the intension of developing Hilog as basis for Logic Programming, paramodulation is introduced in [CKW93] as a derivational calculus for Hilog, and its soundness and completeness with respect to the Hilog semantics is proven. It is also shown that Hilog can be encoded in FOL, and a standard derivational calculus for FOL can be soundly used for the derivation of encoded Hilog formulas.

It should be noted that, despite the second order aspects of Hilog and AL syntax, both are essentially first order theories. In higher order predicate logic, the second order variables range over all relations over the intended domain. That is, in higher order semantics, functions and predicates are identified with their domain. In contrast, in Hilog and AL, the 'second order' variables range over elements of the intended domain. As a result, relation comprehension does not generally hold in AL and Hilog. That is, $\exists p \forall x_1 \ldots \forall x_n (p(x_1, \ldots, x_n) \leftrightarrow \phi(x_1, \ldots, x_n))$ is not generally valid in Hilog and AL. For example, relation comprehension is not true in AL for $\exists z q(x, y, z)$. But, unlike in Hilog, both $\exists p \forall x (p(x) \leftrightarrow \neg q(x))$ and $\exists p \forall x (p(x) \leftrightarrow \exists u (q((u))(x))$ are valid in AL.

A consequence of the identification of functions and predicates with their extensions in higher-order logic, is that the undecidability of this extensional equality carries over to the unification problem. Both for AL and Hilog however, unification is decidable.

2.8 Conclusion

The results reported show that, with minor modifications, basic proof theoretic results for first order predicate logic also go through for Ambivalent Logic. In particular, unification for AL is decidable, and both a standard derivational calculus and resolution are sound and complete inference methods for AL. A conservativity result shows that AL should be considered as a (syntactic) extension of first order predicate logic. All of the results reported relativise to subsystems of Ambivalent Logic that are frequently used in practice, in particular Prolog syntax and the syntax(es) used in Vanilla meta-programming and data bases. In addition, various properties of AL, such as the optional opaqueness with respect to equality and the flexibility of its syntax, suggest that AL may provide an interesting format for the representation of knowledge and belief.

Acknowledgements

This work reported here was originally inspired by ideas of Bob Kowalski and has benefited from many fruitful discussions with Barry Richards. Discussions with Bern Martens, Krzysztof Apt, and John Lloyd have always been very constructive. Recent comments and suggestions by Pat Hill, Danny De Schreye, Marco Schaerf, Sten-Ake Tarnlund, and Johan van Benthem, helped to improve this paper. We especially want to thank the editors of this volume for their support.

Marianne Kalsbeek was supported by the Dutch Organisation for Scientific Research (N.W.O.).

References

[AT95]   K.R. Apt and F. Teusink. Comparing negation in logic programming and in Prolog. This volume.

[CKW93] W. Chen, M. Kifer, and D.S. Warren. Hilog: A foundation for higher-order logic programming. *Journal of Logic Programming*, 15(3):187–230, 1993.

[Doe94] H. C. Doets. *From Logic to Logic Programming*. MIT Press, 1994.

[Gab92] D. Gabbay. Metalevel features in the object level: modal and temporal logic programming III. In L.Farinas del Cerro and M. Penttonen, editors, *Intensional logics for programming*, pages 85–124. Clarendon Press, 1992.

[Jia94a] Y. Jiang. Ambivalent logic as the semantic basis for metalogic programming: I. In P. Van Hentenryck, editor, *Proceedings of the International Conference on Logic Programming*, pages 387–401. MIT Press, June 1994.

[Jia94b] Y. Jiang. Ambivalent logic as the semantic basis of metalogic programming: Theory and practice. Technical report, Imperial College, Dept. of Computing, 1994.

[Kal95] M. Kalsbeek. Correctness of the Vanilla meta-interpreter and ambivalent syntax. This volume.

[KK91]   R.A. Kowalski and J. Kim. A metalogic programming approach to multi-agent knowledge and belief. In V. Lifschitz, editor, *Artificial Intelligence and Mathematical Theory of Computation*, pages 231–246. Academic Press, 1991.

[Ric74] B. Richards. A point of reference. *Synthese*, 28:431–445, 1974.

[Sch89] U. Schöning. *Logic for Computer Scientists*, volume 8 of *Progress in Computer Science and Applied Logic*. Birkhauser, 1989.

3 Two Semantics for Definite Meta-programs, using the Non-ground Representation

Bern Martens and Danny De Schreye

Abstract

We study the semantics of definite vanilla meta-programs in the context of a Prolog-style non-typed approach, using a non-ground representation for object level variables. The two best known semantics for definite Logic Programs are considered: least Herbrand semantics and S-semantics. In both cases we study the relation between the semantics of the object program and the corresponding vanilla meta-program. We also extend the results to various enhanced vanilla meta-programs, multi-level meta-programs, and limited forms of amalgamation. We discuss the benefits and limitations of these semantics, both with respect to each other and with respect to other approaches taken in the literature.

3.1 Introduction

Meta-programming has become increasingly important in logic programming and deductive databases. Applications in knowledge representation and reasoning, program transformation, synthesis and analysis, debugging and expert systems, the modeling of evaluation strategies, the specification and implementation of sophisticated optimisation techniques, the description of integrity constraint checking, etc. are constituting a significantly large part of the recent work in the field (see e.g. Bowen [Bow85], Bowen & Kowalski [BK82], Kowalski [Kow90], Gallaire & Lasserre [GL82], Takeuchi & Furukawa [TF86], Gallagher [Gal86], Sterling & Shapiro [SS86], Sterling & Beer [SB89], Bry [Bry90], Träff & Prestwich [TP92], Bry et al. [BDM88]).

This practical success has more recently been complemented by thorough studies of the theoretical foundations for meta-programming in logic programming. Some important contributions here are Richards [Ric74], Subrahmanian [Sub89], Hill & Lloyd [HL89], Hill & Lloyd [HL88], Lloyd [Llo88], Sato [Sat92], Bonatti [Bon92], Jiang [Jia94], Levi & Ramundo [LR93], Chen et al. [CKW93] and Kalsbeek [Kal95]. Most of these approaches adapt the standard logic programming language and/or its semantics to appropriately deal with the meta-programming issues. In this paper, we investigate to which extent the standard least Herbrand semantics can already provide a sensible meaning for meta-programs written in a Prolog-style, non-typed approach, and using a non-ground representation for object level variables. In order to simplify some technicalities (Martens & De Schreye [MD95]), we hereby restrict the attention to *definite* logic programs.

The main problem with using the least Herbrand semantics to assign meaning to a non-typed meta-program with a non-ground representation, is illustrated in the following example. Consider the object program P:

$p(x) \leftarrow$
$q(a) \leftarrow$

Let M denote the standard *solve* interpreter:

$solve(empty) \leftarrow$
$solve(x \& y) \leftarrow solve(x), solve(y)$
$solve(x) \leftarrow clause(x, y), solve(y)$

In addition, let M_P denote the program M augmented with the following facts:

$clause(p(x), empty) \leftarrow$
$clause(q(a), empty) \leftarrow$

Although the least Herbrand model of our object program is $\{p(a), q(a)\}$, the least Herbrand model of the meta-program M_P contains completely unrelated atoms, such as $solve(p(empty))$, $solve(p(q(a)))$, etc..

This is certainly undesirable, since we, in general, would like at least that the atoms of the form $solve(p(\bar{t}))$ in the least Herbrand model of M_P correspond in a one-to-one way with the atoms of the form $p(\bar{t})$ in the least Herbrand model of P.

In this paper, we introduce the notion of *language independence* of a program and show that it generalises range restrictedness. As one of our main results, we prove that for definite, language independent programs, the least Herbrand model of the program corresponds in a one-to-one way with a natural subset of the least Herbrand model of the corresponding vanilla theory. In addition, we show how this approach can be extended to provide a semantics for a limited form of amalgamation. This extension is rather interesting, since it reflects one of the main advantages the untyped approach may have over the typed one.

Nevertheless, the limitation to language independent programs does not completely correspond to our intuitive expectations. On the level of computed answer substitutions, practical experience with untyped meta-programs reveals no such restriction. This observation motivates a second part of our study, in which the meaning of the programs is defined through S-semantics. We show that in this case the precondition of language independence can be dropped, although some care remains necessary for enhanced vanilla meta-programs.

The paper is organised as follows. In the next section, we present our main results for definite, language independent object programs and their straightforward untyped vanilla meta-version. In Section 3.3, we briefly discuss overloading of symbols and its effect on the semantics of amalgamated programs. We propose a semantics for a limited form of amalgamation in Section 3.4, using the foundations laid out in the previous two sections. Then, in Section 3.5, we repeat the study using the S-semantics. Section 3.6 contains a discussion on the extension of the results for meta-programs with an explicit theory argument and for the *demo* predicate. We end with a discussion of related work and with a short evaluation of the proposed semantics.

A preliminary version of this paper appeared as De Schreye & Martens [DM92]. Martens & De Schreye [MD95] contains a more technically involved extension of the results for the standard Herbrand semantics to the case of *normal* logic programs (also called *general* in e.g. Apt [Apt90]). We briefly comment on this extension in Section 3.7. Our main theorem of Section 3.5 was independently proved in Levi & Ramundo [LR93].

3.2 A Semantics for Untyped Vanilla Meta-programming

As mentioned in the introduction, in this paper, we only consider definite logic programs. We suppose the reader to be familiar with the basic concepts of predicate logic (see e.g. Fitting [Fit90]) and logic programming (see e.g. Lloyd [Llo87]).

3.2.1 Language Independent Programs

We start by introducing the concept of a *language independent* definite program.

Definition 1 Let $\mathcal{L}$ be a (first order) language and $\mathcal{R}$, $\mathcal{F}$ and $\mathcal{C}$ its sets of predicate, function and constant symbols respectively. We call a language $\mathcal{L}'$, determined by $\mathcal{R}'$, $\mathcal{F}'$ and $\mathcal{C}'$ an *extension* of $\mathcal{L}$ iff $\mathcal{R} \subseteq \mathcal{R}'$, $\mathcal{F} \subseteq \mathcal{F}'$ and $\mathcal{C} \subseteq \mathcal{C}'$.

When considering a logic program, it is customary to speak about the *language underlying the program*. In this language, $\mathcal{R}$, $\mathcal{F}$ and $\mathcal{C}$ are the sets of predicate, function and constant symbols occurring *in the program*. Notice that this implies that we assume the set of constants in the underlying language to be empty if the program does not contain any constants. (We return to this below.) Although this is not imposed as a limitation in e.g. Lloyd [Llo87], Herbrand interpretations of the program are usually constructed with this underlying language in mind. For our purposes in this paper, however, we need more flexibility. Therefore, the language in which the interpretations are constructed, is made explicit in the following definition.

Definition 2 Let P be a definite program with underlying language $\mathcal{L}$. A Herbrand interpretation of P in a language $\mathcal{L}'$, extension of $\mathcal{L}$, is called an $\mathcal{L}'$-*Herbrand interpretation* of P.

We are now in a position to introduce the notion of *language independence*.

Definition 3 A definite program P with underlying language $\mathcal{L}$ is called *language independent* iff for any extension $\mathcal{L}'$ of $\mathcal{L}$, its least $\mathcal{L}'$-Herbrand model is equal to its least $\mathcal{L}$-Herbrand model.

Notice that this definition immediately entails that no atom in any least Herbrand model can contain any terms with function and/or constant symbols not occurring in P. In particular, propositions are the only possible elements of any least Herbrand model for a language independent program without constant symbols. This justifies our choice of a language without constant symbols as the one underlying such programs, instead of the more usual approach of picking one with a single arbitrary constant symbol. Martens & De Schreye [MD95] contains an alternative formalisation, following the latter approach.

Before concluding this subsection, we point out that the notion of language independence generalises the well-known concept of *range restriction* for definite logic programs. We first repeat the definition of the latter.

Definition 4 A clause in a definite program P is called *range restricted* iff any variable that appears in its head also appears in its body.
A definite program P is called *range restricted* iff all its clauses are range restricted.

Range restriction has been defined for more general formulas and/or programs and was used in other contexts. See e.g. Nicolas [Nic82] and Bry et al. [BMM92] for its use in the context of integrity checking in relational and deductive databases. Two equivalent notions are *safety*, used by Ullman in Ullman [Ull88] and *allowedness* (at the clause level), defined in Lloyd [Llo87].

The limitation to range restricted programs is natural in many contexts. Moreover, it can be noted that Manthey & Bry [MB88] includes a general method to transform a non range restricted program P into a range restricted program P' through the addition of $dom(X)$ calls to the bodies of clauses for every variable X which is not "restricted". *dom* itself is defined to hold for every ground term in the Herbrand universe of the given language via a range restricted definite program. There is a one-to-one correspondence between Herbrand models of P and P' such that they coincide for all predicates in P.

The following proposition shows that this important class of logic programs is a subclass of the language independent ones.

Proposition 5 *A range restricted definite program is language independent.*

Proof: Let P be a range restricted definite program with underlying language $\mathcal{L}$ and let $\mathcal{L}'$ be an extension of $\mathcal{L}$. Let T_P be the immediate consequence operator applied in the context of $\mathcal{L}$ and T_P' the corresponding operator applied in the context of $\mathcal{L}'$. We prove that for each $n \geq 0 : T_P\uparrow n = T_P'\uparrow n$.

The proof is through induction on n. The base case is trivial. Furthermore, if the body literals in a ground instance of a range restricted clause are instantiated with terms in $\mathcal{L}$, then so is the head. The induction step now follows immediately.

The difference between language independence and range restriction can be characterised more precisely, as the following proposition shows[1].

Proposition 6 *Let P be a definite program. Then P is language independent iff all ground ($\mathcal{L}_P$)-instances of every non range restricted clause in P contain at least one body atom, not true in its least Herbrand model.*

Proof: First suppose there is a non range restricted clause that does not satisfy the stated condition. It immediately follows that P is not language independent. This proves the *only-if*-part. The proof of the *if*-part is completely analogous to that of Proposition 5.

We conclude our reflections on definite, language independent programs with the following proposition.

Proposition 7 *Let P be a definite program. Then P is language independent iff for any definite goal G, all (SLD-)computed answers for $P \cup \{G\}$ are ground.*

Proof: We first prove the *only-if*-part. Suppose there is a goal $\leftarrow A_1, \ldots, A_k$ such that θ is a non-ground computed answer for $P \cup \{\leftarrow A_1, \ldots, A_k\}$. Then there must be at least one $A_i, 1 \leq i \leq k$ such that $\forall \bar{x} A_i \theta$ is a logical consequence of P (here $\bar{x}$ represents the *non-empty* set of variables, free in $A_i\theta$). In particular, $\forall \bar{x} A_i \theta$ must be satisfied in any least Herbrand model. This implies that P is *not* language independent.

For the *if*-part, first observe that for any particular goal, computed answers are not affected by language extensions. Suppose now that all computed answers to all goals are ground. This means that in particular all computed answers to any goal G where each argument is an uninstantiated variable, are ground. From the completeness of SLD-resolution (Theorem 8.6 in Lloyd [Llo87]), it now follows that the set of computed answers for $P \cup \{G\}$ is exactly equal to the set of correct answers for $P \cup \{G\}$. We can conclude that P is language independent.

Proposition 7 provides yet another characterisation of the class of definite, language independent programs. We will find occasion to apply it in Section 3.5.

3.2.2 A Meta-programming Semantics

In this subsection, we show that there is a natural correspondence between the least Herbrand model of a language independent definite program and the least Herbrand model of its vanilla meta-program.

Definition 8 Suppose $\mathcal{L}$ is a first order language and $\mathcal{R}$ its finite (or countable) set of predicate symbols. Then we define $\mathcal{F}_\mathcal{R}$ to be a *functorisation* of $\mathcal{R}$ iff $\mathcal{F}_\mathcal{R}$ is a set of

---
[1]This observation is due to an anonymous referee of De Schreye & Martens [DM92]. It is worth noting that both range restriction and language independence can be defined for (stratified) *normal* programs. In that case, Proposition 5 can be generalised, but not Proposition 6. See Martens & De Schreye [MD95].

function symbols such that there is a one-to-one correspondence between elements of $\mathcal{R}$ and $\mathcal{F_R}$ and the arity of corresponding elements is equal.

We introduce the following notation: Whenever A is an atom in a first order language $\mathcal{L}$ with predicate symbol set $\mathcal{R}$, A' denotes the term produced by replacing in A the predicate symbol by its corresponding element in $\mathcal{F_R}$.

Definition 9 Let P be a definite program. Then, M_P, the *vanilla meta-program associated to P*, will be the following definite program:
$solve(empty) \leftarrow$
$solve(x \& y) \leftarrow solve(x), solve(y)$
$solve(x) \leftarrow clause(x, y), solve(y)$
together with a fact of the form
$clause(A', B_1' \& \ldots \& B_n') \leftarrow$
for every clause $A \leftarrow B_1, \ldots, B_n$ in P and a fact of the form
$clause(A', empty) \leftarrow$
for every fact $A \leftarrow$ in P.

Notice that if $\mathcal{L}_P$, the language underlying P, is determined by $\mathcal{R}_P$, $\mathcal{F}_P$ and $\mathcal{C}_P$, then $\mathcal{L}_{M_P}$, the language underlying M_P, is determined by
$\mathcal{R}_{M_P} = \{solve, clause\}$
$\mathcal{F}_{M_P} = \mathcal{F}_P \cup \mathcal{F}_{\mathcal{R}_P} \cup \{\&\}$
$\mathcal{C}_{M_P} = \mathcal{C}_P \cup \{empty\}$
where $\mathcal{F}_{\mathcal{R}_P}$ is a functorisation of $\mathcal{R}_P$.

In the sequel, when we refer to Herbrand interpretations and/or models (of a program P or M_P) and related concepts, this will be in the context of the languages $\mathcal{L}_P$ and $\mathcal{L}_{M_P}$, as defined above, unless stated explicitly otherwise. Before continuing, we introduce the following notation:

1. U_P: the Herbrand universe of a program P

2. $U_P^n = U_P \times \ldots \times U_P$ (n copies)

3. p/r: a predicate symbol with arity r in $\mathcal{R}_P$
   p'/r: its associated function symbol in $\mathcal{F}_{M_P}$

The following proposition, which will implicitly be used in the sequel, is immediate.

Proposition 10 Let P be a definite program and M_P its vanilla meta-program. Then $U_P \subseteq U_{M_P}$.

Two lemmas now lead to one of the main results in this paper.

Lemma 11 Let P be a definite program and M_P its vanilla meta-program. Then the following holds for every $p/r \in \mathcal{R}_P$:
$\forall \bar{t} \in U_P^r, \forall n \in \mathbb{N} : p(\bar{t}) \in T_P \uparrow n \Longrightarrow \exists m \in \mathbb{N} : solve(p'(\bar{t})) \in T_{M_P} \uparrow m$

Proof: The proof is through induction on n. The base case ($n = 0$; $T_P\uparrow 0 = \emptyset$) is trivially satisfied. Now suppose that $p(\bar{t}) \in T_P\uparrow n, n > 0$. Then there must be at least one clause C in P such that $p(\bar{t}) \leftarrow C_1, \ldots, C_k$ ($k \geq 0$) is a ground instance of C and $C_1, \ldots, C_k \in T_P\uparrow(n-1)$. Consider first the case that we have one with $k = 0$. In other words, $p(\bar{t})$ is a ground instance of a fact in P. In that case Definition 9 immediately implies that $solve(p'(\bar{t})) \in T_{M_P}\uparrow 2$.

Suppose now $k \geq 1$. The induction hypothesis guarantees for every C_i the existence of an $m_i \in \mathbb{N}$ such that $solve(C_i') \in T_{M_P}\uparrow m_i$. Let mm denote the maximum of these m_i. It takes only a completely straightforward proof by induction on l to show the following:

$\forall\, 1 \leq l \leq k : solve(C_{k-l}'\&\ldots\&C_k') \in T_{M_P}\uparrow(mm+k-l)$.

From this, it follows that

$solve(C_1'\&\ldots\&C_k') \in T_{M_P}\uparrow(mm+k-1)$

and therefore

$solve(p'(\bar{t})) \in T_{M_P}\uparrow(mm+k)$.

Note that Lemma 11 holds for any definite object program.

Lemma 12 *Let P be a definite, language independent program and M_P its vanilla meta-program. Then the following holds for every $p/r \in \mathcal{R}_P$:*

$\forall \bar{t} \in U_{M_P}{}^r, \forall n \in \mathbb{N} : solve(p'(\bar{t})) \in T_{M_P}\uparrow n \Longrightarrow$
$\bar{t} \in U_P{}^r \,\&\, \exists m \in \mathbb{N} : p(\bar{t}) \in T_P\uparrow m$

Proof: We first define $\mathcal{L}'$ to be the language determined by $\mathcal{R}_P$, $\mathcal{F}_{M_P}$ and $\mathcal{C}_{M_P}$. $\mathcal{L}'$ is an extension of $\mathcal{L}_P$.

The proof proceeds through an induction on n. The base case ($n = 0$; $T_{M_P}\uparrow 0 = \emptyset$) is trivially satisfied. Suppose that $solve(p'(\bar{t})) \in T_{M_P}\uparrow n$ where $n > 0$. Then either there is a *clause*-fact in M_P of which $clause(p'(\bar{t}), empty) \leftarrow$ is a ground instance or this is not the case. Suppose first there is. Then P must contain a fact of which $p(\bar{t}) \leftarrow$ is a ground instance in $\mathcal{L}'$. This means that $p(\bar{t}) \in T_P\uparrow 1$ in $\mathcal{L}'$. But, since P is language independent, this implies that $\bar{t} \in U_P{}^r$ and $p(\bar{t}) \in T_P\uparrow 1$.

If there is no such *clause*-fact in M_P, then there must be one with a ground instance $clause(p'(\bar{t}), C_1'\&\ldots\&C_k')$ where $k \geq 1$, such that $solve(C_1'\&\ldots\&C_k') \in T_{M_P}\uparrow(n-1)$. A simple induction argument on k shows that we obtain the following:

$\forall\, 1 \leq i \leq k : \exists\, n_i < n \in \mathbb{N} : solve(C_i') \in T_{M_P}\uparrow n_i$.

Through the induction hypothesis, we get:

$\forall\, 1 \leq i \leq k : \exists m_i \in \mathbb{N} : C_i \in T_P\uparrow m_i \,\&\, \bar{t_i} \in U_P{}^{r_i}$

(where r_i is the arity of the predicate symbol in C_i). From the above and the fact that P is language independent, the desired result follows.

Theorem 13 *Let P be a definite, language independent program and M_P its vanilla meta-program. Let H_P and H_{M_P} denote the least Herbrand model of P and M_P respectively. Then the following holds for every $p/r \in \mathcal{R}_P$:*

$$\forall \bar{t} \in U_{M_P}{}^r : solve(p'(\bar{t})) \in H_{M_P} \Longleftrightarrow \bar{t} \in U_P{}^r \;\&\; p(\bar{t}) \in H_P$$

Proof: The theorem follows immediately from Lemmas 11 and 12.

In Lemma 12 and (therefore also) Theorem 13, language independence of the object program appears as a *sufficient* condition. We actually conjecture that it is essentially[2] also a *necessary* one.

3.3 A Justification for Overloading

In the next section, we extend some of the results of Section 3.2 to provide a semantics for a limited form of amalgamation. The most trivial example of the programs we will consider is the (textual) combination $P + M_P$ of the clauses of an object program P with the clauses for its associated vanilla program M_P. A more complex example is obtained by further (textually) extending $P + M_P$ with one additional *clause*/2-fact for each of the (three) clauses in M_P defining the *solve*/1-predicate. In the most general case, we allow – in addition – the occurrence of *solve*/1-calls in the bodies of clauses of P. As a final difference with Section 3.2, we here impose the use of one particular functorisation $\mathcal{F}_{\mathcal{R}_P}$, namely the one in which all functors in $\mathcal{F}_{\mathcal{R}_P}$ are identical to their associated predicate symbols in $\mathcal{R}_P$.

Postponing the discussion on the generalisation of our results, we first address the more basic problem with the semantics of these programs, caused by overloading the symbols in the language. Clearly, the predicate symbols of P occur both as predicate symbol and as functor in $P + M_P$ (and in any further extensions). Now – although this was not made explicit in e.g. Lloyd [Llo87] – an underlying assumption of first order logic is that the class of functors and the class of predicate symbols of a first order language $\mathcal{L}$, are disjoint (see e.g. Ramsay [Ram88]). So, if we aim to extend our results to amalgamated programs – without introducing any kind of naming to avoid the overloading – we need to verify whether the constructions, definitions and results on the foundations of logic programming are still valid if the functors and predicate symbols of the language overlap.

We have checked the formalisation and proofs in Lloyd [Llo87] in detail, starting from the assumption that the set of functors and the set of predicate symbols may overlap. We found that none of the results become invalid. Of course, under this assumption, there is in general no way to distinguish well-formed formulas from terms. They as well

---

[2]The case where $\mathcal{F}_{M_P} = \mathcal{F}_P$ and $\mathcal{C}_{M_P} = \mathcal{C}_P$ might provide an (uninteresting) exception.

have a non-empty intersection. But, this causes no problem in the definition of pre-interpretations, variable- and term-assignments and interpretations (see Lloyd [Llo87], p.12). It is clear however, that a same syntactical object can be both term and formula and can therefore be given two different meanings, one under the pre-interpretation and variable-assignment, the other under the corresponding interpretation and variable-assignment. But this causes no confusion on the level of truth-assignment to well-formed formulas under an interpretation and a variable-assignment (Lloyd [Llo87], p.12–13). This definition performs a complete parsing of the well-formed formulas, making sure that the appropriate assignments are applied for each syntactic substructure. In particular, it should be noted that no paradoxes can be formulated in these languages, since each formula obtains a unique truth-value under every interpretation and variable-assignment.

On the level of declarative semantics (for definite programs), the main results – the existence of a least Herbrand model and its characterisation as the least fixpoint of T_P (Lloyd [Llo87], Proposition 6.1, Theorem 6.2, Theorem 6.5) – remain valid in the extended languages. Thus, the amalgamated programs we aim to study can in any case be given a unique semantics.

In [Kal95], Kalsbeek proposes a closely related extension of first order logic syntax: she includes all atoms among the terms of the language. This corresponds to overloading *all* predicate symbols in the language (of the meta-program). The work by Jiang and Kalsbeek contains detailed formal justifications for using (so-called) ambivalent syntax of this and even more general types. Brief further discussions can be found in Section 3.7.

3.4 Amalgamation

In this section, we build on the results of the previous two sections and present a semantics for different kinds of amalgamated programs. We will *fix the choice of the functorisation* for a given set of predicate symbols: It will contain *exactly the same symbols*. A justification for this practice was given above. It leads to an increased flexibility in considering meta-programs with several layers. In fact, we can now deal with an unlimited amount of meta-layers.

The first extension we consider is completely straightforward: We include the object-program in the resulting meta-program.

Definition 14 Let P be a definite program and M_P its associated vanilla meta-program (see Definition 9). Then we call the textual combination $P + M_P$ of P and M_P the *amalgamated vanilla meta-program associated to P*.

It is clear that $P + M_P$ is also a definite program. Notice that $\mathcal{L}_{P+M_P}$ is determined by
$$\mathcal{R}_{P+M_P} = \mathcal{R}_P \cup \{solve, clause\}$$

$\mathcal{F}_{P+M_P} = \mathcal{F}_P \cup \mathcal{F}_{\mathcal{R}_P} \cup \{\&\}$
$\mathcal{C}_{P+M_P} = \mathcal{C}_P \cup \{empty\}$
where $\mathcal{F}_{\mathcal{R}_P}$ is equal to $\mathcal{R}_P$.
We immediately have the following:
$U_{P+M_P} = U_{M_P}$
We can now prove the following theorem through an immediate extension of the proof for Theorem 13.

Theorem 15 *Let P be a definite, language independent program, M_P its vanilla meta-program and $P + M_P$ its amalgamated vanilla meta-program. Let H_P, H_{M_P} and H_{P+M_P} denote their least Herbrand models. Then the following holds for every $p/r \in \mathcal{R}_P$:*

$\forall \bar{t} \in U_{P+M_P}{}^r : solve(p(\bar{t})) \in H_{P+M_P} \iff p(\bar{t}) \in H_{P+M_P}$
$\forall \bar{t} \in U_{P+M_P}{}^r : solve(p(\bar{t})) \in H_{P+M_P} \iff \bar{t} \in U_P{}^r \ \& \ p(\bar{t}) \in H_P$
$\forall \bar{t} \in U_{P+M_P}{}^r : solve(p(\bar{t})) \in H_{P+M_P} \iff \bar{t} \in U_{M_P}{}^r \ \& \ solve(p(\bar{t})) \in H_{M_P}$

Theorem 13 and Theorem 15 are interesting because they provide us with a reasonable semantics for vanilla and amalgamated vanilla meta-programming for a large class of definite programs. However, they also show that we do not seem to *gain* much by this kind of programming. Indeed, (the relevant part of) the meta-semantics can be *identified* with the object semantics. So, why going through the trouble of writing a meta-program in the first place ? The answer lies of course in useful *extensions* of the vanilla interpreter. (See e.g. Sterling & Shapiro [SS86] and further references given there.) The following definition captures many such extensions. (Henceforth, we no longer explicitly comment on the language underlying our meta-programs, but leave its description implicit in the programs' definitions.)

Definition 16 *Let P be a definite program. Let E be a definite program of the following form – called* extended *meta-interpreter – :*

$solve(empty, t_{11}, \ldots, t_{1n}) \leftarrow C_{11}, \ldots, C_{1m_1}$
$solve(x\&y, t_{21}, \ldots, t_{2n}) \leftarrow solve(x, t_{31}, \ldots, t_{3n}), solve(y, t_{41}, \ldots, t_{4n}),$
$\qquad C_{21}, \ldots, C_{2m_2}$
$solve(x, t_{51}, \ldots, t_{5n}) \leftarrow clause(x, y), solve(y, t_{61}, \ldots, t_{6n}), C_{31}, \ldots, C_{3m_3}$

where the t_{ij}-terms are extra arguments of the solve-predicate and the C_{kl}-atoms extra conditions in its body, together with defining clauses for any other *($\neq$ solve) predicates occurring in the C_{kl}. Then we define E_P, the E-extended meta-program associated to P, to be the program E together with the usual* clause*-facts, one for each clause in P (as in Definition 9).*

Example 17 As an example, we include the following program E, adapted from Sterling & Shapiro [SS86]. It builds proof trees of object level queries.

$solve(empty, empty) \leftarrow$

$solve(x\&y, proofx\&proofy) \leftarrow solve(x, proofx), solve(y, proofy)$
$solve(x, x \ if \ proof) \leftarrow clause(x, y), solve(y, proof)$

As is illustrated in Sterling & Shapiro [SS86], the proof trees thus constructed can be used as a basis for explanation facilities in expert systems. Further examples can be found in e.g. Sterling & Beer [SB89].

We have the following proposition:

Proposition 18 *Let P be a definite, language independent program and E_P an E-extended meta-program associated with P. Let H_P and H_{E_P} denote their respective least Herbrand models. Then the following holds for every $p/r \in \mathcal{R}_P$:*

$$\forall \bar{t} \in U_{E_P}{}^r : (\exists \bar{s} \in U_{E_P}{}^n : solve(p'(\bar{t}), \bar{s}) \in H_{E_P}) \implies \bar{t} \in U_P{}^r \ \& \ p(\bar{t}) \in H_P$$

Proof: It follows immediately from an obvious property of definite logic programs that $solve(p'(\bar{t}), \bar{s}) \in H_{E_P}$ implies $solve(p'(\bar{t})) \in \mathcal{L}_{E_P}\text{-}H_{M_P}$ (M_P's least $\mathcal{L}_{E_P}$-Herbrand model). Let $\mathcal{L}'$ be the language determined by $\mathcal{R}_P$, $\mathcal{F}_{E_P}$ and $\mathcal{C}_{E_P}$. $\mathcal{L}'$ is an extension of $\mathcal{L}_P$. The result now follows from the language independence of P.

It can be noted that the right hand side of the implication in Proposition 18 is equivalent with $\bar{t} \in U_{M_P}{}^r \ \& \ solve(p'(\bar{t})) \in H_{M_P}$ (where M_P denotes the vanilla meta-program associated with P), This follows from Theorem 13.

Proposition 18 essentially ensures us that working with extended meta-programs is "safe" for language independent programs. Indeed, no unwanted solutions of the kind mentioned in the introduction are produced. Further research can perhaps determine conditions on E that allow an equivalence in Proposition 18. We know such extended interpreters exist: The proof tree building program in Example 17 above presents an instance.

Definition 16 and Proposition 18 do not involve amalgamation. But, it is of course perfectly possible to give a definition analogous to Definition 14 for extended meta-interpretation. We will not do this explicitly and only illustrate by an example the extra programming power one can gain in this way.

Example 19 In applications based on the proof tree recording program given in the previous example, it may be the case that users are not interested in branches for particular predicates. To reflect this, clauses of the form:

$solve(p(x), some\_info) \leftarrow p(x)$

can be added (combined with extra measures to avoid also using the standard clause for these cases). Notice that analogues of Theorem 15 and Proposition 18 still hold for such meta-programs.

Further extensions of the framework remain possible. A very interesting one is adding *clause*-facts for the *solve*-clauses themselves. The formal definition is as follows.

Definition 20 Let P be a definite program. Let the vanilla meta-interpreter M as before be defined by the following clauses:
$solve(empty) \leftarrow$
$solve(x\&y) \leftarrow solve(x), solve(y)$
$solve(x) \leftarrow clause(x,y), solve(y)$
Then, $M^2{}_P$, the *vanilla meta2-program associated to P*, will be the definite program consisting of M together with the following clause:
$clause(clause(x,y), empty) \leftarrow clause(x,y)$
and a fact of the form
$clause(A, B_1 \& \ldots \& B_n) \leftarrow$
for every clause $A \leftarrow B_1, \ldots, B_n$ in P or in M and a fact of the form
$clause(A, empty) \leftarrow$
for every fact $A \leftarrow$ in P or in M.
If we also include P, we speak about the *amalgamated vanilla meta2-program associated to P*, which we denote by $P + M^2{}_P$.

(Note that $\{clause, solve\} \subset \mathcal{F}_{M^2{}_P}$.)

We have the following theorem:

Theorem 21 Let P be a definite, language independent program and $M^2{}_P$ its vanilla meta2-program. Let H_P and $H_{M^2{}_P}$ denote the least Herbrand model of P and $M^2{}_P$ respectively. Then the following holds:
$\forall \bar{t} \in U_{M^2{}_P} : solve(solve(\bar{t})) \in H_{M^2{}_P} \iff solve(\bar{t}) \in H_{M^2{}_P}$
Moreover, the following holds for every $p/r \in \mathcal{R}_P$:
$\forall \bar{t} \in U_{M^2{}_P}{}^r : solve(p(\bar{t})) \in H_{M^2{}_P} \iff \bar{t} \in U_P{}^r \ \& \ p(\bar{t}) \in H_P$

In other words, for any "layer" of meta-interpretation, we obtain the same correspondence with the object program as before. Formulating and proving a theorem similar to Theorem 15 for amalgamated vanilla meta2-programs is not difficult: We will not do this explicitly.

Yet another step can be taken. We can consider amalgamated meta2-programs in which the "object" clauses contain meta-calls. It is clear that we can in such cases no longer discern between an object- and a meta-level. Results similar to what we obtained before make no sense. But we can, along lines of reasoning analogous to those above, establish the following:

Proposition 22 Let P be a definite, language independent program and $P + M^2{}_P$ its amalgamated vanilla meta2-program. Let PM be a program textually identical to $P + M^2{}_P$, except that an arbitrary number of atoms A in the bodies of clauses in the P-part of it have been replaced by $solve(A)$. Then the following holds:
$H_{P+M^2{}_P} = H_{PM}$

Example 23 In this framework, we can formulate examples similar to some of the more interesting ones in Bowen & Kowalski [BK82]. Consider e.g. the following clause, telling us that a person is convicted for a crime when he is found guilty of it:

$convicted(x, y) \leftarrow person(x), crime(y), solve(guilty(x, y))$

Of course, such possibilities only become really interesting when using extended meta-interpreters – involving e.g. an extra *solve*-argument limiting the resources available for proving a person's guilt.

Further extensions are possible, but we believe that the above sufficiently illustrates the flexibility, elegance and power of our approach. The essence of this section being that in each of these extensions, true atoms of the form $p(\bar{t})$ or $solve(p(\bar{t}), \ldots)$ or even such atoms involving more layers of *solve* correspond to true atoms of the original object theory.

We conclude this section with a brief comment on *reflection*. Basically, the term refers to the transfer of queries/answers from one level of reasoning to another. In this way, in a meta-programming system, we obtain additional inference rules: When something is shown to be true at one level, conclude the truth of the corresponding statement at the other level. Translation of a meta-level goal to the object level is generally termed *downward* reflection, while proceeding in the other direction is known as *upward* reflection. *Reflection rules* were first introduced to artificial intelligence in Weyhrauch [Wey80]. Bowen and Kowalski mentioned them in the context of logic programming in Bowen & Kowalski [BK82]. An example of a meta-programming approach in logic programming that relies heavily on reflection, can be found in Costantini & Lanzarone [CL89]. A number of papers addressing the issue of reflection, together with further references, can be found in Maes & Nardi [MN88]. In terms of our framework, we can point out that the meta-programs of Proposition 18 provide a basis for considering amalgamated programs involving *downward* reflection (as the one in Example 19), while meta2-programs are a starting point for the incorporation of *upward* reflection (see Theorem 21 and Proposition 22). Of course, in this context, these terms do not really refer to inference rules, but rather to the embodiment of the underlying reasoning in actual program clauses.

3.5 S-semantics for Meta-interpreters

We believe that in many applications, especially in the context of deductive databases, the conditions we imposed on the object programs in the past few sections, are very naturally satisfied. However, it is a fact that our basic results no longer hold for classes of object programs significantly surpassing the limitation of language independence. Nevertheless, the actual *practice* of meta-programming reaches beyond this boundary, without experiencing much trouble. The underlying reason for this phenomenon is the fact that least Herbrand semantics does not really accurately reflect the operational behaviour of many logic programs. Indeed, Herbrand models contain only ground atoms, while logic program execution often produces *non-ground answer substitutions*.

In Falaschi et al. [FLMP89] and Falaschi et al. [FLMP93], non-ground Herbrand models are introduced to bridge the gap between declarative semantics and operational behaviour. In this section, we briefly summarise how many of our results can be generalised beyond language independence in the context of the so-called *S-semantics*. The interested reader is referred to Martens & De Schreye [MD95] for a more detailed treatment.

We first recapitulate some relevant basic notions and results concerning the S-semantics for definite logic programs, as it was introduced in Falaschi et al. [FLMP89] and Falaschi et al. [FLMP93] (see also Bossi et al. [BGLM94]).

For atoms A and A', we define $A \leq A'$ (A is *more general* than A') iff there exists a substitution θ such that $A\theta = A'$. The relation $\leq$ is a preorder. Let $\approx$ be the associated equivalence relation (*renaming*). (Similarly for terms.) Then we can define the following.

Definition 24 Let P be a definite program with underlying language $\mathcal{L}_P$. Then its *S-Herbrand universe* $U^S{}_P$ is the quotient set of all terms in $\mathcal{L}_P$ with respect to $\approx$.

So, $U^S{}_P$ basically contains *all* possible terms, not only ground ones. Notice however that terms which are renamings of each other are considered to be one and the same element of $U^S{}_P$. The following definition similarly extends the concept of Herbrand *base*.

Definition 25 Let P be a definite program with underlying language $\mathcal{L}_P$. Then its *S-Herbrand base* $B^S{}_P$ is the quotient set of all atoms in $\mathcal{L}_P$ with respect to $\approx$.

We can now extend the notions of *interpretation*, *truth* and *model*.

Definition 26 Let P be a definite program. Any subset $I^S{}_P$ of $B^S{}_P$ is called an *S-Herbrand interpretation* of P.

Definition 27 Let P be a definite program and $I^S{}_P$ an S-Herbrand interpretation of P.

- A (possibly non-ground) *atom* A in $\mathcal{L}_P$ is *S-true* in $I^S{}_P$ iff there exists an atom A', such that (the equivalence class of) A' belongs to $I^S{}_P$ and $A' \leq A$.
- A definite *clause* $A \leftarrow B_1, \ldots, B_n$ in $\mathcal{L}_P$ is *S-true* in $I^S{}_P$ iff for every $B_1', \ldots, B_n'$ belonging to $I^S{}_P$, if there exists $\theta = mgu((B_1', \ldots, B_n'), (B_1, \ldots, B_n))$, then $A\theta$ belongs to $I^S{}_P$.

Definition 28 Let $I^S{}_P$ be an S-Herbrand interpretation of a definite program P. $I^S{}_P$ is an *S-Herbrand model* of P iff every clause of P is S-true in $I^S{}_P$.

Returning to the semantics of meta-programs, we remind the reader of the fact that we use functorisations which are identical to their associated sets of predicate symbols. As a preliminary result on the level of the universes, we have:

Proposition 29 *Let P be a definite program and M_P its vanilla meta-program. Then $U^S{}_P \subset U^S{}_{M_P}$.*

Proof: Obvious from the definitions.

Our main result on the S-semantics of vanilla meta-interpreters is expressed by the following theorem.

Theorem 30 *Let P be a definite program and M_P its vanilla meta-program. Let $H^S{}_P$ and $H^S{}_{M_P}$ denote the least S-Herbrand model of P and M_P respectively. Then the following holds for every $p/r \in \mathcal{R}_P$:*

$$\forall \bar{t} \in U^S{}_{M_P}{}^r : solve(p(\bar{t})) \in H^S{}_{M_P} \iff \bar{t} \in U^S{}_P{}^r \,\&\, p(\bar{t}) \in H^S{}_P$$

Proof: The proof is similar to that of Theorem 13. It is done by induction, using the generalised T_P operator of the S-semantics and the generalised fixpoint characterisation of the least S-Herbrand model. We omit it.

Having established Theorem 30, the next question that comes to mind is: Can we accomplish a similar feat for Proposition 18? Contrary to our initial expectations, this question has to be answered negatively. Consider the following example. It involves a very simple, definite, non-language independent object program and an equally simple associated extended meta-program.

Example 31

$P: \quad p(x) \leftarrow$

$H^S{}_P = \{p(x)\}$

$E_P: \quad solve(empty) \leftarrow$
$\quad\quad\quad solve(x \& y) \leftarrow solve(x), solve(y)$
$\quad\quad\quad solve(x) \leftarrow clause(x, y), solve(y), inst(x)$
$\quad\quad\quad clause(p(x), empty) \leftarrow$
$\quad\quad\quad inst(p(a)) \leftarrow$

It is easy to see that $solve(p(a)) \in H^S{}_{E_P}$.

The source of the problem is clearly the fact that answers might become further instantiated than is the case in the object program. Since the least S-Herbrand model represents the most general answer substitutions, this means that a straightforward generalisation of Proposition 18 is impossible. One solution is to again impose the restriction of language independence on the object program. Indeed, Proposition 7 ensures us that computed answers for language independent definite programs are ground. It is clear that such answers *can not be further instantiated*. Therefore, language independent programs again prove to be "safe".

In Martens & De Schreye [MD95], we prove the following variant of Proposition 18:

Proposition 32 *Let P be a definite, language independent program and E_P an E-extended meta-program associated with P. Let $H^S{}_P$ and $H^S{}_{E_P}$ denote the least S-Herbrand model of P and E_P respectively. Then the following holds for every $p/r \in \mathcal{R}_P$:*
$$\forall \bar{t} \in U^S{}_{E_P}{}^r : (\exists \bar{s} \in U^S{}_{E_P}{}^n : solve(p(\bar{t}), \bar{s}) \in H^S{}_{E_P}) \Longrightarrow \bar{t} \in U^S{}_P{}^r \ \& \ p(\bar{t}) \in H^S{}_P$$

One should take care not to draw wrong conclusions from Example 31. The example is very typical for the class of programs for which a straightforward reformulation of Proposition 18 to the case of the S-semantics and for non-language independent programs does not hold. More specifically, in any such counterexample, the extended meta-interpreter, E, needs to contain an explicit reference to an object level predicate symbol. In Example 31, this is the case in the clause $inst(p(a)) \leftarrow$. It would therefore be unjust to consider such examples as an indication of a problem with the semantics. It is more reasonable to assume that, in the case that such an explicit reference to an object level predicate results in a weaker correspondence between the object- and the meta-theory's semantics, then, most likely, this difference simply reflects the programmer's intentions.

It is also interesting to compare Proposition 32 with the following conjecture[3]:

Conjecture *Let P be a definite program and E_P an E-extended meta-program associated with P. Let $H^S{}_P$ and $H^S{}_{E_P}$ denote their least S-Herbrand models. Then the following holds for every $p/r \in \mathcal{R}_P$:*
$$\forall \bar{t} \in U^S{}_{E_P}{}^r : (\exists \bar{s} \in U^S{}_{E_P}{}^n : solve(p(\bar{t}), \bar{s}) \in H^S{}_{E_P}) \Longrightarrow$$
$$\exists \bar{t'} \in U^S{}_P{}^r : (\bar{t'} \leq \bar{t} \ \& \ p(\bar{t'}) \in H^S{}_P)$$

Obviously, this result is weaker then Proposition 32, but we conjecture that it holds for *all* (definite) object programs, not just for the language independent ones.

A treatment of amalgamated meta-programs and meta2-programs in the context of S-semantics is straightforward and not particularly enlightening. All the results from Section 3.4 can be generalised in the expected way.

3.6 Reasoning about Theories and Provability

3.6.1 An explicit theory argument

An interesting variant of the meta-programs we considered above, is obtained when *an extra argument, denoting a particular object theory*, is added to the *solve* and *clause* predicates.

The basic definitions are as follows:

Definition 33 We call the following definite program M^t *th-vanilla meta-interpreter:*
$solve(t, empty) \leftarrow$ \hfill (*)

---
[3]suggested by Marianne Kalsbeek

$solve(t, x\&y) \leftarrow solve(t,x), solve(t,y)$
$solve(t,x) \leftarrow clause(t,x,y), solve(t,y)$

Definition 34 Let P be a definite program. Then $M^t{}_P$, the *th-vanilla meta-program associated with* P, is the definite program consisting of M^t together with a fact of the form
$clause(P, A, B_1 \& \ldots \& B_n) \leftarrow$
for every clause $A \leftarrow B_1, \ldots, B_n$ in P and a fact of the form
$clause(P, A, empty) \leftarrow$
for every fact $A \leftarrow$ in P.

In the above definition, the constant P in the *clause*-facts indicates that they correspond to clauses in the object program P.

Obviously, all earlier results can straightforwardly be generalised to programs of this form. Notice that clause (*) in Definition 33 does not provide a range for its theory argument. Thus, $H_{M^t{}_P}$ will contain numerous irrelevant ground atoms, where the theory argument is instantiated to some other term than the constant P. This can be avoided by either introducing a *theory* range predicate for theories, or typing (see Section 3.7 for some related comments). Of course, an S-semantics approach is likewise free from such inconveniences.

The meta-program in Definition 34 deals with a single object program. Extending it to cope with several such object programs is straightforward: Simply introduce more constants referring to object theories and annotate the *clause* facts correspondingly. More interesting are meta-programs surpassing the vanilla context by defining interactions between the different object theories.

Consider the following example, adapted from Brogi et al. [BMPT90]. It consists of a meta-interpretative definition of (definite) logic program intersection.

Example 35
$solve(t, empty) \leftarrow$
$solve(t, x\&y) \leftarrow solve(t,x), solve(t,y)$
$solve(t,x) \leftarrow clause(t,x,y), solve(t,y)$
$clause(t \cap t', x, y\&y') \leftarrow clause(t,x,y), clause(t',x,y')$ (1)
And, of course, a number of facts of the form
$clause(P_1, A, B_1 \& \ldots \& B_n)$
$\vdots$
$clause(P_k, C, D_1 \& \ldots \& D_n)$
to represent the object programs P_1 to P_k, respectively.

Program union can be defined similarly, as shown in Brogi et al. [BMPT90].

Example 36 We replace clause (1) in Example 35, by the following two clauses:
$clause(t \cup t', x, y) \leftarrow clause(t, x, y)$
$clause(t \cup t', x, y) \leftarrow clause(t', x, y)$

Notice the latter two clauses again do not provide a (full) range for their theory argument.

3.6.2 The *demo* predicate

In many meta-programming applications in logic programming, not the *solve* meta-interpreter, but a related program is used. It involves the so-called *demo* or *demonstrate* predicate. This name refers to a somewhat different origin and/or underlying motivation of most of this work. Indeed, while the *solve* interpreter refers to (possibly modified) query answering in logic programming, the *demo* program is situated in the context of formalising provability. The difference is obviously rather subtle, but, in general, one can perhaps say that *solve* has a slightly more procedural flavour than *demo*.

The *demo* predicate was originally introduced in chapter 12 of Kowalski [Kow79] and further elaborated upon in among others Bowen & Kowalski [BK82] and Eshghi [Esh86]. In Kowalski [Kow90], we find the following definition of a *demo* for definite propositional programs.

Definition 37
$demo(t, true) \leftarrow$
$demo(t, p\&q) \leftarrow demo(t, p), demo(t, q)$
$demo(t, p) \leftarrow demo(t, p \leftarrow q), demo(t, q)$

The same meta-interpreter was used in Brogi et al. [BMPT92] as the basis for a reformulation of the work in Brogi et al. [BMPT90].

The most striking difference with the M^t program in Definition 33 above, is *the absence of a clause predicate*. Kowalski prefers the formulation in Definition 37 because of its greater generality and its similarity with a modal logic approach. However, in the context of a meta extension/simulation of an object program, the formulation with *clause* seems more natural to us. Indeed, the distinction between data in the program and results derivable from the program remains more clear.

A related issue is the choice between *a term- and a constant-based representation of the object theory*. In Definition 34 and in Examples 35 and 36, we introduced an extra argument in the *clause*-facts and -clauses, and named object theories through meta-level constants. The same technique can be used with the *demo* predicate. In that case, Definition 37 needs to be extended with *demo*-facts, representing the object theory, in exactly the same way as the *clause*-facts above. An alternative possibility is representing the object theory as a term in the meta-goal. We adapt the following simple example from Kowalski [Kow90].

Example 38 Execution of the object level program and query

$p \leftarrow q, r$
$q \leftarrow s$
$r \leftarrow$
$s \leftarrow$
$\leftarrow p$

can be simulated by executing the meta-query

$\leftarrow demo([p{\leftarrow}q\&r, q{\leftarrow}s, r{\leftarrow}true, s{\leftarrow}true], p)$

on the condition that the program in Definition 37 is extended with the following clause:

$demo(t, p) \leftarrow member(p, t)$

and a definition of the *member* predicate.

This works well for propositional object theories, but can produce inconvenient variable bindings when used for object programs that contain variables. Moreover, since the object theory is no longer represented in the meta-program itself, comparing least Herbrand models does not immediately bring us anything. Even an extended least Herbrand model, somehow incorporating the information contained in the query, does not really refer to the object theory, but to ground instances of that theory. Therefore, in these cases, a ground representation of the object theory (using meta-level constants to refer to object-level variables) might be preferable. We briefly return to this issue in Section 3.7.

Further applications and comments, including a number of interesting references, can be found in Kowalski [Kow90]. Finally, notice that Kowalski & Kim [KK91] as well as Kowalski [Kow90] use symbol overloading as addressed in Section 3.3.

3.7 Discussion and Conclusions

This section contains some further comments on our approach to meta-semantics and its results, and a discussion of some related work. Since there is a vast literature on meta-logic, its semantics, possible applications, advantages and disadvantages, we do not strive for completeness. Instead, we only consider some closely related papers. For a discussion on the relation with others, such as Sato [Sat92], Bonatti [Bon92], Chen et al. [CKW93] and Subrahmanian [Sub89], we refer to the discussion in Martens & De Schreye [MD95].

A work very closely related to what we presented here is Hill & Lloyd [HL89]. In the first part of that paper, it is shown that through the use of appropriate typing, vanilla meta-programs can be given a suitable declarative (the well-known Clark's completion semantics is used) and procedural semantics. Moreover, in a second part, a ground representation for object level terms at the meta-level is considered, and it is shown how a number of problematic Prolog built-ins (static and of the type called "first-order", i.e. not referring to clauses or goals, in Apt et al. [AMP92]) can be given a declarative meaning in this setting.

Addressing the latter topic first, it should be noted that such Prolog meta-predicates, of which $var/1$ and $nonvar/1$ are prototypical examples, are not included in our language. This certainly puts some limitation on the obtained expressiveness. Observe, however, that in the typed non-ground representation proposed in Hill & Lloyd [HL89], no alternative for $var/1$ was introduced either. Apt et al. [AMP92] proposes a declarative semantics for such predicates. We conjecture that, if one so desires, our basic methodology can be adapted to the semantics described there, thus enabling the inclusion of such built-ins in our language. A, perhaps superior, alternative for providing the latter kind of facilities are *delay* control annotations as in Gödel.

For the $assert/1$ and $retract/1$ Prolog built-in predicates, the solution of Hill & Lloyd [HL88], to represent dynamic theories as terms in the meta-program, can as well be applied in our approach. However, as was noted in Section 3.6 above, this requires special care with variable bindings, and leads to some inconveniences in the context of a ground Herbrand model approach to semantics. A thorough discussion of the problems related to these predicates is given in Hill & Lloyd [HL88] and Lloyd [Llo88]. As a final remark about Prolog built-ins, we would like to mention that our use of overloading largely eliminates the need for a $call/1$ predicate, as Example 19 illustrates.

Next, observe that the condition of range restriction, which is the practical, verifiable, sufficient condition for language independence, is strongly related to typing. Indeed, typing can in principle be converted into additional atoms that are added in the bodies of clauses, expressing the range of each variable. See e.g. Enderton [End72] and Lloyd [Llo87]. In this context, it is interesting to note the apparent duality between range restriction at the object level and typing at the meta-level. If one "hardwires" types into the code of the object program through range restriction, typing (or range restriction) at the meta-level is no longer required for a sensible declarative semantics. Gallagher & de Waal [GdW93] can also be mentioned here. It presents a program transformation technique that enables to minimise run-time type checking in systems which represent types as (unary) predicates, thus largely eliminating one of the main advantages types might offer in comparison with ranges.

Finally, Hill & Lloyd [HL89] does not address amalgamation. And Gödel (see Hill & Lloyd [HL94]), the programming language whose extensive meta-programming facilities are largely based upon the foundations laid out in Hill & Lloyd [HL89] and Hill & Lloyd [HL88] does not allow it. While dealing with amalgamated programs of the kind introduced in Definition 14 and Example 19 seems relatively straightforward, a generalisation of the typed approach to meta2-programs is probably not immediate.

With respect to the extension to amalgamated programs, we would like to repeat that our use of overloading is related to the use of ambivalent logics, blurring the distinction between terms and formulas. A first such logic has been proposed (in a different context) by Richards [Ric74]. Indeed, in his analysis of the problems connected with reference and modality, Richards considers logical languages that contain their *well-formed formulas as terms*. He interprets these languages on specially devised models whose domains are a union of *the constants and the sentences* (i.e. closed formulas) in the language.

As pointed out in Section 3.3, Kalsbeek [Kal95] studies the Herbrand semantics of the vanilla meta-interpreter in a language that features a certain degree of ambivalence, allowing atoms as terms. In the logic framework thus constructed, soundness and completeness of definite vanilla meta-programs with respect to their definite object program, restricted to terms in the *object* level language, is obtained as the main result.

Jiang [Jia94] proposes a much more ambivalent language, allowing arbitrary formulas as terms as well as *terms as formulas*. It is shown that a number of interesting properties (Herbrand theorem, completeness), lost in Richards' logic, are recovered. Vanilla meta-programs are considered, leading to similar results as obtained by others. The framework is however more powerful and particularly suitable for addressing quantified object level statements and full amalgamation. To this end, a sophisticated treatment of variables, not syntactically distinguishing between variables and their names is proposed. This allows to consider meta-theories dealing with full first order object statements.

Finally, we can also mention Kalsbeek & Jiang [KJ95]. Starting from first principles, this paper contains a detailed formal treatment of ambivalent logic, its syntax and semantics, main logic properties and unification theory.

Summarising, our technique of overloading function and predicate symbols is probably less powerful than approaches using more fully ambivalent syntax, but it requires no modification of the familiar notion of Herbrand interpretations and models. For a study of programs with negation as failure and their ground Herbrand semantics, we refer the interested reader to Martens & De Schreye [MD95]. In that paper, we generalise the notion of language independence to stratified normal programs. We show that non-ground vanilla-like meta-programs associated with stratified object programs are weakly stratified and investigate correspondences between the perfect object- and the weakly perfect meta-model. Appropriate versions of Theorems 13, 15 and 21 are shown to hold, but generalising Proposition 18 requires more care. Finally, no generally accepted extension of S-semantics to the class of normal logic programs has been proposed yet. An account of recent developments can be found in Sections 6.3 and 6.4 of Bossi et al. [BGLM94].

To conclude, in this paper, we have considered definite untyped non-ground meta-programs. We have studied rather extensively the semantic properties of vanilla meta-interpreters of this kind. And we have looked at interesting extensions and variants involving (a limited form of) amalgamation. It turned out that in most of these cases, the least Herbrand semantics is well-behaved for definite *language independent* object programs. This is an interesting result since these semantics are widely accepted as good declarative semantics for logic programs. Moreover, we believe that our methodology can often also be applied when considering untyped, non-ground meta-programs that do not immediately fall within one of the categories we explicitly considered. So, contrary to what was generally assumed, untyped non-ground meta-programming often does not really present semantic problems.

We have also shown how, for non-extended meta-programs, the restriction of *language independence* can be lifted in the context of a declarative semantics that more closely reflects the procedural behaviour of logic programs. These results explain why the language independence condition almost never surfaces in logic programming practice. Next, we addressed some issues which arise in the context of meta-programs with an explicit theory argument. And we briefly discussed the formalisation of provability in the related *demo* predicate, including an indication of some limitations inherent to our current work. Finally, we compared our approach with some relevant other work.

The question remains for which type of applications the proposed semantics is adequate. Applications in the area of compositionality of theories, such as discussed in Subsection 3.6.1, definitely fall within the scope of our approach. Brogi & Turini [BT94] gives a detailed account of this type of application and Brogi & Contiero [BC94] discusses the alternative of using the ground representation. Expert systems written as meta-programs (extending Example 17) form another application area. The examples in Sterling & Beer [SB89] can typically be correctly understood on the basis of our semantics. An exception in this respect are meta-programs aimed at controlling the procedural behaviour of the object program, such as those presented in Gallaire & Lasserre [GL82]. Such programs need the ability to inspect the instantiation of object level variables, which is currently not supported in our semantics (see the comments concerning $var/1$ and $nonvar/1$ above). As mentioned before, the semantics is also suited for dealing with limited forms of amalgamated knowledge, although we are unaware of fully developed applications of this type.

On the negative side, it is our current understanding that for applications in program development and verification, such as program transformation, program analysis (e.g. abstract interpretation), program synthesis and termination analysis, the ground representation seems more appropriate. The problem with our semantics for these applications is similar to the problem pointed out in Subsection 3.6.2 regarding the *demo* predicate with a term representation of the object theory. Typical for most program development and verification applications is the need to be able to reason about SLD(NF)-trees, instead of merely individual SLD(NF)-derivations. As a result, in the meta-program, several different bindings for a same object level variable may need to be considered. In a non-ground representation, this requires a *copying* functionality, the (declarative) semantics of which is very unclear. Note that, although the Prolog *assert*/1 provides this functionality, its declarative counterpart, using term representations of dynamic theories, does not.

Acknowledgements

We thank Antonio Brogi, François Bry, Michael Codish, Marc Denecker, Wlodzimierz Drabent, Michael Gelfond, Pat Hill, Yuejun Jiang, Marianne Kalsbeek, Robert Kowalski, John Lloyd, Rodney Topor and Frank van Harmelen for interesting discussions and/or comments. We also appreciated helpful observations from anonymous referees. Bern Martens is supported by the Belgian GOA "Non-standard applications of abstract interpretation" and by the Belgian National Fund for Scientific Research. Danny De Schreye is a senior research associate of the Belgian National Fund for Scientific Research.

References

[AMP92] K. R. Apt, E. Marchiori, and C. Palamidessi. A theory of first-order built-in's ot Prolog. In H. Kirchner and G. Levi, editors, *Proceedings of the 3rd International Conference on Algebraic and Logic Programming*, pages 69–83. Springer-Verlag, LNCS 632, 1992.

[Apt90] K. R. Apt. Logic programming. In J. van Leeuwen, editor, *Handbook of Theoretical Computer Science, Volume B, Formal Models and Semantics*, pages 493–574. Elsevier Science Publishers B.V., 1990.

[BC94] A. Brogi and S. Contiero. Gödel as a meta language for composing logic programs. In A. Turini, editor, *Proceedings Meta'94*. University of Pisa, 1994.

[BDM88] F. Bry, H. Decker, and R. Manthey. A uniform approach to constraint satisfaction and constraint satisfiability in deductive databases. In *Proceedings EDBT'88*, March 1988.

[BGLM94] A. Bossi, M. Gabbrielli, G. Levi, and M. Martelli. The S-semantics approach: Theory and applications. *Journal of Logic Programming*, 19/20:149–197, 1994.

[BK82] K. A. Bowen and R. A. Kowalski. Amalgamating language and metalanguage in logic programming. In K. L. Clark and S.-Å. Tärnlund, editors, *Logic Programming*, pages 153–172. Academic Press, 1982.

[BMM92] F. Bry, R. Manthey, and B. Martens. Integrity verification in knowledge bases. In A. Voronkov, editor, *Proceedings 1st and 2nd Russian Conference on Logic Programming*, pages 114–139. Springer-Verlag, LNAI 592, 1992.

[BMPT90]   A. Brogi, P. Mancarella, D. Pedreschi, and F. Turini. Composition operators for logic theories. In J. W. Lloyd, editor, *Proceedings of the Esprit Symposium on Computational Logic*, pages 117–134. Springer-Verlag, November 1990.

[BMPT92]   A. Brogi, P. Mancarella, D. Pedreschi, and F. Turini. Meta for modularising logic programming. In A. Pettorossi, editor, *Proceedings Meta'92*, pages 105–119. Springer-Verlag, LNCS 649, 1992.

[Bon92]   P. A. Bonatti. Model theoretic semantics for Demo. In A. Pettorossi, editor, *Proceedings Meta'92*, pages 220–234. Springer-Verlag, LNCS 649, 1992.

[Bow85]   K. A. Bowen. Meta-level programming and knowledge representation. *New Generation Computing*, 3(4):359–383, 1985.

[Bry90]   F. Bry. Query evaluation in recursive databases: Bottom-up and top-down reconciled. *Data & Knowledge Engineering*, 5(4):289–312, 1990.

[BT94]   A. Brogi and F. Turini. Semantics of meta-logic in an algebra of programs. In S. Abramski, editor, *Proceedings Ninth Annual IEEE Symposium on Logic in Computer Science, LICS'94*. IEEE Society Press, 1994.

[CKW93]   W. Chen, M. Kifer, and D. S. Warren. HiLog: A foundation for higher-order logic programming. *Journal of Logic Programming*, 15(3):187–230, 1993.

[CL89]   S. Costantini and G. A. Lanzarone. A metalogic programming language. In G. Levi and M. Martelli, editors, *Proceedings ICLP'89*, pages 218–233, Lisbon, Portugal, June 1989. MIT Press.

[DM92]   D. De Schreye and B. Martens. A sensible least Herbrand semantics for untyped vanilla meta-programming and its extension to a limited form of amalgamation. In A. Pettorossi, editor, *Proceedings Meta'92*, pages 192–204. Springer-Verlag, LNCS 649, 1992.

[End72]   H. B. Enderton. *A Mathematical Introduction to Logic*. Academic Press, 1972.

[Esh86]   K. Eshghi. *Meta-Language in Logic Programming*. PhD thesis, Department of Computing, Imperial College, London, U.K., 1986.

[Fit90]   M. Fitting. *First-Order Logic and Automated Theorem Proving*. Springer-Verlag, 1990.

[FLMP89]   M. Falaschi, G. Levi, M. Martelli, and C. Palamidessi. Declarative modeling of the operational behaviour of logic programs. *Theoretical Computer Science*, 69:289–318, 1989.

[FLMP93]   M. Falaschi, G. Levi, M. Martelli, and C. Palamidessi. A model-theoretic reconstruction of the operational semantics of logic programs. *Information and Computation*, 103(1):86–113, 1993.

[Gal86]   J. Gallagher. Transforming logic programs by specialising interpreters. In *Proceedings ECAI'86*, pages 109–122, 1986.

[GdW93]   J. Gallagher and D. A. de Waal. Deletion of redundant unary type predicates from logic programs. In K.-K. Lau and T. Clement, editors, *Proceedings LOPSTR'92*, pages 151–167. Springer-Verlag, Workshops in Computing Series, 1993.

[GL82]   H. Gallaire and C. Lasserre. Metalevel control for logic programs. In K. L. Clark and S.-Å. Tärnlund, editors, *Logic Programming*, pages 173–185. Academic Press, 1982.

[HL88]   P. M. Hill and J. W. Lloyd. Meta-programming for dynamic knowledge bases. Technical Report CS-88-18, Computer Science Department, University of Bristol, U.K., 1988.

[HL89]   P. M. Hill and J. W. Lloyd. Analysis of meta-programs. In H. D. Abramson and M. H. Rogers, editors, *Proceedings Meta'88*, pages 23–51. MIT Press, 1989.

[HL94]   P. Hill and J. Lloyd. *The Gödel Programming Language*. MIT Press, 1994.

[Jia94]   Y. Jiang. Ambivalent logic as the semantic basis of metalogic programming: I. In P. Van Hentenryck, editor, *Proceedings ICLP'94*, pages 387–401. MIT Press, June 1994.

[Kal95]   M. Kalsbeek. Correctness of the Vanilla meta-interpreter and ambivalent syntax. This volume.

[KJ95]   M. Kalsbeek and Y. Jiang. A vademecum of ambivalent logic. This volume.

[KK91]   R. A. Kowalski and J.-S. Kim. A metalogic programming approach to multi-agent knowledge and belief. In V. Lifschitz, editor, *Artificial Intelligence and Mathematical Theory of Computation*, pages 231–246. Academic Press, 1991.

[Kow79]   R. A. Kowalski. *Logic for Problem Solving*. North-Holland, 1979.

[Kow90]   R. A. Kowalski. Problems and promises of computational logic. In J. W. Lloyd, editor, *Proceedings of the Esprit Symposium on Computational Logic*, pages 1–36. Springer-Verlag, November 1990.

[Llo87]   J. W. Lloyd. *Foundations of Logic Programming*. Springer-Verlag, 1987.

[Llo88]   J. W. Lloyd. Directions for meta-programming. In *Proceedings FGCS'88*, pages 609–617. ICOT, 1988.

[LR93]   G. Levi and D. Ramundo. A formalization of metaprogramming for real. In D. S. Warren, editor, *Proceedings ICLP'93*, pages 354–373, Budapest, June 1993. MIT Press.

[MB88]   R. Manthey and F. Bry. SATCHMO: a theorem prover implemented in Prolog. In E. Lusk and R. Overbeek, editors, *Proceedings CADE'88*, pages 415–434. Springer-Verlag, LNCS 310, May 1988.

[MD95]   B. Martens and D. De Schreye. Why untyped non-ground meta-programming is not (much of) a problem. *Journal of Logic Programming*, 22(1):47–99, 1995.

[MN88]   P. Maes and D. Nardi, editors. *Meta-Level Architectures and Reflection*. North-Holland, 1988.

[Nic82]   J.-M. Nicolas. Logic for improving integrity checking in relational databases. *Acta Informatica*, 18(3):227–253, 1982.

[Ram88]   A. Ramsay. *Formal Methods in Artificial Intelligence*. Cambridge University Press, 1988.

[Ric74]   B. Richards. A point of reference. *Synthese*, 28:361–454, 1974.

[Sat92]   T. Sato. Meta-programming through a truth predicate. In K. Apt, editor, *Proceedings JICSLP'92*, pages 526–540, Washington, November 1992. MIT Press.

[SB89]   L. Sterling and R. D. Beer. Meta interpreters for expert system construction. *Journal of Logic Programming*, 6(1&2):163–178, 1989.

[SS86]   L. Sterling and E. Shapiro. *The Art of Prolog*. MIT Press, 1986.

[Sub89]   V. S. Subrahmanian. A simple formulation of the theory of metalogic programming. In H. D. Abramson and M. H. Rogers, editors, *Proceedings Meta'88*, pages 65–101. MIT Press, 1989.

[TF86]   A. Takeuchi and K. Furukawa. Partial evaluation of Prolog programs and its application to metaprogramming. In H.-J. Kugler, editor, *Information Processing 86*, pages 415–420, 1986.

[TP92]   J. L. Träff and S. D. Prestwich. Meta-programming for reordering literals in deductive databases. In A. Pettorossi, editor, *Proceedings Meta'92*, pages 280–293. Springer-Verlag, LNCS 649, 1992.

[Ull88]   J. D. Ullman. *Database and Knowledge-Base Systems, Volume I*. Computer Science Press, 1988.

[Wey80]   R. W. Weyhrauch. Prolegomena to a theory of mechanized formal reasoning. *Artificial Intelligence*, 13(1&2):133–170, 1980.

4 Meta-Logic for Program Composition: Semantics Issues

Antonio Brogi and Franco Turini

Abstract

Meta-programming is a powerful technique for extending and modifying the semantics of an existing object language. Along with the expressiveness, however, meta-programming puts forth some subtle semantics problems, among which the most critical is bound to the representation of object programs at the meta-level. We propose a semantic justification for a simple representation technique in the field of a generalised notion of meta-programming in logic. The generalisation consists in specifying the meta-programs with respect to object programs defined via program expressions. The expressions are defined by means of a rich suite of operations on logic programs. The technique allows one to build straightforward and concise meta-programs via the representation of object level variables by meta-level variables.

4.1 Introduction

Meta-logic can be fruitfully employed for composing separate object level programs in various ways. Such a use of meta-logic permits the evaluation of queries with respect to arbitrary program expressions rather than with respect to a single object program. Roughly, a program expression defines a combination of object programs through a set of basic composition operations. The provability of a query with respect to a composition of programs can be defined by meta-axioms specifying the intended meaning of the various composition operations.

We consider a set of basic composition operations that make up an algebra of logic programs with interesting properties for reasoning about programs and program compositions, originally introduced by Mancarella and Pedreschi [MP88] and by Brogi [Bro93]. The usefulness of the suite of proposed operations for the rational reconstruction of knowledge representation and software engineering techniques has been widely illustrated by Brogi, Mancarella, Pedreschi and Turini [Bro93, BMPT90, BMPT94, BT90]. As an example, the use of program composition operations for modularising logic programming is discussed in this paper. (A thorough description of this application is reported in [BMPT94]).

The semantics of meta-logic is one of the most debated issues in the logic programming field. Inconsistencies may arise, for instance, when object language and meta-language are amalgamated in the same language. In other chapters of this volume, Jiang [JK95], Jiang and Kalsbeek [Kal95], and Martens and DeSchreye [MD95] thoroughly discuss the problems of representing object level constructs within an amalgamated logic language, in which an ambivalent syntax is employed. Another way of referring to the use of ambivalent syntax is to say that we are using a *non ground* representation, in which object level variables are represented by meta-level variables.

In this paper, we aim at showing that there is a sensible set of conditions that makes the use of the non-ground representation semantically sound for a meta-logic extended in order to deal with program expressions. The main result of the paper is a theorem that justifies the use of a flexible and straightforward representation of object programs in the extended meta-logic. The soundness of the meta-logical axioms that define our suite of operations is proved with respect to a semantics based on the immediate consequence operator. Brogi and Turini [BT95] showed how this choice is the proper one by proving that the semantics enjoys the properties of compositionality and full abstraction.

The original definition of the semantics of the operations was given in terms of the standard immediate consequence operator as defined over *ground* interpretations. Here, to make the proof of soundness, that is the correspondence between the expected semantics and the answers provided by the meta-interpreter, more stringent, we lift our semantic definitions to *non-ground* interpretations, as provided in the S-semantics by Falaschi et al. [FLMP89]. It is worth noting that our result extends the results presented by Martens and DeSchreye in [MD92] to a meta-logical language which allows program expressions.

The plan of the paper is the following. In section 4.2 we present an extension to logic programming consisting of a set of program composition operations that allow one to compute a goal with respect to a *program expression* rather than with respect to a plain program, as it is usual. The abstract meaning of the operations used to build program expressions is given in terms of the *ground* immediate consequence operator. In section 4.2.2 we show a possible use of this extension by discussing a modularisation mechanism for logic programming based on the operations. In section 4.2.3, we lift the semantic definitions of the operations to *non-ground* interpretations and we state the correctness of the lifting. In section 4.3 we introduce the classic approach to meta-logic and how the related semantics problems have been tackled so far. Then we show how this standard approach can be extended to capture program expressions. section 4.3.3 contains the principal result, that is, the correctness of the meta-logical axioms with respect to the abstract semantics of program expressions. A preliminary version of this paper appeared in [BT94].

4.2 An Algebra of Logic Programs

This section is devoted to introduce the set of composition operations devising the algebra of logic programs, originally defined by Brogi, Mancarella, Pedreschi and Turini in [Bro93, BMPT94, MP88].

4.2.1 Program composition operations

Four basic operations for composing logic programs are introduced: Encapsulation (denoted by *), union (∪), intersection (∩), and import (◁). The operations are defined in a semantics-driven style, following the intuition that if the meaning of a program P is denoted by the corresponding *immediate consequence operator* $T(P)$ then such a meaning is a homomorphism for several interesting operations on programs. In contrast, the standard least Herbrand model semantics of logic programming is not appropriate to model compositions of programs, as it does not enjoy the compositionality requirement. As an example, the least Herbrand model of a program cannot be obtained, in general, from the least Herbrand models of its clauses. Each program P is therefore denoted by the corresponding $T(P)$.

Recall that, for a logic program P, the immediate consequence operator $T(P)$ is a continuous mapping over Herbrand interpretations defined as follows by van Emden and Kowalski [vEK76]. For any Herbrand interpretation I:

$$A \in T(P)(I) \iff (\exists \bar{B} : A \leftarrow \bar{B} \in ground(P) \land \bar{B} \subseteq I)$$

where $\bar{B}$ is a (possibly empty) conjunction of atoms. The powers of $T(P)$ are defined as usual:

$$\begin{aligned} T^0(P)(I) &= I \\ T^{n+1}(P)(I) &= T(P)\,(T^n(P)(I)) \\ T^\omega(P)(I) &= \bigcup_{n<\omega} T^n(P)(I) \end{aligned}$$

and $T^\alpha(P)(\emptyset)$ is abbreviated to $T^\alpha(P)$.

The semantics of program compositions is given in a compositional way by extending the definition of T with respect to the first argument. For any Herbrand interpretation I:

$$\begin{aligned} T(P^*)(I) &= T^\omega(P) \\ T(P \cup Q)(I) &= T(P)(I) \cup T(Q)(I) \end{aligned}$$

$$T(P \cap Q)(I) = T(P)(I) \cap T(Q)(I)$$
$$T(P \triangleleft Q)(I) = T(P)(I \cup T^\omega(Q))$$

The above definition generalises the notion of immediate consequence operator from programs to compositions of programs. In particular, for any interpretation I, the formulae that may be derived in an encapsulated program P^* are precisely the formulae that may be derived from P in an arbitrary (finite) number of steps. The operations of union and intersection of programs directly relate to their set-theoretic equivalent. The set of immediate consequences of the union (resp. intersection) of two programs is the set-theoretic union (resp. intersection) of the set of immediate consequences of the separate programs. Finally, for any interpretation I, the immediate consequences of the import of two programs $P \triangleleft Q$ are the set of formulae that may be derived in the importing program P in a single deduction step from I and from the set of formulae that may be derived in the imported program Q in an arbitrary (finite) number of steps.

The operations $*$, $\cup$, $\cap$ and $\triangleleft$ satisfy a number of algebraic properties such as associativity, commutativity and distributivity. The resulting algebra studied by Brogi [Bro93] extends the algebra presented by Mancarella and Pedreschi [MP88] and offers a formal basis for proving properties of program compositions. For instance, syntactically different program compositions may be compared and simplified by means of the properties of program composition operations.

4.2.2 A motivating example

We show how the previously introduced operations can be used to address one of the hot issues in the logic programming field, that is how to provide logic programming with a useful and, at the same time, logic-based modular structure. A more thorough presentation of these ideas can be found in [Bro93, BMPT94].

Traditional modular programming requires that a module can import from another one only its functionality without caring of the implementation. In the logic programming setting, this means that the clauses of the imported module can be used for proving a goal but they are not visible to and usable directly by the clauses of the importing module. For example, if a module needs a *sort* operation over lists, the modular style requires the importation of a *sort* relation from another module, which is free to implement it according to any of the sorting algorithms.

This kind of behaviour can be easily realised in logic programming by exploiting the operators $\cup$ and $*$. If P is the "main" program and S is the module that implements the *sort* relation, the combined program is expected to behave as:

$$P \cup S^*.$$

The usefulness of the operator $\cap$ for knowledge representation and reasoning has been shown by Brogi, Mancarella, Pedreschi and Turini elsewhere [BMPT90, BT90]. Here we show that, in the context of modularising logic programs, it can be used to restrict import/export operations to a set of pre-selected predicates. Given a set of predicates π, let π itself denote the program:

$$\{p(x) \leftarrow | \ p \in \pi\}.$$

Now, the importation of only the extensional definitions of predicates in π from a module Q to a module P is easily defined as:

$$P \cup (\pi \cap Q^*).$$

Yet, the very nature of logic programming allows a more refined definition of module composition. Indeed, while in traditional languages the objects that can be imported/exported are either data or functions/procedures, in logic programming we can distinguish between either importing/exporting a whole predicate definition or part of it, that is, some of its clauses. Even more drastically, we can think that a predicate definition is in general spread over several different modules.

By exploiting this possibility, we obtain several forms of information hiding, based on the observation that the knowledge about a procedure is available at two different levels: the *intensional* and the *extensional* level. The former is (an abstraction of) the code of a predicate (i.e., the clauses defining it). The latter is the set of atomic formulae provable for that predicate. The exportation at the extensional level is the one that is supported by conventional modular languages, and the operations $\cup$, $\cap$ and $*$ provide suitable mechanisms for dealing with it.

The operation $\triangleleft$ allows us to exploit the potential flexibility of logic programming with respect to importing/exporting knowledge. In its basic usage, the $\triangleleft$ operation builds a module $P \triangleleft Q$ out of a pair of modules P and Q, which play the role of the *visible* and *hidden* part of the module, respectively. Several forms of information hiding/exporting can be accommodated by exploiting the operation $\triangleleft$:

- *Full hiding*: a predicate is defined only in the hidden part Q of a module $P \triangleleft Q$. In such a way, the predicate is exported neither at the extensional nor at the intensional level.

- *Implementation hiding*: a predicate is defined in the visible part P of a module $P \triangleleft Q$, but its actual implementation is entirely in the hidden part Q. In such a way the predicate is exported only at the extensional level.

- *Partial visibility*: a predicate is defined partially in the visible part P and partially in the hidden part Q of a module $P \triangleleft Q$. In such a way the predicate is partially exported at the intensional level.
- *Full visibility*: a predicate is defined in the visible part P of a module $P \triangleleft Q$. In such a way the predicate is exported both at the intensional and at the extensional level.

It is worth stressing that the adoption of the $T(P)$ semantics accounts for both the intensional and the extensional levels. In fact, $T(P)$ itself is the semantic counterpart of the intensional level, whereas its least fixpoint $T^\omega(P)$ is the semantic counterpart of the extensional level.

As an example of use of $\triangleleft$, consider the following simple situation. Suppose that a company is supported by an expert system that simulates the behaviour of a "Wall Street guru" suggesting possible investments for the company. One of the procedures of the expert system involves judging the quality of investments, and this may involve some confidential information that should be kept hidden. This can be achieved by implementing the expert system by means of a module expression in which the intension of the critical predicates is partly hidden. In what follows the programs *Rules*, *Public* and *Private* represent the basic expert system rules, the public domain database information and the critical expert system knowledge, respectively.

Rules
$invest\_on(amount, stock) \leftarrow$
    $good\_investment(stock),$
    $determine(amount)$

$good\_investment(stock) \leftarrow$
    $high\_risk(stock),$
    $good\_trend(stock)$

$determine(amount) \leftarrow$
    $capital(total),$
    $percentage(total, amount)$

...

Public
$high\_risk(ibm) \leftarrow$
$good\_trend(sun) \leftarrow$

...

Private
$good\_investment(stock) \leftarrow$
$\qquad suggests(x, stock),$
$\qquad insider(x, stock)$
$good\_investment(stock) \leftarrow$
$\qquad suggests(x, stock),$
$\qquad wizard(x)$
$suggests(john, ibm) \leftarrow$
$suggests(x, olivetti) \leftarrow$
$insider(john, ibm) \leftarrow$
$insider(bob, fujitsu) \leftarrow$
$wizard(henry) \leftarrow$
$capital(\ldots) \leftarrow$
...

The overall expert system can then be obtained by the module

$$GrantCredits = (Rules \triangleleft Private) \cup Public.$$

Notice that in this way neither the extension nor the intension of the *Private* component is visible to the external world.

4.2.3 Non-ground semantics of program compositions

In section 4.2.1 we have introduced an algebra of logic programs by defining the semantics of each composition operation of the algebra. The semantics of program compositions has been defined in terms of the immediate consequence operator $T(P)$ that gives a denotational characterisation of the set of ground formulae provable in a program. Our main objective, however, is to show that the execution of an extended vanilla meta-interpreter on a query yields only the correct answers with respect to the abstract semantics. To establish such a result, we first define a denotational semantics of program expressions that is more adherent to the operational behaviour of program expressions than the ground definition presented in section 4.2.1.

Falaschi et al. [FLMP89] introduced a new declarative semantics for logic programs, called *S-semantics*, which is based on Herbrand interpretations containing (possibly) non-ground atoms. The main motivation for the introduction of the S-semantics was to reduce the gap between the least Herbrand model semantics (and the equivalent least fixpoint semantics) and the operational semantics for logic programs. While the former characterises only ground formulae, the latter characterises also computed answers that are not necessarily ground. Formally speaking, the objective was to obtain stronger soundness and completeness results for SLD resolution.

Let us briefly recall the definition of the S-semantics (for a complete description see [FLMP89]). The starting point is to consider an extended Herbrand universe that contains possibly non-ground terms rather than the standard Herbrand universe that contains necessarily ground terms. The extended Herbrand universe $\mathcal{U}$ is the quotient set of all terms with respect to the renaming relation $\approx$. Similarly, the extended Herbrand base $\mathcal{B}$ is the quotient set of all atoms with respect to the renaming relation $\approx$. In this context, the immediate consequence operator is a mapping over extended Herbrand interpretations, that is, over subsets of the extended Herbrand base.

Let us recall from [FLMP89] the definition of the non-ground immediate consequence operator, denoted here by $\mathcal{T}$. For any non-ground Herbrand interpretation $I \subseteq \mathcal{B}$:

$$\begin{aligned} A \in \mathcal{T}(P)(I) \iff (\exists B, \bar{C}, \bar{D}, \vartheta : \\ B \leftarrow \bar{C} \in P \quad \wedge \\ \bar{D} \subseteq I \quad \wedge \\ \vartheta = mgu(\bar{C}, \bar{D}) \; \wedge \\ A = B\vartheta). \end{aligned}$$

Notice that the non-ground operator $\mathcal{T}$ is the natural extension of the ground operator T in order to deal with the unification of non-ground atoms. The powers of $\mathcal{T}$ are defined as usual.

The non-ground semantics of program compositions can be given by extending the definition of $\mathcal{T}$ with respect to the first argument, in the same style of section 4.2. It is worth noting that the $\mathcal{T}$-based definition of the operations may be obtained as a natural lifting of the T-based definition, by virtue of the compositionality of the latter. Intuitively, the lifting of the definition reduces to lifting the operations on interpretations in order to deal with non-ground sets of atoms. In orther to capture the meaning of the intersection operator, however, it is necessary to define an auxiliary operation $\bowtie$ on interpretations. The operation $\bowtie$ is the natural extension of set-theoretic intersection in order to deal with unification (between sets of possibly non-ground atoms), and it is defined as follows. Let I and J be two sets of possibly non-ground atoms, then:

$$I \bowtie J = \{A\vartheta \mid \exists A, B, \vartheta : \; A \in I \wedge B \in J \wedge \vartheta = mgu(A, B)\}.$$

Notice that $\bowtie$ coincides with set-theoretic intersection when applied to ground interpretations. The following Lemma establishes a natural property of $\bowtie$, which will be used in the proof of the correctness of the lifting.

Lemma 1 *For any non-ground interpretations I and J:*

$$ground(I \bowtie J) = ground(I) \cap ground(J).$$

Proof.

$\quad\quad A \in ground(I \bowtie J)$
$\iff \quad \{\text{definition of ground}\}$
$\quad\quad \exists D, \vartheta : \ D \in I \bowtie J \ \wedge \ A = D\vartheta \ \wedge \ A \text{ is ground}$
$\iff \quad \{\text{definition of } \bowtie\}$
$\quad\quad \exists D, E, F, \vartheta, \gamma : \ E \in I \ \wedge \ F \in J \ \wedge \ \gamma = mgu(E, F) \ \wedge \ D = E\gamma = F\gamma \ \wedge$
$\quad\quad\quad\quad A = D\vartheta \ \wedge \ A \text{ is ground}$
$\iff \quad \{\text{definition of substitution}\}$
$\quad\quad \exists E, F, \delta : \ E \in I \ \wedge \ F \in J \ \wedge \ A = E\delta = F\delta \ \wedge \ A \text{ is ground}$
$\iff \quad \{\text{definition of ground}\}$
$\quad\quad A \in ground(I) \ \wedge \ A \in ground(J)$
$\iff \quad \{\text{definition of set-theoretic intersection}\}$
$\quad\quad A \in ground(I) \cap ground(J)$

$\square$

Now we have that for any non-ground Herbrand interpretation I:

$$\begin{aligned}
\mathcal{T}(P^*)(I) &= T^\omega(P) \\
\mathcal{T}(P \cup Q)(I) &= \mathcal{T}(P)(I) \cup \mathcal{T}(Q)(I) \\
\mathcal{T}(P \cap Q)(I) &= \mathcal{T}(P)(I) \bowtie \mathcal{T}(Q)(I) \\
\mathcal{T}(P \triangleleft Q)(I) &= \mathcal{T}(P)\,(I \cup T^\omega(Q))
\end{aligned}$$

The definition of $*$, $\cup$ and $\triangleleft$ is straightforward. The only modification with respect to the ground case is that the set-theoretic union on the right hand side of the definitions must be modulo renaming. More precisely, the union of two extended interpretations I and J is the quotient set of the set-theoretic union of I and J with respect to the renaming relation $\approx$.

The correctness of the lifting of the definition of program compositions is established by showing the equivalence between the T-based definition of the operations of section 4.2.1 and the new $\mathcal{T}$-based definition. More precisely, for any program expression P, the least fixpoint of the ground immediate consequence operator $T(P)$ coincides with the set of ground instances of the least fixpoint of the non-ground immediate consequence operator $\mathcal{T}(P)$. The proposition exends Proposition 6.12(b), proved by Falaschi et al. [FLMP89] which establishes the result for plain programs.

Proposition 2 *For any program expression P:*

$$ground(\mathcal{T}^\omega(P)) = T^\omega(P).$$

Proof.
We prove the following stronger statement.
For any program expression P and for any non-ground Herbrand interpretation I:

(i) $ground(\mathcal{T}(P)(I)) = T(P)(ground(I))$

and

(ii) $ground(\mathcal{T}^\omega(P)) = T^\omega(P).$

The proof is organised as follows. We first establish that $(i) \Longrightarrow (ii)$. Then we prove (i) by structural induction on P. In this second proof, we exploit the statement $(i) \Longrightarrow (ii)$. Finally, we obtain (ii) from (i) and $(i) \Longrightarrow (ii)$.

Proof of $(i) \Longrightarrow (ii)$.
We first prove by induction on n that:

$$(i) \implies (\forall n : ground(\mathcal{T}^n(P)) = T^n(P))$$

(Base case)
Trivial since $\mathcal{T}^0(P) = T^0(P) = \emptyset$.
(Inductive case)
Assume that

$$(i) \implies (ground(\mathcal{T}^n(P)) = T^n(P))$$

Now assume (i). Then

$\quad ground(\mathcal{T}^{n+1}(P))$
= $\quad$ {definition of powers of $\mathcal{T}$}
$\quad ground(\mathcal{T}(P)(\mathcal{T}^n(P)))$
= $\quad$ **{assuming (i)}**
$\quad T(P)(ground(\mathcal{T}^n(P)))$
= $\quad$ **{assuming (i)}**, inductive hypothesis on n}
$\quad T(P)(T^n(P))$
= $\quad$ {definition of powers of T}
$\quad T^{n+1}(P)$

Thus

$$(i) \implies (ground(\mathcal{T}^{n+1}(P)) = T^{n+1}(P)).$$

Since, by definition of *ground*, $ground(\bigcup_{n<\omega} \mathcal{T}^n(P)) = \bigcup_{n<\omega} ground(\mathcal{T}^n(P))$ we can conclude that $(i) \Longrightarrow (ii)$.

Meta-Logic for Program Composition: Semantics Issues

Proof of (i).
We now prove that for any program expression P and any non-ground Herbrand interpretation I:
$$ground(\mathcal{T}(P)(I)) = T(P)(ground(I)).$$
The proof is by structural induction on P.
(P plain program).
The base case of the proof, in which P is a plain program, is proved in [FLMP89].
($P = Q \cup R$).

$\qquad A \in ground(\mathcal{T}(Q \cup R)(I))$
$\Longleftrightarrow \qquad$ {non-ground definition of $\cup$}
$\qquad A \in ground(\mathcal{T}(Q)(I) \cup \mathcal{T}(R)(I))$
$\Longleftrightarrow \qquad$ {definition of set-theoretic union}
$\qquad A \in ground(\mathcal{T}(Q)(I)) \cup ground(\mathcal{T}(R)(I))$
$\Longleftrightarrow \qquad$ {inductive hypothesis on Q and R}
$\qquad A \in T(Q)(ground(I)) \cup T(R)(ground(I))$
$\Longleftrightarrow \qquad$ {ground definition of $\cup$}
$\qquad A \in T(Q \cup R)(ground(I))$

($P = Q \cap R$).

$\qquad A \in ground(\mathcal{T}(Q \cap R)(I))$
$\Longleftrightarrow \qquad$ {non-ground definition of $\cap$}
$\qquad A \in ground(\mathcal{T}(Q)(I) \bowtie \mathcal{T}(R)(I))$
$\Longleftrightarrow \qquad$ {by the above observation}
$\qquad A \in ground(\mathcal{T}(Q)(I)) \cap ground(\mathcal{T}(R)(I))$
$\Longleftrightarrow \qquad$ {inductive hypothesis on Q and R}
$\qquad A \in T(Q)(ground(I)) \cap T(R)(ground(I))$
$\Longleftrightarrow \qquad$ {ground definition of $\cap$}
$\qquad A \in T(Q \cap R)(ground(I))$

($P = Q^*$).

$\qquad ground(\mathcal{T}(Q^*)(I))$
$\Longleftrightarrow \qquad$ {non-ground definition of $*$}
$\qquad ground(\mathcal{T}^\omega(Q))$
$\Longleftrightarrow \qquad$ {inductive hypothesis on Q, (i) $\Longrightarrow$ (ii)}
$\qquad T^\omega(Q)$
$\Longleftrightarrow \qquad$ {ground definition of $*$}
$\qquad T(Q^*)(ground(I))$

($P = Q \triangleleft R$).

$$\begin{aligned}
&\quad A \in ground(\mathcal{T}(Q \triangleleft R)(I))\\
&\Longleftrightarrow \quad \{\text{non-ground definition of } \triangleleft\}\\
&\quad A \in ground(\mathcal{T}(Q)(I \cup T^\omega(R)))\\
&\Longleftrightarrow \quad \{\text{inductive hypothesis on } Q\}\\
&\quad A \in \mathcal{T}(Q)(ground(I \cup T^\omega(R)))\\
&\Longleftrightarrow \quad \{\text{definition of set-theoretic union}\}\\
&\quad A \in \mathcal{T}(Q)(ground(I) \cup ground(T^\omega(R)))\\
&\Longleftrightarrow \quad \{\text{inductive hypothesis on } R, \text{ (i)} \Longrightarrow \text{(ii)}\}\\
&\quad A \in \mathcal{T}(Q)(ground(I) \cup T^\omega(R))\\
&\Longleftrightarrow \quad \{\text{ground definition of } \triangleleft\}\\
&\quad A \in \mathcal{T}(Q \triangleleft R)(ground(I))
\end{aligned}$$

□

4.3 Meta-logic and Program Composition Operations

This section introduces the meta-logical definition of program expressions. First we discuss the basic definition of meta-logic, which is based on the notion of the so-called *vanilla meta-interpreter*. Then we extend this basic definition to handle program expressions and finally we establish the correctness of the meta-logic with respect to the $\mathcal{T}$-based definition of the operators.

4.3.1 Vanilla meta-programs

We consider a large class of meta-programs, which are based on the so-called *vanilla meta-interpreter* [SS86]. The vanilla meta-interpreter consists of the following three clauses:

$$\begin{aligned}
solve(empty) &\leftarrow \\
solve((x,y)) &\leftarrow solve(x),\\
&\quad solve(y)\\
solve(x) &\leftarrow clause(x \leftarrow y),\\
&\quad solve(y)
\end{aligned}$$

The vanilla meta-program is used to interpret an object logic program, which is represented by means of the *clause* predicate. The predicate *solve* is used to represent the provability relation, so that $solve(x)$ states that the formula x is provable in the interpreted object program.

Declaratively, the vanilla meta-interpreter reads as follows. The constant *empty* is always true. The conjunction (x,y) is true if x is true and y is true. A goal x is true if there is a clause $x \leftarrow y$ in the interpreted program such that y is true. As any other logic program, the vanilla meta-interpreter has also a procedural reading. The first unit clause states that the empty goal, represented by the constant symbol *empty* is always solved. The second clause deals with conjunctive goals. It reads: "To solve a conjunction (x, y), solve x and y". Finally, the third clause deals with the general case of goal reduction. To solve a goal, choose a clause from the program whose head unifies with the goal, and recursively solve the body of the clause.

Object level programs are represented by means of the *clause* predicate. For example, the object level program:

$$nat(s(x)) \leftarrow nat(x)$$
$$nat(zero) \leftarrow$$

is represented at the meta-level by the clauses:

$$clause(nat(s(x)) \leftarrow nat(x)) \leftarrow$$
$$clause(nat(zero) \leftarrow empty) \leftarrow$$

where object level variables are represented by themselves at the meta-level.

Hill and Lloyd [HL89] observed that the declarative meaning of the vanilla meta-interpreter is unclear. The problem is that variables in the definition of *clause* and variables in the definition of *solve* intuitively range over different domains. Informally speaking, the variables in *clause* range over elements of the intended interpretation domain of the object program, while the variables in *solve* range over conjunctions of instances of atoms. Therefore they concluded that the intended meaning of the program is simply not a model of the program.

On the other hand, Kowalski showed [Kow90] that the vanilla meta-interpreter employing the non-ground representation can be used for programs containing variables as well as for variable-free programs, although it is, strictly speaking, incorrect. In fact, the representation of object level variables by meta-level variables is incorrect because the implicit quantifiers are treated incorrectly.

Recently, De Schreye and Martens [MD92, DM92] made the commendable effort to try to clarify this oddity. They showed that it is possible to provide a sensible semantics for a class of vanilla meta-programs using the incorrect representation. The approach basically consists in relating the semantics of an object program P to the semantics of the corresponding vanilla meta-program V_P. In [DM92] such a correspondence is stated for a restricted class of object programs, called *language independent* programs. This restriction has been then relaxed in [MD92] for definite object programs. Roughly, for any definite object program P, if q is predicate symbol of P and t a (possibly non-ground) term they prove the following. A formula $solve(p(t))$ is provable in the vanilla metaprogram associated to P if and only if $p(t)$ is provable in the object program. A formal statement of this result will be given as a special case of the general theorem presented in section 4.3.3. Levi and Ramundo [LR93] independently presented the same result as a generalisation of the work by De Schreye and Martens [DM92]. They applied the same construction to an enhanced meta-interpreter defining some inheritance mechanisms on logic programs.

4.3.2 Meta-logical definition of program composition operations

Program composition operations can be naturally provided with a meta-logical definition. More precisely, the set of operations of the algebra can be implemented by a metainterpreter that is a simple extension of the vanilla meta-interpreter.

We consider a more general form of the vanilla meta-interpreter introduced in section 4.3.1. Following Bowen and Kowalski [BK82], we employ the two-argument proof predicate *demo* rather than *solve* to express provability. The extra argument of *demo* is used to denote explicitly the interpreted object program. Namely, $demo(x,y)$ means that the formula y is provable in the object program x. The definition of the vanilla meta-interpreter for *demo* properly extends the definition of the *solve* meta-interpreter.

$$demo(x, empty) \leftarrow \tag{1}$$

$$demo(x, (y, z)) \leftarrow demo(x, y),$$
$$demo(x, z) \tag{2}$$

$$demo(x, y) \leftarrow demo(x, y \leftarrow z),$$
$$demo(x, z) \tag{3}$$

Notice that the *solve* meta-interpreter of section 4.3.1 uses different predicates to represent clauses that can be proved because they are axioms (the predicate *clause*) and clauses which can be proved by one or more steps of inference (the predicate *solve*). According to Kowalski [Kow90], we prefer to use only one predicate (*demo*) to represent both object level clauses and provable formulae.

The *demo* vanilla meta-interpreter has the following operational reading. The unit clause (1) states that the empty goal, represented by the constant symbol *empty*, is solved in any program x. Clause (2) deals with conjunctive goals. It states that a conjunction (y, z) is solved in the program x if y is solved in x and z is solved in x. Finally, clause (3) deals with the case of atomic goal reduction. To solve an atomic goal y, choose a clause from the program x and recursively solve the body of the clause in x.

Object level programs are named by constant symbols, denoted by capital letters such as P and Q. As in the *solve* meta-interpreter, object level expressions are represented by themselves at the meta-level. In particular, object level variables are denoted by meta-level variables, according to the non-ground representation. The meta-level representation of an object program P contains a clause of the form

$$demo(P, A \leftarrow \bar{B}) \leftarrow$$

if and only if the clause

$$A \leftarrow \bar{B}$$

belongs to the object program P. For instance, the program P representing the natural numbers

$$nat(s(x)) \leftarrow nat(x)$$
$$nat(zero) \leftarrow$$

is represented at the meta-level by the clauses:

$$demo(P, nat(s(x)) \leftarrow nat(x)) \leftarrow$$
$$demo(P, nat(zero) \leftarrow empty) \leftarrow$$

Program composition operations can be provided with a meta-logical definition in a simple and concise way. Each program composition operation is represented at the meta-level by a functor. The meaning of each functor is defined by new clauses to be added to the vanilla meta-interpreter.

The meta-logical definition of the operations $*$, $\cup$, $\cap$ and $\triangleleft$ is given by extending the vanilla meta-interpreter with the following clauses.

$$demo(x \cup y, z \leftarrow w) \leftarrow demo(x, z \leftarrow w) \qquad (4)$$
$$demo(x \cup y, z \leftarrow w) \leftarrow demo(y, z \leftarrow w) \qquad (5)$$
$$demo(x \cap y, (z \leftarrow u, v)) \leftarrow demo(x, z \leftarrow u),$$
$$demo(y, z \leftarrow v) \qquad (6)$$
$$demo(x^*, y \leftarrow) \leftarrow demo(x, y) \qquad (7)$$
$$demo(x \triangleleft y, z \leftarrow w) \leftarrow demo(x, (z \leftarrow w, v)),$$
$$demo(y, v) \qquad (8)$$

The first argument of *demo* is used to represent a composition of programs rather than a single program as happens in the *solve* vanilla meta-interpreter. The meaning of the clauses (4)—(8) is straightforward. For instance, clauses (4) and (5) define the meta-level implementation of the operation $\cup$. A clause $z \leftarrow w$ belongs to the meta-level representation of the composition $P \cup Q$ if it belongs either to the meta-level representation of P or to the meta-level representation of Q. Notice that the adoption of the non-ground representation of object programs allows clause (6) to exploit the basic unification mechanism of logic programming. The meta-level representation of an encapsulated program x^* (clause (7)) consists of assertions of the form $demo(x^*, y \leftarrow)$ where y is provable in x. In this way, the code of x is hidden to other programs, which may only refer to the set of sentences that are provable in x. Finally, clause (8) defines the import operation $\triangleleft$. The clauses in the visible part of $x \triangleleft y$ are obtained from the clauses of x by dropping the calls to hidden predicates in the original clause body, provided that they are provable in the private part y, and possibly instantiating the public calls. It is worth observing the intuitive correspondence of the meta-level definition of $x \triangleleft y$ with its abstract specification. First, the call $demo(y, v)$ in (8) mirrors that only the least fixpoint of the semantics of program y is considered. Second, the call $demo(x, (z \leftarrow w, v))$ establishes that only the predicates in x are made visible.

4.3.3 Correctness of the extended meta-logic

We now state the principal result of the paper that establishes the correctness of the meta-logical implementation of program composition operations. Namely, we prove that the meta-logical provability relation defined by clauses (1)—(8) faithfully realises the abstract semantics of program expressions given in the previous section.

We first relate, for an arbitrary program expression P, the (possibly non-ground) object formulae in the least fixpoint of the immediate consequence operator $\mathcal{T}(P)$ to the object formulae that are provable at the meta-level. For any program expression P, let M denote the vanilla meta-interpreter of clauses (1)—(8) extended with the meta-level representation of the object programs occurring in P.

Theorem 3 *For any program expression P and for any object level atom A:*

$$demo(P, A) \in \mathcal{T}^\omega(M) \iff A \in \mathcal{T}^\omega(P).$$

Proof.
The proof is carried on for the following stronger statement.
For any program expression P and any object level interpretation I:
(i) $A \in \mathcal{T}(P)(I) \iff \exists B, \bar{C}, \bar{D}, \vartheta : \ demo(P, B \leftarrow \bar{C}) \in \mathcal{T}^\omega(M) \land$
$\bar{D} \subseteq I \land$
$\vartheta = mgu(\bar{C}, \bar{D}) \land$
$A = B\vartheta$

and
(ii) $A \in T^\omega(P) \iff demo(P, A) \in T^\omega(M)$

The proof is organised as follows. We first establish that $(i) \implies (ii)$. Then we prove (i) by structural induction on P. In this second proof, we exploit the statement $(i) \implies (ii)$. Finally, we obtain (ii) from (i) and $(i) \implies (ii)$.

Proof of $(i) \implies (ii)$.
We first show that:

$$(i) \implies (demo(P, A) \in T^\omega(M) \implies A \in T^\omega(P))$$

by proving that:

$$(i) \implies (\forall m : demo(P, A) \in T^m(M) \implies (\exists n : A \in T^n(P))).$$

and by exploiting the continuity of T.
The proof is by induction on m.
(Base case).
Trivial since $T^0(M) = \emptyset$.
(Inductive case).
Assume that

$$(i) \implies (demo(P, A) \in T^m(M) \implies (\exists n : A \in T^n(P)))$$

then

$$(i) \implies (demo(P, A) \in T^{m+1}(M) \implies (\exists n : A \in T^n(P)))$$

since

$$demo(P, A) \in T^{m+1}(M)$$
$\iff$ {definition of powers of T}
$$demo(P, A) \in T(M)(T^m(M))$$
$\iff$ {definition of T}
$$\exists B, \bar{C}, \bar{D}, \vartheta : \quad B \leftarrow \bar{C} \in M \land$$
$$\bar{D} \subseteq T^m(M) \land$$
$$\vartheta = mgu(\bar{C}, \bar{D}) \land$$
$$demo(P, A) = B\vartheta$$
$\implies$ {only clause (3) of vanilla (suitably standardised apart) applies since A is atomic}
$$\exists X, Y, Z, P_1, P_2, E, \bar{F}, \bar{G}, \vartheta :$$
$$demo(X, Y) \leftarrow demo(X, Y \leftarrow Z), demo(X, Z) \in M \land$$

$$demo(P_1, E \leftarrow \bar{F}) \in \mathcal{T}^m(M) \wedge$$
$$demo(P_2, \bar{G}) \in \mathcal{T}^m(M) \wedge$$
$$\vartheta = mgu((X, Y, Z, X, Z), (P_1, E, \bar{F}, P_2, \bar{G})) \wedge$$
$$demo(P, A) = (demo(X, Y))\vartheta$$

$\Longrightarrow$ {the existence of ϑ implies the existence of
$\gamma = mgu((X, Y, Z, X, Z, X, Y), (P_1, E, \bar{F}, P_2, \bar{G}, P, A))$.
Since P is ground then $P = P_1 = P_2$, and by standardisation apart,
$\gamma = \{X/P, \delta\}$ for some δ, where $X \notin dom(\delta)$.
In particular, $\delta = mgu((Y, Z, Z), E, \bar{F}, \bar{G}))$ and $A = A\delta = E\delta$. }

$\exists Y, Z, E, \bar{F}, \bar{G}, \delta :$
$$demo(P, E \leftarrow \bar{F}) \in \mathcal{T}^m(M) \wedge$$
$$demo(P, \bar{G}) \in \mathcal{T}^m(M) \wedge$$
$$\delta = mgu((Y, Z, Z), (E, \bar{F}, \bar{G})) \wedge$$
$$A = Y\delta$$

$\Longrightarrow$ {definition of mgu, Y and Z do not occur in $E, \bar{F}, \bar{G}$)}

$\exists E, \bar{F}, \bar{G}, \delta :$
$$demo(P, E \leftarrow \bar{F}) \in \mathcal{T}^m(M) \wedge$$
$$demo(P, \bar{G}) \in \mathcal{T}^m(M) \wedge$$
$$\delta = mgu(\bar{F}, \bar{G}) \wedge$$
$$A = E\delta$$

$\Longrightarrow$ {let $G = (G_1, \ldots, G_h)$}

$\exists E, \bar{F}, G_1, \ldots, G_h, \delta :$
$$demo(P, E \leftarrow \bar{F}) \in \mathcal{T}^m(M) \wedge$$
$$demo(P, (G_1, \ldots, G_h)) \in \mathcal{T}^m(M) \wedge$$
$$\delta = mgu(\bar{F}, (G_1, \ldots, G_h)) \wedge$$
$$A = E\delta$$

$\Longrightarrow$ {clause (2) of vanilla, definition of $\mathcal{T}$}

$\exists E, \bar{F}, G_1, \ldots, G_h, \delta :$
$$demo(P, E \leftarrow \bar{F}) \in \mathcal{T}^m(M) \wedge$$
$$\{demo(P, G_1), \ldots, demo(P, G_h)\} \subseteq \mathcal{T}^m(M) \wedge$$
$$\delta = mgu(\bar{F}, (G_1, \ldots, G_h)) \wedge$$
$$A = E\delta$$

$\Longrightarrow$ {**assuming (i)**, inductive hypothesis on $G_1, \ldots, G_h$}

$\exists E, \bar{F}, G_1, \ldots, G_h, \delta, n :$
$$demo(P, E \leftarrow \bar{F}) \in \mathcal{T}^m(M) \wedge$$
$$\{G_1, \ldots, G_h\} \subseteq T^n(P) \wedge$$
$$\delta = mgu(\bar{F}, (G_1, \ldots, G_h)) \wedge$$
$$A = E\delta$$

$\Longrightarrow$ {assuming (i)}
$\exists n: \ A \in \mathcal{T}(P)(\mathcal{T}^n(P))$

We now show that:
$$(i) \implies (A \in \mathcal{T}^\omega(P) \implies demo(P,A) \in \mathcal{T}^\omega(M))$$

by proving that:
$$(i) \implies (\forall n: \ A \in \mathcal{T}^n(P)) \implies (\exists m: demo(P,A) \in \mathcal{T}^m(M)).$$

and by exploiting the continuity of $\mathcal{T}$.
The proof is by induction on n.
(Base case).
Trivial since $\mathcal{T}^0(M) = \emptyset$.
(Inductive case).
Assume that
$$(i) \implies (A \in \mathcal{T}^n(P)) \implies (\exists m: demo(P,A) \in \mathcal{T}^m(M)).$$

then
$$(i) \implies (A \in \mathcal{T}^{n+1}(P)) \implies (\exists m: demo(P,A) \in \mathcal{T}^m(M)).$$

since
$A \in \mathcal{T}(P)(\mathcal{T}^n(P))$
$\Longleftrightarrow$ {assuming (i)}
$\exists B, \bar{C}, \bar{D}, \vartheta :$
  $demo(P, B \leftarrow \bar{C}) \in \mathcal{T}^\omega(M) \land$
  $\bar{D} \subseteq \mathcal{T}^n(P) \land$
  $\vartheta = mgu(\bar{C}, \bar{D}) \land$
  $A = B\vartheta$
$\Longleftrightarrow$ $\{\bar{D} = D_1, \ldots, D_h\}$
$\exists B, \bar{C}, D_1, \ldots, D_h, \vartheta :$
  $demo(P, B \leftarrow \bar{C}) \in \mathcal{T}^\omega(M) \land$
  $\{D_1, \ldots, D_h\} \subseteq \mathcal{T}^n(P) \land$
  $\vartheta = mgu(\bar{C}, (D_1, \ldots, D_h)) \land$
  $A = B\vartheta$
$\Longrightarrow$ {assuming (i), by inductive hypothesis on $D_1, \ldots, D_h$}
$\exists B, \bar{C}, D_1, \ldots, D_h, \vartheta, m' :$
  $demo(P, B \leftarrow \bar{C}) in \mathcal{T}^\omega(M) \land$
  $\{demo(P, D_1), \ldots, demo(P, D_h)\} \subseteq \mathcal{T}^{m'}(M) \land$

$$\vartheta = mgu(\bar{C}, (D_1, \ldots, D_h)) \wedge$$
$$A = B\vartheta$$
$\implies$ { clause (2) of vanilla, definition of $\mathcal{T}$ }
$$\exists B, \bar{C}, D_1, \ldots, D_h, \vartheta, m', k :$$
$$demo(P, B \leftarrow \bar{C}) \in \mathcal{T}^k(M) \wedge$$
$$\{demo(P, D_1, \ldots, D_h)\} \subseteq \mathcal{T}^{m'+h-1}(M) \wedge$$
$$\vartheta = mgu(\bar{C}, (D_1, \ldots, D_h)) \wedge$$
$$A = B\vartheta$$
$\implies$ $\{m = max\{k, (m'+h-1)\} + 1$, clause (3) of vanilla, definition of $\mathcal{T}\}$
$\exists m :$ $\quad demo(P, A) in \mathcal{T}^m(M)$

Proof of (i).
The proof is by structural induction on P.
(P is a plain program).
$\quad A \in \mathcal{T}(P)(I)$
$\iff$ {definition of $\mathcal{T}$}
$\quad \exists B, \bar{C}, \bar{D}, \vartheta : B \leftarrow \bar{C} \in P \wedge \bar{D} \subseteq I \wedge \vartheta = mgu(\bar{C}, \bar{D}) \wedge A = B\vartheta$
$\iff$ {non-ground representation of object programs}
$\quad \exists B, \bar{C}, \bar{D}, \vartheta : demo(P, B \leftarrow \bar{C}) \leftarrow \in M \wedge \bar{D} \subseteq I \wedge \vartheta = mgu(\bar{C}, \bar{D}) \wedge A = B\vartheta$
$\iff$ {definition of $\mathcal{T}$}
$\quad \exists B, \bar{C}, \bar{D}, \vartheta : demo(P, B \leftarrow \bar{C}) \in \mathcal{T}^\omega(M) \wedge \bar{D} \subseteq I \wedge \vartheta = mgu(\bar{C}, \bar{D}) \wedge A = B\vartheta$

($P = Q \cup R$).
$\quad A \in \mathcal{T}(Q \cup R)(I)$
$\iff$ {definition of $\cup$}
$\quad A \in \mathcal{T}(Q)(I) \vee A \in \mathcal{T}(R)(I)$
$\iff$ {inductive hypothesis on Q and R}
$\quad \exists B, \bar{C}, \bar{D}, \vartheta :$
$\quad (demo(Q, B \leftarrow \bar{C}) \in \mathcal{T}^\omega(M) \wedge \bar{D} \subseteq I \wedge \vartheta = mgu(\bar{C}, \bar{D}) \wedge A = B\vartheta)$
$\quad \vee$
$\quad (demo(R, B \leftarrow \bar{C}) \in \mathcal{T}^\omega(M) \wedge \bar{D} \subseteq I \wedge \vartheta = mgu(\bar{C}, \bar{D}) \wedge A = B\vartheta)$
$\iff$ {by clauses (4) and (5) of vanilla, definition of $\mathcal{T}$}
$\quad \exists B, \bar{C}, \bar{D}, \vartheta :$
$\quad (demo(Q \cup R, B \leftarrow \bar{C}) \in \mathcal{T}^\omega(M) \wedge \bar{D} \subseteq I \wedge \vartheta = mgu(\bar{C}, \bar{D}) \wedge A = B\vartheta)$

($P = Q \cap R$).
$\quad A \in \mathcal{T}(Q \cap R)(I)$
$\iff$ {definition of $\cap$}

$A \in \mathcal{T}(Q)(I) \bowtie \mathcal{T}(R)(I)$
$\iff$ {definition of $\bowtie$}
$\exists E, F, \vartheta : E \in \mathcal{T}(Q)(I) \wedge F \in \mathcal{T}(R)(I) \wedge \vartheta = mgu(E, F) \wedge A = E\vartheta = F\vartheta$
$\iff$ {inductive hypothesis on Q and R}
$\exists E, F, \vartheta, A_1, \bar{C}_1, \bar{D}_1, \vartheta_1, A_2, \bar{C}_2, \bar{D}_2, \vartheta_2 :$
$\quad demo(Q, A_1 \leftarrow \bar{C}_1) \in \mathcal{T}^\omega(M) \wedge \bar{D}_1 \subseteq I \wedge \vartheta_1 = mgu(\bar{C}_1, \bar{D}_1) \wedge E = A_1\vartheta_1 \wedge$
$\quad demo(R, A_2 \leftarrow \bar{C}_2) \in \mathcal{T}^\omega(M) \wedge \bar{D}_2 \subseteq I \wedge \vartheta_2 = mgu(\bar{C}_2, \bar{D}_2) \wedge E = A_2\vartheta_2 \wedge$
$\quad \vartheta = mgu(E, F) \wedge A = E\vartheta = F\vartheta$
$\iff$ {clause (6) of vanilla, basic properties of substitutions [FLMP89]}
$\exists \delta, \gamma, A_1, \bar{C}_1, \bar{D}_1, A_2, \bar{C}_2, \bar{D}_2 :$
$\quad demo(Q \cap R, (A_1 \leftarrow \bar{C}_1, \bar{C}_2)\gamma) \in \mathcal{T}^\omega(M) \wedge$
$\quad \gamma = mgu(A_1, A_2) \wedge$
$\quad \{\bar{D}_1, \bar{D}_2\} \subseteq I \wedge$
$\quad \delta = mgu((\bar{C}_1, \bar{C}_2)\gamma, (\bar{D}_1, \bar{D}_2)) \wedge A = A_1\gamma\delta$

$(P = Q^*)$.

$A \in \mathcal{T}(Q^*)(I)$
$\iff$ {definition of $*$}
$A \in \mathcal{T}^\omega(Q)$
$\iff$ {inductive hypothesis on Q, (i) $\implies$ (ii)}
$demo(Q, A) \in \mathcal{T}^\omega(M)$
$\iff$ {vanilla (7)}
$demo(Q, A \leftarrow) \in \mathcal{T}^\omega(M)$

$(P = Q \triangleleft R)$.

$A \in \mathcal{T}(Q \triangleleft R)(I)$
$\iff$ {definition of $\triangleleft$}
$A \in \mathcal{T}(Q)(I \cup \mathcal{T}^\omega(R))$
$\iff$ {inductive hypothesis on Q}
$\exists B, \bar{C}, \bar{D}, \vartheta :$
$\quad demo(Q, B \leftarrow \bar{C}) \in \mathcal{T}^\omega(M) \wedge$
$\quad \bar{D} \subseteq (I \cup \mathcal{T}^\omega(R)) \wedge$
$\quad \vartheta = mgu(\bar{C}, \bar{D}) \wedge A = B\vartheta$
$\iff$ {putting $\bar{C} = (\bar{C}_1, \bar{C}_2)$ and $\bar{D} = (\bar{D}_1, \bar{D}_2)$}
$\exists B, \bar{C}_1, \bar{C}_2, \bar{D}_1, \bar{D}_2, \vartheta :$
$\quad demo(Q, B \leftarrow (\bar{C}_1, \bar{C}_2)) \in \mathcal{T}^\omega(M) \wedge$
$\quad \bar{D}_1 \subseteq I \wedge \bar{D}_2 \subseteq \mathcal{T}^\omega(R) \wedge$
$\quad \vartheta = mgu((\bar{C}_1, \bar{C}_2), (\bar{D}_1, \bar{D}_2)) \wedge A = B\vartheta$

$\iff$ { for each atom in $\bar{D}_2$: inductive hypothesis on R, (i) $\implies$ (ii) }
$\exists B, \bar{C}_1, \bar{C}_2, \bar{D}_1, \bar{D}_2, \vartheta :$
$\quad demo(Q, B \leftarrow (\bar{C}_1, \bar{C}_2)) \in \mathcal{T}^\omega(M) \wedge$
$\quad \bar{D}_1 \subseteq I \wedge demo(R, \bar{D}_2) \in \mathcal{T}^\omega(M) \wedge$
$\quad \vartheta = mgu((\bar{C}_1, \bar{C}_2), (\bar{D}_1, \bar{D}_2)) \wedge A = B\vartheta$
$\iff$ {clause (8) of vanilla}
$\exists B, \bar{C}_1, \bar{D}_1, \vartheta :$
$\quad demo(Q \triangleleft R, (B \leftarrow \bar{C}_1)\vartheta) \in \mathcal{T}^\omega(M) \wedge$
$\quad \bar{D}_1 \subseteq I \wedge \vartheta = mgu(\bar{C}_1, \bar{D}_1) \wedge A = B\vartheta$

□

The above theorem states that an object formula A is a (possibly non-ground) immediate consequence of a program expression P if and only if $demo(P, A)$ can be proved at the meta-level. Intuitively, this means that the vanilla meta-interpreter can prove all the non-ground immediate consequences of a program expression P. Notice that, when P is simply a plain program, theorem 3 states that an object formula A is provable in P if and only if $demo(P, A)$ is provable in the corresponding meta-program.

Theorem 3, however, states only a partial correctness result for the use of meta-logic to prove object level queries. In fact, we would like that only "sensible" formulae are provable at the meta-level. For a single object program P, for instance, we would ultimately like a formula of the kind $demo(P, F)$ to be provable only if the corresponding object formula F is provable in the object program P. Obviously this is not true in general (even for a single object program [MD92]) because, for instance, the constant *empty* is provable in any program P by virtue of clause (1) of the vanilla meta-interpreter.

An intriguing question therefore is under which conditions it is guaranteed that only "sensible" formulae are provable at the meta-level. For a single object program P, for instance, one is actually interested in querying object level relations by means of meta-level queries of the form $demo(P, q(x))$ where q is a predicate name in the object program P. The expected guarantee therefore is that if the formula $demo(P, q(t))$ (where t is a generic, possibly non-ground term) is provable at the meta-level then the corresponding object formula $q(t)$ is provable at the object level as well. By theorem 3, this is equivalent to say that if $demo(P, q(t))$ is provable at the meta-level then t is necessarily an object level term. The following theorem states that such a condition is satisfied even in the general case in which P is an arbitrary program expression.

Theorem 4 *For any program expression P and for any object level predicate symbol q:*

$$demo(P, q(t)) \in \mathcal{T}^\omega(M) \implies t \text{ is an object level term.}$$

Proof.
We prove the following stronger statement.
For any program expression P and for any object level predicate symbols $q, q_1, \ldots q_h$:

(i) $demo(P, q(t)) \in T^\omega(M) \implies t$ is an object level term
and
(ii) $demo(P, q(t) \leftarrow q_1(t_1), \ldots, q_h(t_h)) \in T^\omega(M) \implies t, t_1, \ldots t_h$ are object level terms.

The proof of (i) and (ii) is by simultaneous induction on the powers of T.

Proof of (i)
(Base case)
Trivial since $T^0(M) = \emptyset$.
(Inductive case)

$$\begin{aligned}
& demo(P, q(t)) \in T^{n+1}(M) \\
\iff & \quad \{\text{definition of powers of } T\} \\
& demo(P, q(t)) \in T(M)(T^n(M)) \\
\iff & \quad \{\text{definition of } T\} \\
& \exists B, \bar{C}, \bar{D}, \vartheta : \\
& \quad B \leftarrow \bar{C} \in M \land \\
& \quad \bar{D} \subseteq T^n(M) \land \\
& \quad \vartheta = mgu(\bar{C}, \bar{D}) \land \\
& \quad demo(P, q(t)) = B\vartheta \\
\implies & \quad \{\text{only clause (3) of vanilla applies}\} \\
& \exists X, Y, Z, q_1, \ldots, q_h, u, u_1, \ldots, u_h, v_1, \ldots, v_h, \vartheta : \\
& \quad demo(X, Y) \leftarrow demo(X, Y \leftarrow Z), demo(X, Z) \in M \land \\
& \quad demo(P, q(u) \leftarrow q_1(u_1), \ldots, q_h(u_h)) \in T^n(M) \land \\
& \quad demo(P, q_1(v_1), \ldots, q_h(v_h)) \in T^n(M) \land \\
& \quad \vartheta = mgu((X, Y, Z, Z), (P, q(u), (q_1(u_1), \ldots, q_h(u_h)), (q_1(v_1), \ldots, q_h(v_h)))) \land \\
& \quad q(t) = q(u)\vartheta \\
\implies & \quad \{\text{inductive hypotheses (i) and (ii)}\} \\
& \quad t \text{ is an object-level term}
\end{aligned}$$

Proof of (ii)
(Base case)
Trivial since $T^0(M) = \emptyset$.
(Inductive case)
The proof is by induction on the structure of P. We show below the proof for $P = Q \cup R$. The proofs of the other cases rely on similar arguments.

$\iff$ $demo(Q \cup R, q(t) \leftarrow q_1(t_1), \ldots, q_h(t_h)) \in T^{n+1}(M)$
         {definition of powers of T}
$\iff$ $demo(Q \cup R, q(t) \leftarrow q_1(t_1), \ldots, q_h(t_h)) \in T(M)T^n(M)$
         {definition of T}
$\exists H, \bar{C}, \bar{D}, \vartheta :$
     $H \leftarrow \bar{C} \in M \land$
     $\bar{D} \subseteq T^n(M) \land$
     $\vartheta = mgu(\bar{C}, \bar{D}) \land$
     $(demo(Q \cup R, q(t) \leftarrow q_1(t_1), \ldots, q_h(t_h))) = H\vartheta$
$\iff$ {definition of clause (4) and clause (5) of vanilla}
$\exists X, Y, Z, W, u, u_1, \ldots, u_h, u', u'_1, \ldots, u'_h, \gamma, \gamma' :$
     $(demo(X \cup Y, Z \leftarrow W) \leftarrow demo(X, Z \leftarrow W) \in M \land$
     $demo(Q, q(u) \leftarrow q_1(u_1), \ldots, q_h(u_h)) \in T^n(M) \land$
     $\gamma = mgu((X, Z, W), (Q, q(u), (q_1(u_1)), \ldots, q_h(u_h)))$
     $q(t) = q(u)\gamma \land$
     $q_i(t_i) = q_i(u_i)\gamma \;\; \forall i \in \{1, \ldots, h\})$
     $\lor$
     $(demo(X \cup Y, Z \leftarrow W) \leftarrow demo(Y, Z \leftarrow W) \in M \land$
     $demo(R, q(u') \leftarrow q_1(u'_1), \ldots, q_h(u'_h)) \in T^n(M) \land$
     $\gamma' = mgu((Y, Z, W), (Q, q(u'), (q_1(u'_1)), \ldots, q_h(u'_h)))$
     $q(t) = q(u')\gamma' \land$
     $q_i(t_i) = q_i(u'_i)\gamma' \;\; \forall i \in \{1, \ldots, h\})$
$\implies$ {inductive hypothesis on Q and R}
$t, t_1, \ldots, t_h$ are object level terms.

$\square$

The above theorem gives a formal justification to the use of meta-logic for querying compositions of programs. Actually, it states that meta-logic can be "safely" used to prove non-ground object level queries in arbitrary composition of programs, with the guarantee that only sensible answers will be generated. The relevance of this result may be outlined by the following corollary, which follows immediately from theorems 3 and 4 by the strong soundness and completeness results of the S-semantics [FLMP89]. Let $R \vdash_\vartheta G$ denote that there exists an SLD refutation for G in the program R with computed answer substitution ϑ.

Corollary 5 *For any program expression P and for any object level predicate symbol q:*

$$M \vdash_\vartheta demo(P, q(x)) \iff (q(x))\vartheta \in T^\omega(P).$$

Proof.

Straightforward by theorem 4 and by strong soundness and strong completeness of S-semantics [FLMP89] (Theorems 7.1 and 7.7). □

4.4 Conclusions

We have presented an extension to logic programming which allows one to prove a goal with respect to a program expression rather than with respect to a simple program. Such an extension offers, in our opinion, a simple way for re-constructing useful programming idioms and reasoning styles in the logic programming framework. In this chapter we have proved that a correct way of providing an implementation for our extension is to employ a generalized meta-interpreter, in which an ambivalent syntax is used.

Although the generalized meta-interpreter provides us with an effective implementation of the extension, it does not solve the problem of an *efficient* implementation of it. Brogi et al. discuss in [BCM+94] other ways of implementing the extension. Besides meta-logic, two other approaches are used: a transformational approach, based upon the compilation of the extended language into plain logic programming, and an extended WAM. The latter approach can be seen as an efficient implementation of the extended meta-interpreter in that program expressions are still interpreted, although at a much lower level. The extended WAM approach has proved to be quite effective, allowing us to speed up the execution of the extended program by a factor of 20 with respect to the meta-interpreted version. The drawback is the fact that this implementation makes our implementation not portable to other Prolog systems.

We are currently pursuing another approach: keeping the extended meta-interpreter and applying partial compilation techniques. This has been done in the context of using Gödel as a programming language along with its partial evaluator. The results of this experiment are reported by Brogi and Contiero in [BC95].

Acknowledgements

This work was partly supported also by Progetto Finalizzato Sistemi Informatici e Calcolo Parallelo of C.N.R. under grant n. 92.01564.PF69. Our gratitude goes to the anonymous referee, who provided us with sharp and extremely helpful advice on how to restructure and improve the presentation.

References

[BK82]   K.A. Bowen and R.A. Kowalski. Amalgamating Language and Metalanguage in Logic Programming. In K.L. Clark and S.A. Tarnlund, editors, *Logic Programming*, pages 153–173. Academic Press, 1982.

[Bro93]   A. Brogi. *Program Construction in Computational Logic*. PhD thesis, University of Pisa, March 1993.

[BCM+94]   A. Brogi, A. Chiarelli, P. Mancarella, V. Mazzotta, D. Pedreschi, C. Renso, and F. Turini. Implementation of Program Composition Operations. In M. Hermenegildo and J. Penjam, editors, *Programming Language Implementation and Logic Programming: Proceedings of the 6th International Symposium*, number 844 in LNCS, pages 292–307. Springer-Verlag, 1994.

[BC95]   A. Brogi and S. Contiero. Composing Logic Programs by Meta-programming in Gödel. This volume.

[BMPT90]   A. Brogi, P. Mancarella, D. Pedreschi, and F. Turini. Composition Operators for Logic Theories. In J.W. Lloyd, editor, *Computational Logic, Symposium Proceedings*, pages 117–134. Springer-Verlag, 1990.

[BMPT94]   A. Brogi, P. Mancarella, D. Pedreschi, and F. Turini. Modular Logic Programming. *ACM Transactions on Programming Languages and Systems*, 16(4):1361–1398, 1994.

[BT90]   A. Brogi and F. Turini. Metalogic for Knowledge Representation. In J.A. Allen, R. Fikes, and E. Sandewall, editors, *Principles of Knowledge Representation and Reasoning: Proceedings of the Second International Conference*, pages 100–106. Morgan Kaufmann, 1990.

[BT94]   A. Brogi and F. Turini. Semantics of Meta-logic in an Algebra of Programs. In S. Abramsky, editor, *Ninth Annual IEEE Symposium on Logic in Computer Science*. IEEE Computer Society, 1994.

[BT95]   A. Brogi and F. Turini. Fully Abstract Compositional Semantics for an Algebra of Logic Programs. *Theoretical Computer Science*, 1995.

[DM92]   D. De Schreye and B. Martens. A Sensible Least Herbrand Semantics for Untyped Vanilla Meta-Programming and its Extension to a Limited Form of Amalgamation. In A. Pettorossi, editor, *Proceedings of the Third Workshop on Meta-programming in Logic*, pages 127–141, 1992.

[FLMP89]   M. Falaschi, G. Levi, M. Martelli, and C. Palamidessi. Declarative modeling of the operational behavior of logic languages. *Theoretical Computer Science*, 69(3):289–318, 1989.

[HL89]   P.M. Hill and J.W. Lloyd. Analysis of metaprograms. In H.D. Abramson and M.H. Rogers, editors, *Metaprogramming in Logic Programming*, pages 23–52. The MIT Press, 1989.

[JK95]   Y. Jiang and M. Kalsbeek. Ambivalent Logic as the Semantical Basis of Meta-logic Programming. This volume.

[Kal95]   M. Kalsbeek. Correctness of the Vanilla meta-interpreter and ambivalent syntax. This volume.

[Kow90]   R.A. Kowalski. Problems and Promises of Computational Logic. In J.W. Lloyd, editor, *Computational Logic, Symposium Proceedings*, pages 1–36. Springer-Verlag, 1990.

[LR93]   G. Levi and D. Ramundo. A Formalization of Metaprogramming for Real. In D.S. Warren, editor, *Proceedings Tenth International Conference on Logic Programming*, pages 354–373. The MIT Press, 1993.

[MD92]   B. Martens and D. De Schreye. Why untyped non-ground meta-programming is not (much of) a problem. Technical Report CW159, Departement Computerwetenschappen, K.U.Leuven, Belgium, December 1992. Revised November 1993, Abridged version to appear in The Journal of Logic Programming.

[MD95]   B. Martens and D. De Schreye. Two Semantics for Definite Meta-Programs using the Non-Ground Representation. This volume.

[MP88]   P. Mancarella and D. Pedreschi. An algebra of logic programs. In R. A. Kowalski and K. A. Bowen, editors, *Proceedings Fifth International Conference on Logic Programming*, pages 1006–1023. The MIT Press, 1988.

[SS86]   L. Sterling and E. Shapiro. *The Art of Prolog*. The MIT Press, 1986.
[vEK76]   M. H. van Emden and R. A. Kowalski. The semantics of predicate logic as a programming language. *Journal of the ACM*, 23(4):733–742, 1976.

5 Comparing Negation in Logic Programming and in Prolog

Krzysztof R. Apt and Frank Teusink

Abstract

We compare here two uses of negation – in logic programming and in Prolog. As in Prolog negation is defined by means of meta-programming facilities and the cut operator, this requires a careful reexamination of the assumptions about the underlying syntax and a precise definition of the computational processes involved.

After taking care of these matters we establish a formal result showing an equivalence in appropriate sense between these two uses of negation. This result allows us to argue about correctness of various known Prolog programs which use negation by reasoning about the corresponding general logic programs.

5.1 Introduction

During the last 15 years, a lot of attention was devoted to the study of negation in logic programming. No less than seven survey articles on this subject were published. Just to mention two most recent ones: Dix [Dix93] and Apt and Bol [AB94].

The main reason for this interest is that in the logic programming setting negative literals can be used to model non-monotonic reasoning. The computation process of logic programming provides then a readily available computational interpretation. This is not the case with other approaches to non-monotonic reasoning. This computation process is called SLDNF-resolution and was proposed by Clark [Cla78]. Negation is interpreted in it using the "negation as finite failure" rule. Intuitively, this rule works as follows: for a ground atom A,

$\neg A$ succeeds iff A finitely fails,

$\neg A$ finitely fails iff A succeeds,

where "finitely fails" means that the corresponding evaluation tree is finite and all its leaves are marked as failed.

However, SLDNF-resolution is not a practical way of computing and usually one resorts to Prolog when seeking for a computational interpretation. But in Prolog negation is implemented in a different way, namely by the predicate (or synonymously relation symbol) neg defined internally by the following two clauses:

$$\text{neg}(X) \leftarrow X, !, \text{fail}. \qquad (1)$$

$$\text{neg}(X) \leftarrow . \qquad (2)$$

where "!" is the cut operator and fail is a Prolog built-in with the empty definition.

The intuition behind this definition is perhaps best revealed by first introducing the if_then_else predicate defined as follows:

   if_then_else(P, Q, R) ← P,!,Q.
   if_then_else(P, Q, R) ← R.

if_then_else is intended to model within Prolog the customary **if** P **then** Q **else** R construct of imperative programming languages. Then neg can be equivalently defined by

   neg(X) ← if_then_else(X, fail, □).

where □ is the empty query which immediately succeeds. So intuitively, neg(X) can be interpreted as "if X succeeds then fail else succeed".

It is usually tacitly assumed that logic programming and Prolog ways of dealing with negation are "equivalent", in the sense that SLDNF-resolution combined with the leftmost selection rule (henceforth called LDNF-resolution) properly reflects Prolog's way of handling negation. Upon closer scrutiny this assumption is far from being obvious. The above definition of the neg predicate and its use in programs calls upon a number of features which are present in Prolog, but absent in logic programming, and for which a formal treatment is lacking. These are:

- the use of meta-variables, that is variables which occur in an atom position, like X in the first clause,

- the use of meta-programming facilities that arise when applying this definition of neg, so in constructs of the form neg(A) where A is an atom, or a query in general.

Additionally, two better understood, though not necessarily simpler to handle, features of Prolog need to be taken care of, namely:

- the ordering of the program clauses,

- the use of the cut operator "!".

The aim of this paper is to relate precisely these two uses of negation: in logic programming and in Prolog. To do this we appropriately tune the definition of the SLDNF-resolution given in Apt and Doets [AD94] to our present needs and formally define "Prolog trees" in the presence of the cut operator. Then we prove a result that shows an appropriate equivalence between these two definitions of negation.

The outcome of this study is that we can now interpret various results about correctness of general logic programs executed by means of the LDNF-resolution (see e.g. Apt [Apt95]) as correctness results about the corresponding Prolog programs that use negation.

5.2 Syntactic Matters

5.2.1 General Logic Programs

To relate general logic programs to Prolog programs we have to be precise about the syntax. Fix a first-order language $\mathcal{L}$. To make this comparison possible we assume that

- a general program is a *sequence* and not a *set* of general clauses,
- the predicates !, neg and fail are not present in the language $\mathcal{L}$.

A *general clause* is defined in the usual way (see e.g. Lloyd [Llo87]), so as a construct of the form $A \leftarrow L_1, \ldots, L_n$, where A is an atom and $L_1, \ldots, L_n$ are literals, i.e. atoms or their negations, all in the language $\mathcal{L}$. And a *query* is a finite sequence of literals. In the context of logic programming the negation connective is written as "$\neg$".

5.2.2 Prolog Programs

Prolog programs here considered are intended to be the programs that allow us to model the negation by means of the predicate neg defined by the clauses (1) and (2). However, the syntax of clause (1) creates a number of problems, even if we ignore the cut operator "!".

First of all, the use of the meta-variable X in clause (1) violates the syntax of the first-order logic. This use of X in the resolution process leads to further complications. Take an n-ary function symbol p in the language $\mathcal{L}$ and let $s_1, \ldots, s_n$ be some terms. Consider now the query neg(p($s_1, \ldots, s_n$)). During Prolog computation process it resolves using the clause (1) to the query p($s_1, \ldots, s_n$),!,fail. Now in the first query p occurs in a position of a function symbol, whereas in the second one p occurs in a position of a relation symbol. So every function symbol needs also to be accepted as a relation symbol.

Also conversely: take an n-ary relation symbol p with some terms $s_1, \ldots, s_n$, and consider the general clause $p(s_1, \ldots, s_n) \leftarrow \neg p(s_1, \ldots, s_n)$. Its desired translation into a Prolog clause is p($s_1, \ldots, s_n$) $\leftarrow$ neg(p($s_1, \ldots, s_n$)). In the head of the latter clause p occurs in a position of a relation symbol, whereas in its body in the position of a function symbol.

As in both cases p was arbitrarily chosen, we conclude that to render the resolution process meaningful we need to accept that the classes of function symbols and of relation symbols in the underlying language coincide.

This is clearly in violation with the (usually tacit) assumption that in the first-order language, say $\mathcal{L}$, fixed above, the classes F_m and R_n of, respectively, its function symbols of arity m and its relation symbols of arity n are pairwise disjoint for $m, n \geq 0$. In short, the use of the clause (1) cannot be properly accounted for by just referring to the first-order logic.

A simple solution to the above mentioned two problems is to modify the syntax of the language $\mathcal{L}$ by allowing

- *meta-variables*, so variables that can occur in atoms positions, both in the queries and in the clause bodies,

- *ambivalent syntax*, so – in this case – by assuming that the classes of function and relation symbols coincide.

The latter can be achieved by extending $\mathcal{L}$ to a language in which for each $m \geq 0$ $F_m \cup R_m$ are the classes of both its function symbols and relation symbols. Thus in this language terms and atoms coincide.

Additionally, we assume that

- the predicates !, `neg` and `fail` are present in the underlying language,

- ! is a built-in 0-ary predicate (with a meaning to be explained later), and no clause uses it in its head,

- `neg` is a built-in predicate defined by the clauses (1) and (2), so no other clause uses it in its head,

- `fail` is a built-in 0-ary predicate with the empty definition, so no clause uses it in its head.

The last two assumptions ensure that `neg` and `fail` are indeed defined internally in the desired way. For the purposes of syntax the cut operator "!" is viewed here as a 0-ary predicate with the empty definition. This might suggest that its meaning coincides with that of `fail`. However, this is not the case. Its real, operational, "meaning" will be defined in Section 5.4 by means external to the resolution process.

So in the resulting language, apart of the customary atoms, also !, `fail` and meta-variables are admitted as atoms (henceforth called *special atoms*).

Now, a *Prolog program* is defined as a sequence of Prolog clauses preceded by the clauses (1) and (2). In turn a *Prolog clause* is a construct of the form $A \leftarrow B_1, \ldots, B_n$, where $A, B_1, \ldots, B_n$ are atoms in the language $\mathcal{L}$, and A is not a special atom. And a *Prolog query* is a finite sequence of atoms. For brevity, in the examples of Prolog programs, we drop the listing of the clauses (1) and (2). Finally, we denote sequences of atoms or literals by bold capital letters.

Note that at this stage we use two notions of an atom – one within the language $\mathcal{L}$ and another in its ambivalent extension just defined. From the context it will be always clear to which of these two languages we refer.

5.2.3 Restricted Prolog Programs

The translation of a general program to a Prolog program is now straightforward and as expected: we just replace everywhere a logic programming literal $\neg A$ by Prolog's atom neg(A) and prefix the resulting program with the clauses (1) and (2). In short, the logic programming negation connective "$\neg$" is traded for the built-in predicate neg. Similarly, a general query is translated to a Prolog query by replacing everywhere $\neg A$ by neg(A).

This translation process maps every general program (resp. general query) onto a Prolog program. However, not every Prolog program (resp. Prolog query) is the result of translating a general program (resp. general query). Indeed, in general the cut operator "!" can be used in any Prolog clause, not only (1).

Let us now characterize the Prolog programs (resp. Prolog queries) which are the result of the above translation of general programs (resp. general queries). We call them *restricted Prolog programs* (resp. *restricted Prolog queries*). To this we translate "back" every Prolog program (resp. Prolog query) onto a general program (resp. general query) by replacing everywhere neg(A) by $\neg A$, and omitting the clauses (1) and (2) that define the neg predicate. Then a Prolog program (resp. Prolog query) is restricted if the outcome of this reverse translation is a syntactically legal general program (resp. general query). For example the Prolog query neg(q),q is restricted because its reverse translation is $\neg q, q$, whereas neither neg(q(neg(a))) nor p(q),q is restricted because their respective reverse translations violate the syntactic assumptions concerning general programs.

Of course, it is possible to define the class of restricted Prolog programs and queries directly, though the resulting definition is rather tedious.

We now define a *resolvent* of a Prolog query as follows.

Definition 1 Consider a non-empty Prolog query $A, \mathbf{M}$ and a Prolog clause c. Let $H \leftarrow \mathbf{L}$ be a variant of c variable disjoint with $A, \mathbf{M}$ and let θ be an mgu of A and H. Then $(\mathbf{L}, \mathbf{M})\theta$ is called a *resolvent* of $A, \mathbf{M}$ and c with an mgu θ. □

The only unusual feature in the present setting is, that now the mgu's also bind the meta-variables. Also, note that the selected literal is always the leftmost literal.

It is worthwhile to mention that a resolvent of a restricted Prolog query w.r.t. a restricted Prolog program is not necessarily a restricted Prolog query. This is due to the use of clause (1), which introduces a cut atom. Thus, the Prolog queries generated in a computation of a restricted Prolog query are not necessarily restricted Prolog queries. However, the Prolog queries so generated do have one important property: they do not contain meta-variables. To prove this fact we need a stronger property.

Definition 2
- An atom A is called *unsafe* if one of the following holds:

- A is a meta-variable,
- A is `neg(X)` where `X` is a variable,
- A is `neg(neg(s))` where `s` is a term.

- A Prolog query is called *meta-safe* if none of its atoms is unsafe. □

For example, the Prolog query `X, p(X)` is not meta-safe because its leftmost atom is a meta-variable, `neg(X)` is not meta-safe because the argument of `neg` is a meta-variable, and `neg(neg(p(X)))` is not meta-safe because it is of the form `neg(neg(s))`.

Note that restricted Prolog queries and bodies of the restricted Prolog clauses are meta-safe.

Lemma 3 *Let Q be a meta-safe Prolog query and P a restricted Prolog program. Then all resolvents of Q are meta-safe.*

Proof: Let Q be of the form $A, \mathbf{L}$, and let $(\mathbf{M}, \mathbf{L})\theta$ be a resolvent of Q, with an input clause c and mgu θ. As Q is meta-safe, we know that $\mathbf{L}\theta$ is meta-safe. We prove that $\mathbf{M}\theta$ is meta-safe as well. Three cases arise.

Case 1 : c is clause (1).
  Then $\mathbf{M}\theta$ is of the form $B, !, \texttt{fail}$, where A is of the form $\text{neg}(B)$. But Q is meta-safe, so B is neither a meta-variable nor of the form $\text{neg}(B')$. So $\mathbf{M}\theta$ is meta-safe.

Case 2 : c is clause (2).
  Then $\mathbf{M}\theta$ is the empty query, so obviously meta-safe.

Case 3 : c is different from clauses (1) and (2).
  Then the body of c is meta-safe, and consequently so is $\mathbf{M}\theta$.

This proves that $(\mathbf{M}, \mathbf{L})\theta$ is meta-safe. □

Corollary 4 *All Prolog queries generated in a computation of a restricted Prolog query and a restricted Prolog program are meta-safe.* □

In Prolog, if the selected atom is a meta-variable, an *error* arises. The above result thus shows that no errors arise in Prolog computations for queries and programs that are obtained by a translation of a general query and a general program.

5.3 Computing with General Logic Programs: LDNF-resolution

As the next step we define the LDNF-resolution that allows us to compute with general logic programs. The definition of LDNF-resolution given here is derived in a straightforward way from that of the SLDNF-resolution given in Apt and Doets [AD94]. Apart of the fact that we view in this paper a general program as a finite sequence and not as a finite set of general clauses, the differences are that:

- the leftmost selection rule is used,
- *floundering*, so –in this context– an abnormal termination due to selection of a non-ground literal is ignored.

In this way we bring the procedural interpretation of general programs closer to that of the corresponding Prolog programs and make the subsequent comparison possible. Recall from Clark [Cla78] and Lloyd [Llo87] that floundering is a problem that arises only when dealing with the semantic aspects of the SLDNF-resolution, which are irrelevant here.

Before giving the definition of LDNF-resolution, we recall the definitions of *resolvent* and *pseudo-derivation*.

Definition 5 Consider a non-empty general query $L, \mathbf{M}$ and a general clause c.

- Suppose L is a positive literal.
  Let $H \leftarrow \mathbf{L}$ be a variant of c variable disjoint with $L, \mathbf{M}$ and let θ be an mgu of L and H. Then $(\mathbf{L}, \mathbf{M})\theta$ is called a *resolvent* of $L, \mathbf{M}$ and c w.r.t. L, with an mgu θ. We write then $L, \mathbf{M} \stackrel{\theta}{\underset{c}{\Longrightarrow}} (\mathbf{L}, \mathbf{M})\theta$, and call it a *positive derivation step*. We call $H \leftarrow \mathbf{L}$ the *input clause* of the derivation step.
- Suppose L is a negative literal. Then $\mathbf{M}$ is called a *resolvent* of $L, \mathbf{M}$ with the identity substitution ϵ w.r.t. L.
  We write then $L, \mathbf{M} \stackrel{\epsilon}{\underset{\emptyset}{\Longrightarrow}} \mathbf{M}$, and call it a *negative derivation step*.
- A general clause c is called *applicable* to an atom if it has a variant the head of which unifies with the atom. □

Fix, until the end of this section, a general program P.

Definition 6 A (finite or infinite) sequence $Q_0 \stackrel{\theta_1}{\underset{c_1}{\Longrightarrow}} Q_1 \cdots Q_n \stackrel{\theta_{n+1}}{\underset{c_{n+1}}{\Longrightarrow}} Q_{n+1} \cdots$ of derivation steps is called a *pseudo derivation of* $P \cup \{Q_0\}$ if

- $Q_0, \ldots, Q_n, \ldots$ are general queries,
- $\theta_1, \ldots, \theta_n, \ldots$ are substitutions,
- $c_1, \ldots, c_n, \ldots$ are general clauses of P, or $\emptyset$,

and for every step involving selection of a positive literal the following condition holds:
Standardization apart: the input clause employed is variable disjoint from the initial general query Q_0 and from the substitutions and input clauses used at earlier steps. □

Intuitively, an LDNF-derivation is a pseudo derivation in which the deletion of every negative literal is justified by means of a subsidiary (finitely failed LDNF-) tree. This brings us to consider special types of trees, called *forests*.

Definition 7 A *forest* is a system $\mathcal{F} = (\mathcal{F}, T, subs)$ where

- $\mathcal{F}$ is a set of trees,
- T is an element of $\mathcal{F}$ called the *main tree*, and

- *subs* is a function assigning to some nodes of trees in $\mathcal{F}$ a ("subsidiary") tree from $\mathcal{F}$.

By a *path* in $\mathcal{F}$ we mean a sequence of nodes $N_0, \ldots, N_i, \ldots$ such that for all i, N_{i+1} is either an immediate descendant of N_i in some tree in $\mathcal{F}$, or the root of the tree $subs(N_i)$. The *depth* of $\mathcal{F}$ is the length of the longest path in $\mathcal{F}$. □

Thus a forest is a special directed graph with two types of edges – the "usual" ones stemming from the tree structures, and the ones connecting a node with the root of a subsidiary tree. An LDNF-tree is a special type of forest, built as a limit of certain finite forests: *pre-LDNF trees*.

Definition 8 A *pre-LDNF-tree* (relative to P) is a forest whose nodes are queries. Leaves can be unmarked, or can be marked as either *success* or *failure*. The class of pre-LDNF-trees is defined inductively:

- For every general query Q, the forest consisting of the main tree which has the single unmarked node Q is a pre-LDNF-tree (an *initial* pre-LDNF-tree),
- If $\mathcal{T}$ is a pre-LDNF-tree, then any *extension* of $\mathcal{T}$ is a pre-LDNF-tree.

Before defining the notion of an *extension* of a pre-LDNF-tree, we need to define the notion of *successful* and *finitely failed* trees: for $T \in \mathcal{T}$,

- T is called *successful*, if one of its leaves is marked as *success*, and
- T is called *finitely failed*, if it is finite and all its leaves are marked as *failure*.

Now, an *extension* of a pre-LDNF-tree $\mathcal{T}$ is defined by performing the following actions for every non-empty general query Q (with leftmost literal L) which is an unmarked leaf in some tree $T \in \mathcal{T}$:

- Suppose that L is a positive literal.
  - If Q has no resolvents w.r.t. L and a clause from P:
    Mark Q as *failure*.
  - If Q has such resolvents:
    For every clause c from P which is applicable to L, choose one resolvent Q' of Q w.r.t. L and c, with an mgu θ, and add this as an immediate descendant of Q in T. Choose the input clauses in such a way that all branches of T remain pseudo derivations.
- Suppose that L is a negative literal, say $\neg A$.
  - If $subs(Q)$ is undefined:
    Add a new tree T', consisting of the single node A, to $\mathcal{T}$, and let $subs(Q) = T'$.
  - If $subs(Q)$ is defined and successful:
    Mark Q as *failure*.

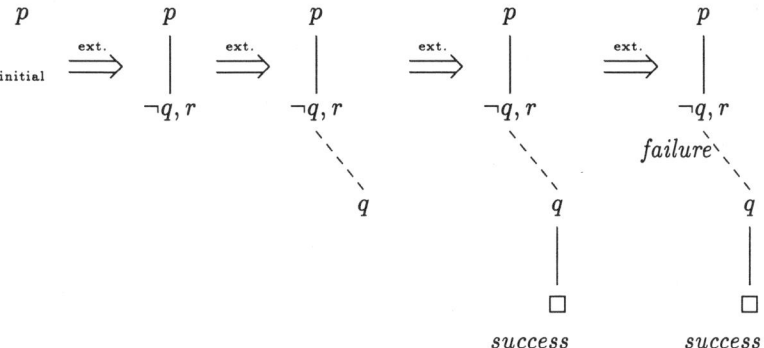

Figure 5.1
Step-by-step construction of an LDNF-tree for the query p w.r.t. the general program $p \leftarrow \neg q, r \quad q \leftarrow$.

- If $subs(Q)$ is defined and finitely failed:
   Add the resolvent $Q - \{L\}$ of Q as the only immediate descendant of Q in T.

Additionally, all empty queries are marked as *success*. □

Note that, if no tree in T has unmarked leaves, then trivially T is an extension of itself, and the extension process becomes stationary.

Next, we define LDNF-trees as the limit of sequences of pre-LDNF-trees. Every pre-LDNF-tree is a tree with two types of edges between possibly marked nodes, so the concepts of *inclusion* between such trees and of *limit* of a growing sequence of such trees have a clear meaning.

Definition 9

- An *LDNF-tree* is a limit of a sequence $T_0, \ldots, T_\alpha, \ldots$ such that T_0 is an initial pre-LDNF-tree, and for all i T_{i+1} is an extension of T_i.
- An *LDNF-tree for* Q is an LDNF-tree in which Q is the root of the main tree.
- A (pre-)LDNF-tree is called *successful* (resp. *finitely failed*) if the main tree is successful (resp. finitely failed).
- An LDNF-tree is called *finite* if no infinite path exists in it (cf. Definition 7). □

In Figure 5.1, we show how the notions of initial pre-LDNF-trees and extensions of pre-LDNF-trees are used to construct a P-tree.

Finally, we recall the notion of a computed answer substitution.

Definition 10 Consider a branch in the main tree of a (pre-)LDNF-tree for Q which ends with the empty query. Let $\alpha_1, \ldots, \alpha_n$ be the consecutive substitutions along this branch.

Then the restriction $(\alpha_1 \cdots \alpha_n)|Q$ of the composition $\alpha_1 \cdots \alpha_n$ to the variables of Q is called a *computed answer substitution* (*c.a.s.* for short) of Q. □

5.4 Computing with Prolog Programs: P-resolution

In this section, we define the computation process used in Prolog to find answers to queries, which we call *P-resolution*. To this end we proceed in two steps.

First, we restrict the LDNF-resolution to logic programs, so general logic programs without negation, by simply disregarding the selection of a negative literal. We call the resulting computation process *LD-resolution*.

Then, we extend the LD-resolution to Prolog programs by allowing the choice of a meta-variable or of a cut atom as a selected atom. In the first case an error is reported, and in the second case the computation tree constructed so far is appropriately pruned.

To better understand the issues involved in defining the effect of the cut operator, let us consider the definition of a predicate p:

$$p(s_1) \leftarrow L_1.$$
$$\ldots$$
$$p(s_i) \leftarrow M, !, N.$$
$$\ldots$$
$$p(s_k) \leftarrow L_k.$$

Here, the i-th clause contains a cut atom (there could be others, either in the same clause, or in other clauses). Now, suppose that during the execution of a query, some atom p(t) is resolved using (a variant of) the i-th clause, and that later on, the cut atom thus introduced becomes the leftmost atom. Then, according to the customary definition of the cut operator "!", once the indicated occurrence of ! is selected:

1. all other ways of resolving **M** are discarded, and

2. all derivations using (variants of) the $i+1$-th to k-th clause for p are discarded.

Note that this operational definition of the behaviour of the cut operator depends on the leftmost selection rule, and on viewing a program as a sequence of clauses, instead of a set of clauses.

To model this operational behaviour of the cut operator in P-resolution, we have to define it in terms of a pruning operator on LD-trees, but first, let us give an example of the behaviour of the cut operator. Consider the following Prolog program:

$$p \leftarrow q,!,t. \quad q \leftarrow r,!,t. \quad r \leftarrow s. \quad s \leftarrow .$$
$$p \leftarrow . \quad\quad q \leftarrow . \quad\quad\quad r \leftarrow .$$

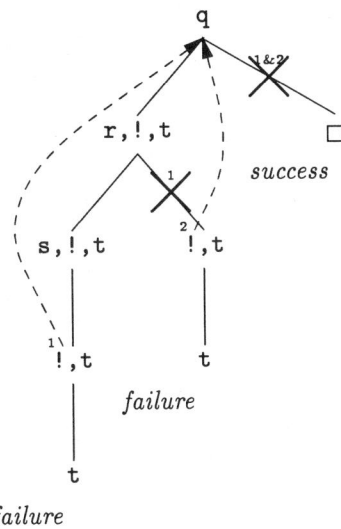

Figure 5.2
A computation tree for the query q

In Figure 5.2, an LD-tree for the query q is shown. In this tree, there are two nodes with a cut atom as the leftmost atom. Both of these cut atoms are introduced by resolving q in the root node of the tree. We say that their *origin* is the root node. In the figure, we use dashed arrows to to point from a selected cut atom to its origin. The two cut atoms that appear as leftmost atoms are marked as 1 and 2 respectively. Now, consider the cut atom marked as 1. Execution of this cut atom results in pruning: the middle branch has to be pruned according to rule 1, and the rightmost branch has to be pruned following rule 2. Execution of the cut atom marked as 2 also leads to a pruning of the rightmost branch (using rule 1 for the cut operator). In the figure, the pruned branches are marked using a cross. The label on the cross refers to the cut atoms that where responsible for the pruning of that branch.

Now, we can restate the behaviour of the cut operator as a pruning operator on LD-trees. Consider an LD-tree T. Let Q be a node in T with a cut atom as the selected atom and let Q' be the origin of this cut atom (i.e. the node that introduced this cut atom). Then, execution of this cut atom results in pruning all branches that are to the right of Q, contain Q', and do not contain Q.

In the tree of Figure 5.2, the order in which selected cut atoms where processed, was not important. However, in general, the order *is* important. Consider the LD-tree for p

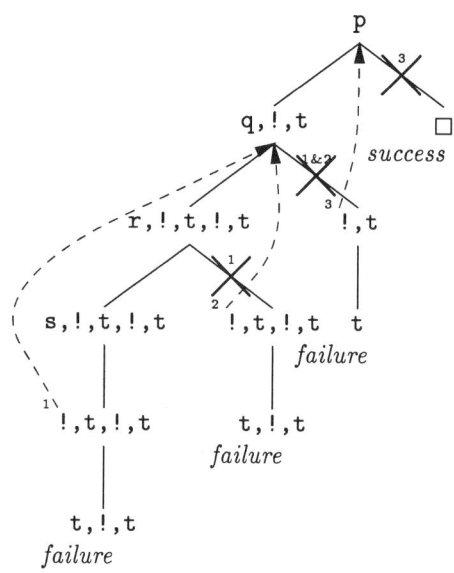

Figure 5.3
A computation tree for the query p

in Figure 5.3. Here, there are three nodes with a cut atom as leftmost atom, marked as 1, 2 and 3, respectively. Suppose we would process them from right to left. First, the cut atom marked as 3 would prune the rightmost branch. Then, the cut atoms marked as 2 and 1 would prune the third and the second branch from the left, respectively. The resulting tree would consist of the leftmost branch only. On the other hand, when processed from left to right, the cut atom marked as 1 would prune the middle two branches. As a result, the cut atoms marked as 2 and 3 would disappear, which would prevent the rightmost branch from being pruned. Thus, the resulting tree would consists of the leftmost branch *and* the rightmost branch.

In Prolog, answers are computed using a left to right depth-first strategy. In particular, Prolog processes the cut atoms in the tree from left to right. On the other hand, LD-resolution is defined in a breadth-first manner: the process of extending a pre-tree consists of extending all unmarked leaves of that tree simultaneously. To solve this problem, we have to refine LD-resolution so that the depth-first strategy is used instead of the breadth-first strategy. At first sight it seems that to this end we have to implement the backtracking mechanism used by Prolog. Fortunately, it is not so. A simpler alternative is to generate at each stage all direct successors of the *leftmost* unmarked leaf only. In this way the backtracking process is taken care of automatically.

Having discussed the modifications of the LD-resolution we now model the computation process of Prolog, by providing a formal definition of P-resolution. The central notion in this definition is that of a *P-tree*. We define them as the limit of a sequence of *pre-P-trees*, which in turn are a subclass of a class of ordered trees called *semi-P-trees*.

Definition 11 A *semi-P-tree* (relative to P) is an ordered tree whose nodes contain queries, possibly marked with *success*, *failure*, or *error*. □

In an ordered tree, by definition for every node there is a strict total order on its children. To define the behaviour of the cut operator, we use these total orders to define a partial order on the nodes of an ordered tree.

Definition 12 Let m, n be two nodes in an ordered tree. We say that n is *to the right of* m if for some predecessors m' and n' of m and n, respectively,

- m' and n' are siblings,
- m' is strictly smaller than n' in the total order on the children of a node. □

The first step in defining pre-P-trees is to define the effect of the cut operator.

Definition 13 Let $\mathcal{B}$ be a branch in a semi-P-tree, and let Q be a node in this branch with a cut atom as the leftmost atom. Then, the *origin* of this cut atom is the first predecessor of Q in $\mathcal{B}$ that contains less cut atoms than Q. □

To see that this definition properly captures the informal meaning of the origin note that, when following a branch from top to bottom, the cut atoms are introduced and removed in a First-In Last-Out manner.

Definition 14 Let $\mathcal{T}$ be a semi-P-tree, Q a query in $\mathcal{T}$ which has a cut atom as the leftmost atom, and Q' be the origin of this cut atom. Then, the operator $cut(\mathcal{T}, Q)$ removes from $\mathcal{T}$ all the nodes that are descendants of Q' and lie to the right of Q. □

In Figure 5.4, we illustrate the effect of $cut(\mathcal{T}, Q)$.

Definition 15 The class of *pre-P-trees* is defined as follows:

- For every query Q, the tree consisting of the single unmarked node Q is a pre-P-tree (an *initial* pre-P-tree).
- If $\mathcal{T}$ is a pre-P-tree, then any *extension* of $\mathcal{T}$ is a pre-P-tree.

An *extension* of a pre-P-tree $\mathcal{T}$ is defined as follows:

Let Q be the leftmost unmarked leaf in $\mathcal{T}$. If Q is the empty query, mark Q as *successful*. Otherwise, let Q be of the form $A, \mathbf{M}$.

- Suppose A is an ordinary atom (i.e. not a special atom).
    - If Q has no resolvents w.r.t. a clause from P:
      Mark Q as *failure*.

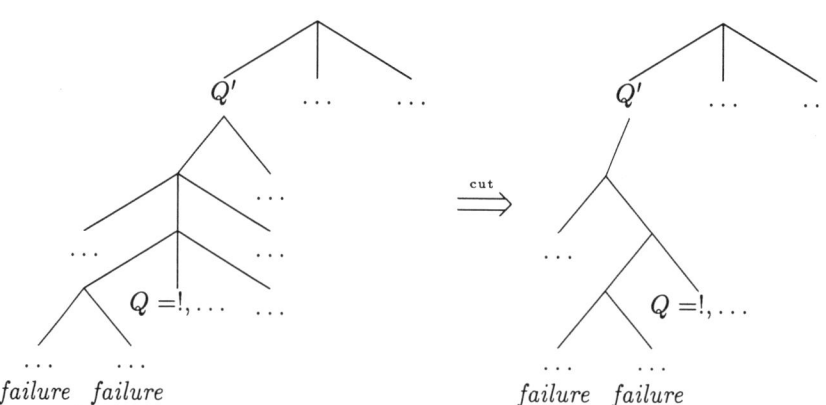

Figure 5.4
The effect of the operator $cut(T, Q)$

- If Q has such resolvents:

    For every clause c from P which are applicable to A, choose one resolvent Q' of Q w.r.t. c and add this as a child of Q in T. Choose the input clauses in such a way that all branches of T remain pseudo derivations. Order these children according to the the order in which their input-clauses appear in P.

- Suppose A is a cut atom.
  Apply the operation $cut(T, Q)$.
  Provide Q with a single child **M**.
- Suppose A is a meta-variable.
  Mark Q as *error*. □

We now define P-trees as the limit of sequences of pre-P-trees. In Figure 5.5, we show how the notions of initial pre-P-trees and extensions of pre-P-trees can be used to construct a P-tree (the program used in the figure is the translation of the program used in Figure 5.1). Note that in this Figure, the result of the 'cut step' (that is, the fifth tree) is not itself part of the sequence of extensions; it was added to clarify the use of the cut operator in the construction of P-trees.

To be able to define the limit of a sequence of pre-P-trees, we have to define a notion of an *inclusion* between pre-P-trees, and of the *limit* of a growing sequence of pre-P-trees. For pre-LD-trees and pre-LDNF-trees, these notions were obvious. In the case of pre-P-trees, the pruning that takes place when extending a pre-P-tree, complicates the matters a bit.

Definition 16 Let T and T' be pre-P-trees. T is said to be *included* in T' if T' can be constructed from T by means of one of the following two operations:

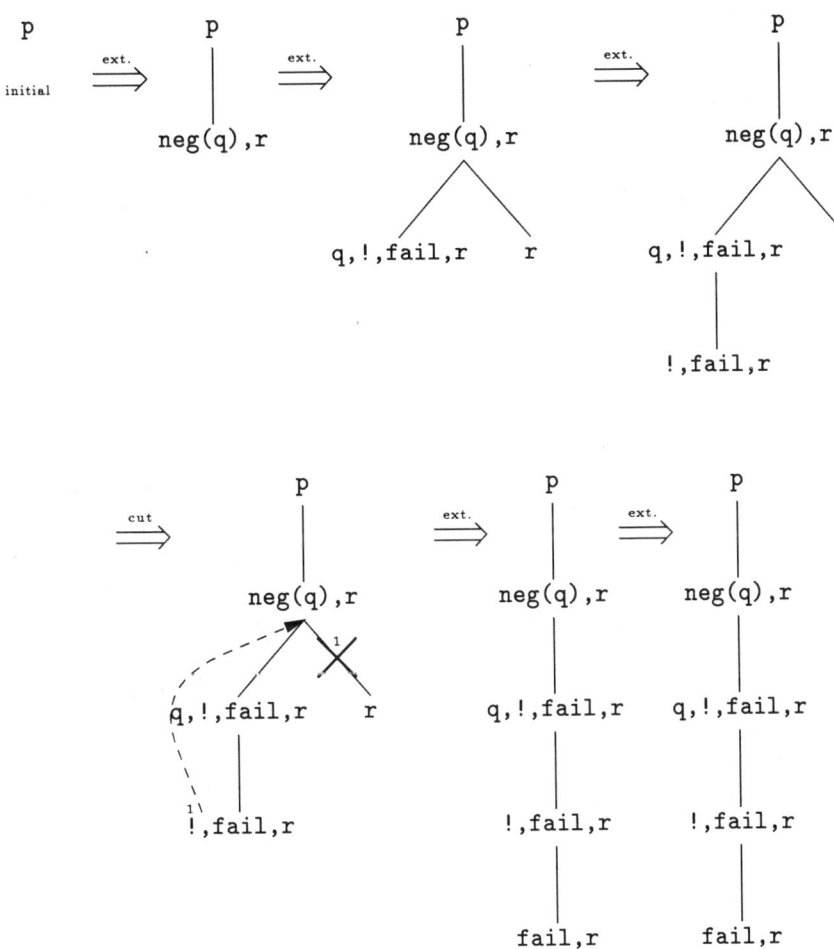

Figure 5.5
Step-by-step construction of a P-tree for the Prolog query p w.r.t. the Prolog program
p ← neg(q), r. q ← . .

1. adding some children to a leaf of T.
2. removing a single subtree from T, provided its root is not a single child in T.

We say that T is *properly included* in T', if T is included in T' and T' is not included in T. We use $\subset$ to denote the transitive closure of the relation "T is properly included in T'" and define $T \subseteq T'$ as $(T \subset T') \vee (T = T')$. □

Note that operation (2) never turns an internal node into a leaf.

Lemma 17 *The relation $\subset$ is a strict partial order on pre-P-trees.*

Proof: We have to prove that the conditions for a strict partial order hold.

1. $T \not\subset T$

   Suppose by contradiction that $T \subset T$. Then, there exists a T' such that T is properly included in T', and $T' \subseteq T$. There are two cases:

   - T' is constructed by adding children to a leaf of T.

     But then, some node Q that is a leaf in T, is an internal node in T'. By definition of inclusion, and the fact that $T' \subseteq T$, Q is an internal node in T. This is in contradiction with the fact that Q is a leaf T.

   - T' is constructed by pruning a single subtree from T.

     By definition of inclusion, the parent of the pruned subtree has at least two children in T, and therefore, it has at least one child in T'. Moreover, new nodes can only "grow" from leaves. Thus subtrees pruned from T can never be "regenerated", to reconstruct T out of T'. Therefore, $T' \not\subseteq T$, which leads to a contradiction.

2. $T \subset T'$ and $T' \subset T''$ imply $T \subset T''$.

   Straightforward by the definition of $\subset$. □

Corollary 18 *The relation $\subseteq$ is a partial order on pre-P-trees.* □

Clearly, with this notion of inclusion, we have that if T extends T' in the sense of Definition 15, then $T' \subseteq T$, so we can use this notion of extension to construct monotonously growing chains of pre-P-trees.

Definition 19

- A *P-tree* is a limit of a sequence $T_0, \ldots, T_i, \ldots$ such that T_0 is an initial pre-P-tree, and for all i, T_{i+1} is an extension of T_i.
- A *P-tree for* Q is a P-tree whose root is the query Q.
- An P-tree is called *finite* if no infinite branch exists in it. □

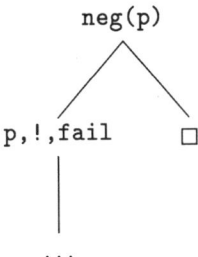

Figure 5.6
A P-tree for the query neg(p) w.r.t. p ← p.

Formally, this definition is justified by the fact that every countable partial order with the least element (here the relation $\subseteq$ on pre-P-trees with the initial pre-P-tree as least element) can be canonically extended to a countable cpo (see e.g. Gierz [GHK+80]).

Next, we define the concepts of *successful* and *finitely failed* P-trees.

Definition 20

- A P-tree is called *successful* if one of its leaves is marked as *success*.
- A (pre-)P-tree is called *finitely failed*, if it is finite, and all its leaves are marked as *failure*. □

Note that in P-trees, in contrast to LDNF-trees, some leaves can be unmarked. Whenever this is the case, the P-tree will contain exactly one infinite branch to the left of all these unmarked leaves. Such unmarked leaves represent the resolvents the Prolog computation process did not reach, because it got "trapped" in an infinite derivation (the infinite branch). For example, take the program p ← p., and the query neg(p). Its P-tree is shown in Figure 5.6. This tree contains a branch ending with a leaf containing the empty query. However, this leaf is never reached by the Prolog computation process (and therefore never marked) because there is an infinite branch to the left of it.

Finally, it is clear how to define the notion of a computed answer substitution.

Definition 21 Consider a successful derivation in a pre-P-tree for Q. Let $\alpha_1, \ldots, \alpha_n$ be the consecutive substitutions along this branch.

Then the restriction $(\alpha_1 \cdots \alpha_n)|Q$ of the composition $\alpha_1 \cdots \alpha_n$ to the variables of Q is called a *computed answer substitution* (*c.a.s.* for short) of Q. □

5.5 Correspondence between LDNF-trees and P-trees

In this section, we prove that there is a close correspondence between (computed answers of) LDNF-trees and P-trees. More precisely, we prove that termination results on general programs w.r.t. LDNF-resolution translate directly into termination of their translated Prolog programs w.r.t. Prolog computation. For this purpose, we start by examining finite LDNF-trees, and their corresponding P-trees.

Theorem 22 *Let T_L be a finite LDNF-tree for a general query Q. Then, there exists a finite P-tree T_P for Q such that T_L and T_P have the same set of computed answers.*

Proof: We prove the claim by induction on the depth of LDNF-trees (cf. Definition 7). Assume that the claim holds for all LDNF-trees of depth less than r. We have to prove the claim for LDNF-trees of depth r.

Let T_L be an LDNF-tree for Q of some finite depth r. In the remainder of this proof, we identify a general query with its translation into a Prolog query. From the context it will always be clear whether we refer to a general query, or a Prolog query. Two cases arise.

- Suppose that Q is of the form $A, \mathbf{L}$.

  Let $Q_1, \ldots, Q_k$ ($k \geq 0$) be the children of Q in T_L. Let, for $i \in [1..k]$, T_L^i denote the subtree of T_L starting at Q_i.

  As, for $i \in [1..k]$, T_L^i is finite and of depth less than r, by induction hypothesis there exists a P-tree T_P^i for Q_i such that T_P^i contains the same computed answers as T_L^i. Now consider the semi-P-tree T_P with root Q, children $Q_1, \ldots, Q_k$ (ordered according to the order of their input clauses in P) and, for $i \in [1..k]$, T_P^i as the subtree starting at Q_i, as depicted by the following diagram:

  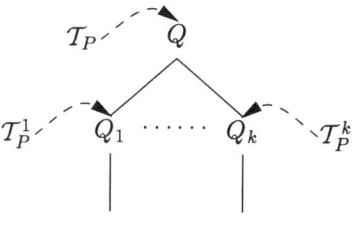

  To prove that T_P is a P-tree for Q, it is sufficient to show that all pruning caused by selection of cut atoms is guaranteed to be local to the respective subtrees T_P^i (for $i \in [1..k]$). Neither Q, nor its children $Q_1, \ldots, Q_k$ in T_P, contain a cut atom, so no atom in T_P has Q as its origin. It follows from the definition of the cut operator that all pruning is indeed local to the respective subtrees T_P^i. Thus T_P is a P-tree for Q. From its construction, it follows that it contains the same computed answers as T_L. Moreover, it is finite.

- Suppose that Q is of the form $\neg A, \mathbf{L}$.

Let $\mathcal{T}_L^1$ be the subtree of $\mathcal{T}_L$ starting at the root of $subs(Q)$. As the LDNF-tree $\mathcal{T}_L^1$ for A is finite and of depth less than r, by induction hypothesis there exists a finite P-tree $\mathcal{T}_P^1$ for A that has the same computed answers as $\mathcal{T}_L^1$. There are two sub-cases.

- Suppose that Q has a child in $\mathcal{T}_L$.

  Then, $\mathcal{T}_L^1$ is finitely failed, and therefore $\mathcal{T}_P^1$ is finitely failed as well. But then, we can construct a finitely failed P-tree $\mathcal{T}_P^{1'}$ for $A,!,\texttt{fail},\mathbf{L}$. In this P-tree, the cut atom introduced at the root will never be reached.

  Let $\mathcal{T}_L^2$ be the subtree of $\mathcal{T}_L$ starting at the single child $\mathbf{L}$ of Q. As the LDNF-tree $\mathcal{T}_L^2$ for $\mathbf{L}$ is finite and of depth less than r, by induction hypothesis there exists a finite P-tree $\mathcal{T}_P^2$ for $\mathbf{L}$ that has the same computed answers as $\mathcal{T}_L^2$.

  Using $\mathcal{T}_P^{1'}$ and $\mathcal{T}_P^2$ we can construct a finite P-tree $\mathcal{T}_P$ for Q that has the same computed answers as $\mathcal{T}_L$. This tree has the following form:

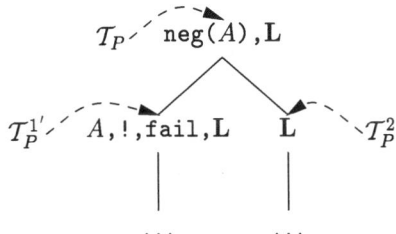

- Suppose that Q has no children in $\mathcal{T}_L$.

  Then, $\mathcal{T}_L^1$ is successful, and therefore $\mathcal{T}_P^1$ is successful as well. But then we can construct a finitely failed P-tree $\mathcal{T}_P^{1'}$ for $A,!,\texttt{fail},\mathbf{L}$, in which the cut atom present in its root is selected at some point.

  Let $\mathcal{T}_P$ be the semi-P-tree such that its root is Q, and the subtree starting at the single child $A,!,\texttt{fail},\mathbf{L}$ of Q is $\mathcal{T}_P^{1'}$. In this tree, the origin of the cut atom that appears in the single child of Q, is Q. This cut atom is the selected atom in some node within $\mathcal{T}_P^{1'}$. Thus $\mathcal{T}_P$ is a P-tree for Q, because the potential second child of Q, that would contain the query $\mathbf{L}$ has been pruned at some stage. Thus $\mathcal{T}_P$ is finitely failed, just as $\mathcal{T}_L$ is. □

Thus if we have a general query Q that terminates w.r.t. a general program P, we know that Prolog computation on that query and that program will terminate, and give the same computed answers as LDNF-resolution.

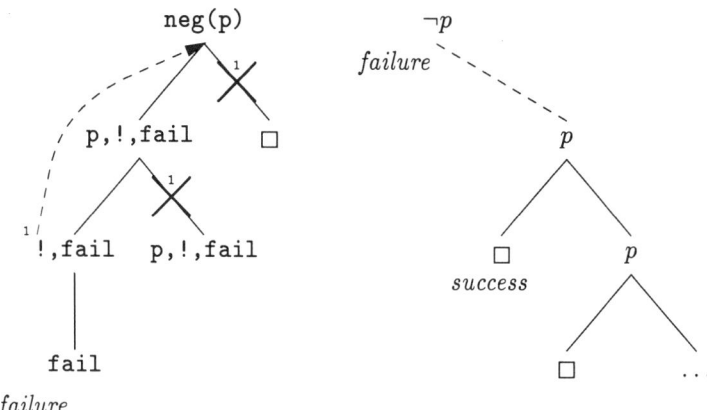

Figure 5.7
A P-tree and an LDNF-tree for neg(p)

Now what if we have a finite P-tree for a restricted Prolog query Q and a restricted Prolog program P? Consider the following restricted Prolog program

$$p \leftarrow$$
$$p \leftarrow p$$

and the restricted Prolog query neg(p). The P-tree and LDNF-tree for this query and this program are shown in Figure 5.7 (note that the pruned branches are not really part of the P-tree for neg(p), but existed at some point during the construction of this P-tree). In this example, the P-tree is finite, because the potentially infinite branch caused by the clause $p \leftarrow p$ is pruned. However, in the LDNF-tree, this branch has been constructed in full, and therefore this LDNF-tree is infinite.

5.6 Applications

Due to the presence of cut in the definition of the predicate neg it is difficult to reason in a declarative way about Prolog programs that use negation. In other words, it is not clear how to prove correctness of such programs using their declarative interpretation.

We now show how this is possible using the results of this paper. The key observation is that Theorem 22 provides a crucial relationship between the computational behaviour of Prolog programs and their translations into general logic programs.

In the subsequent discussion we assume that the variables in the input clauses and the mgu's are chosen in a fixed way. We can then assume that for every Prolog program

P and Prolog query Q there exists exactly one P-tree, and similarly for general logic programs, general queries and LDNF-trees.

So consider a restricted Prolog program P with a restricted query Q and their translation P_L and Q_L onto a general logic program and a general logic query, respectively. To reason about correctness of P with Q it is sufficient to reason about P_L and Q_L. Indeed, suppose that we proved already that all LDNF-derivations of P and Q are finite. Then by Theorem 22 the P-tree for P_L and Q_L is finite, and P_L with Q_L and P with Q have the same set of computed answers.

As an example consider the following well-known Prolog program TRANS about which one claims that it computes the transitive closure a binary relation e:

```
trans(X, Y, E, Avoids)  ←  member([X, Y], E).
trans(X, Z, E, Avoids)  ←
    member([X, Y], E),
    neg(member(Y, Avoids)),
    trans(Y, Z, E, [Y | Avoids]).

member(X, [X | Xs])  ← .
member(X, [Y | Xs])  ←  member(X, Xs).
```

In Apt [Apt95] the following facts about its translation TRANS_L to a general logic program and a binary relation e were established:

- all LDNF-derivations of trans(X, Y, e, []) are finite,
- the computed answer substitutions of trans(X, Y, e, []) determine all pairs of elements which form the transitive closure of e.

Now, by Theorem 22 the same conclusions can be drawn about the original program TRANS.

The fact that above approach to correctness is limited to restricted Prolog programs is in our opinion not serious. In fact, we noticed that practically all "natural" Prolog programs that use negation are restricted.

5.7 Related Work

We conclude by briefly discussing related work.

There is an enormous literature on the subject of negation in logic programming, see, e.g., the references in the surveys cited in the introduction. However, to our knowledge, no work has been done on negation in Prolog.

The use of the ambivalent syntax was first advocated in mathematical logic by Richards [Ric74], in the logic programming setting by Kalsbeek [Kal93] (who actually coined the term), and Jiang [Jia94], and in the programming languages area by Chen, Kifer and Warren[CKW89] in their language proposal HiLog. In each of these references different versions of ambivalence are assumed. For example, in Kalsbeek [Kal93] atoms can appear as terms and in Jiang [Jia94] formulas can appear as terms. Here we use an alternative version of ambivalence, which amounts to identification of atoms and terms, though we also allow meta-variables. For a systematic and thorough overview of various versions of ambivalent syntax see Kalsbeek and Jiang [KJ95].

The definition of the LDNF-resolution given in Section 3 is derived from the definition of the SLDNF-resolution provided in Apt and Doets [AD94]. An alternative definition of SLDNF-resolution was given earlier by Martelli and Tricomi [MT92]. Both definitions overcome problems encountered in the original definition of Clark [Cla78].

In our operational semantics of Prolog programs, given in Section 4, we also provided a meaning to the cut operator. The problem of formalizing the meaning of cut has been studied in a number of publications during the last 10 years. Jones and Mycroft [JM84] defined various semantics for Prolog with cut. This work was pursued by Arbab and Berry [AB87], Debray and Mishra [DM88], and more recently by Lilly and Bryant [LB92].

In the literature several alternatives to the cut operator have been proposed – see e.g. Moss [Mos86] and more recent Hill, Lloyd and Shepherdson [HLS90].

Acknowledgements

We would like to thank Marianne Kalsbeek and two referees for helpful comments. The work of the second author was supported by the Foundation for the Computer Science Research in the Netherlands (SION).

References

[AB87]   B. Arbab and D.M. Berry. Operational and denotational semantics of Prolog. *Journal of Logic Programming*, 4(4):309–329, 1987.

[AB94]   K.R. Apt and R. Bol. Logic programming and negation: a survey. *Journal of Logic Programming*, 19-20:9–71, 1994.

[AD94]   K.R. Apt and K. Doets. A new definition of SLDNF-resolution. *Journal of Logic Programming*, 18(2):177–190, 1994.

[Apt95]   K. R. Apt. Program verification and Prolog. In E. Börger, editor, *Specification and Validation methods for Programming languages and systems*. Oxford University Press, 1994. To appear.

[CKW89]   W. Chen, M. Kifer, and D.S. Warren. Hilog: A first-order semantics for higher-order logic programming constructs. In *Proceedings of the North-American Conference on Logic Programming*, Cleveland, Ohio, October 1989.

[Cla78]   K.L. Clark. Negation as failure. In H. Gallaire and G. Minker, editors, *Logic and Data Bases*, pages 293–322. Plenum Press, 1978.

[Dix93]   J. Dix. Semantics of Logic Programs: Their Intuitions and Formal Properties. An Overview. In Andre Fuhrmann and Hans Rott, editors, *Logic, Action and Information. Proceedings of the Konstanz Colloquium in Logic and Information (LogIn '92)*. DeGruyter, 1993.

[DM88]   S.K. Debray and P. Mishra. Denotational and operational semantics for Prolog. *Journal of Logic Programming*, 5(1):61–91, 1988.

[GHK+80]   G. Gierz, K.H. Hofmann, K. Keimel, J.D. Lawson, M.W. Mislove, and D.S. Scott. *A Compendium of Continuous Lattices*. Springer-Verlag, 1980.

[HLS90]   P.M. Hill, J.W. Lloyd, and J.C. Shepherdson. Properties of a pruning operator. *Journal of Logic and Computation*, 1(1):99–143, 1990.

[Jia94]   Y. Jiang. Ambivalent logic as the semantic basis fo metalogic programming: I. In P. Van Hentenryck, editor, *Proceedings of the International Conference on Logic Programming*, pages 387–401. MIT Press, June 1994.

[JM84]   N.D. Jones and A. Mycroft. Stepwise development of operational and denotational semantics for Prolog. In *International Symposium on Logic Programming*, pages 281–288, 1984.

[Kal93]   M. Kalsbeek. The vanilla meta-interpreter for definite logic programs and ambivalent syntax. Technical Report CT-93-01, Department of Mathematics and Computer Science, University of Amsterdam, The Netherlands, 1993.

[KJ95]   M. Kalsbeek and Y. Jiang. A vademecum of ambivalent logic. This volume.

[LB92]   A. Lilly and B.R. Bryant. A prescribed cut for Prolog that ensures soundness. *Journal of Logic Programming*, 14(4):287–339, 1992.

[Llo87]   J.W. Lloyd. *Foundations of Logic Programming*. Symbolic Computation – Artificial Intelligence. Springer-Verlag, 1987. Second, extended edition.

[Mos86]   C. Moss. Cut & Paste – defining the impure primitives of Prolog. In E. Shapiro, editor, *Proceedings of the International Conference on Logic Programming*, number 225 in Lecture Notes in Computer Science, pages 686–694. Springer Verlag, 1986.

[MT92]   M. Martelli and C. Tricomi. A new SLDNF-tree. *Information Processing Letters*, 43(2):57–62, 1992.

[Ric74]   B. Richards. A point of reference. *Synthese*, 28:431–445, 1974.

II LANGUAGE SUPPORT FOR META-LOGICS

6 Towards Fast and Declarative Meta-Programming

Antony F. Bowers and Corin A. Gurr

Abstract

Meta-programming, the ability to manipulate programs as data, is fundamental to the success of declarative languages. Regardless of the choice of logic, without properly representing object variables as ground terms, declarative meta-programs are severely limited. However, use of the ground representation seems to involve unacceptable programming effort and computational overhead. We examine the underlying reasons for these problems, and investigate techniques for addressing them so that the dream of declarative meta-programming can be realised. Gödel is a programming language based on first order logic which is intended to have better declarative semantics than Prolog. Taking Gödel interpreters as typical meta-programs, we show how awareness of efficiency issues in the interpretation algorithm suggests a particular programming style. The careful design and construction of the Gödel system modules has made writing such interpreters straightforward, and also made them amenable to partial evaluation. By using mutable data structures to represent programs and substitutions, it is possible to optimise these interpreters further until they approach the speed of interpreters using the traditional non-ground representation. The details of these optimisations may remain entirely hidden from the programmer. The efficiency gains apply equally to interpreters performing tasks that require the full power of the ground representation, and so have no declarative non-ground equivalent for comparison.

6.1 Introduction

Declarative meta-programming is vital for the future of computing. If the use of computers is to continue to expand, but the task of programming them is not to occupy the entire working population, we must eventually learn to teach them to create and maintain their own software. Declarative programming languages surely offer the best hope that machines will one day be able to assist us in mastering the complexity of their own programming, because it is declarative programs that are most readily understood and manipulated by other programs. These meta-programs must also take their turn as data, and so should be declarative.

The full potential of a truly declarative language for meta-programming is sometimes overlooked. Hill and Lloyd [HL94] summarise declarative programming as being "much more concerned with writing down *what* should be computed and much less concerned with *how* it should be computed". This approach has two major benefits of particular relevance to meta-programming. The first is that declarative meta-programs, particularly large programs, can be much easier to understand than comparable procedural meta-programs. The second advantage becomes apparent when writing meta-programs which manipulate object programs or theories which are themselves declarative. Such meta-programs often need pay little or no concern to the procedural semantics or non-logical quirks of the object programs which they manipulate. As a consequence these meta-programs can employ far simpler algorithms than would be needed to manipulate comparable non-declarative object programs.

It is well-known that declarative meta-programming requires carefully representing the language of the object program in the language of the meta-program, and that this can be done in two ways, the non-ground and ground representations described by Hill and Lloyd [HL89]. The non-ground representation is easy to implement efficiently and easy to program, but its power to manipulate object programs is extremely limited because object variables cannot be inspected. Regardless of the particular choice of logic, without the representation of object variables by ground terms, declarative meta-programming is not powerful enough to implement sophisticated meta-programs such as program transformers. However, at first sight the management of object variables in the ground representation involves unacceptable programming effort and computational overhead, so in current Prolog practice object variables are usually represented by meta-variables and handled by means of non-logical operations. Consequently the declarative property of meta-programs is lost.

Gödel is a programming language based on first-order logic and intended to have better declarative semantics than Prolog. Gödel places particular emphasis on declarative meta-programming, and provides considerable support for programmers wanting to manipulate ground representations of Gödel object programs. Gödel has a type system based on *polymorphic many-sorted logic*, and a module system. Gödel provides a rich set of system modules, such as **Integers**, **Lists**, **Rationals**, **Floats**, **Strings**, **Sets** and **IO** to support the commonly used data types and common operations on those types. Gödel predicates may have control declarations, called **DELAY** declarations. These are syntactic variants of the **when** declarations of NU-Prolog (Thom and Zobel [TZ88]). It is not our intention to describe the Gödel language here. For a complete technical description and specification the reader is referred to Hill and Lloyd [HL94].

A Gödel system has been implemented at Bristol. This implementation (known as "Bristol Gödel") is based on a compiler that translates the Gödel source into SICStus Prolog. The experiments described in this paper were conducted on this system.

Meta-programs using the ground representation incur a considerable overhead due to the necessity of explicitly managing the ground variables. In this paper we investigate possible techniques to reduce this overhead in Gödel. These techniques include partial evaluation and the use of arrays with destructive assignment to represent substitutions. We take simple interpreters as examples of typical meta-programs, as the performance of interpreters is generally critically dependent on the efficiency with which variables are handled.

The central problem we are struggling with is to find a way to relate the complex and laborious high-level manipulations performed by an interpreter executing SLD-resolution on the ground representation of an object program, to the relatively simple and efficient operations performed by the underlying language implementation, which is essentially doing the same thing. If the interpreter can somehow be made to copy the low-level mechanism without sacrificing the declarative semantics, it can be made efficient. Remember that the declarative semantics of a meta-interpreter is the procedural semantics of the object program. It is therefore not surprising that our optimisations lead to interpreters that resemble compiled Prolog.

The standard Prolog implementation, using the WAM (Warren [War83]), is unsuitable as a basis for interpreting programs in the ground representation as it has too many optimisations which are specific to resolution without the occur check in a depth-first backtracking system. The only interpretation mechanism previously formulated for the ground representation was the explicit *unify-compose-apply* style of resolution presented in the interpreters of first Bowen and Kowalski [BK82] and then Hill and Lloyd [HL89]. The course of this paper follows the historical course taken in the development of the current implementation of the ground representation where successive development steps have led to an implementation which is similar, but not identical, to the WAM.

Of course, an essential feature of meta-programming in Gödel is that all the cunning is hidden from the programmer inside the library modules provided by the system. In fact, the style of meta-interpreter that emerges from these considerations, once it becomes familiar, can be seen to be very much simpler than traditional interpreters for the ground representation, and is analogous to the Vanilla interpreter for the non-ground representation.

In the next section we look briefly at the design of the ground representation in Gödel, and the important issues arising from the simulation of unification in the ground representation. In section 6.3 we consider the design of some simple Gödel meta-interpreters and compare their performance. Finally, in section 6.4 we investigate the performance improvements that can be obtained by partial evaluation and by adding special-purpose low-level support to the implementation. Although performed in a crude prototype implementation, these experiments show a dramatic performance improvement which clearly promises that the techniques we outline have the potential to make meta-programming in the ground representation fully practical.

6.2 The Ground Representation in Gödel

6.2.1 Syntax and programs

Anyone attempting to write a meta-program using a ground representation in a logic programming language will immediately meet the first difficulty: the labour involved. A lot of routine decisions have to be made concerning the details of the representation, and a lot of routine programming work is required to implement basic operations on the representation such as unification.

To solve this problem, Gödel provides a fixed representation for Gödel programs, in the form of a collection of abstract data types so that programmers need not be concerned with the details of the representation. Two of Gödel's system modules, Syntax and Programs, export an extensive library of basic operations on these abstract data types. There are disadvantages in a fixed representation, as argued by van Harmelen [vH92], but the facilities provided are intended to be sufficient to cover most meta-programming tasks involving manipulations of Gödel object programs, and to assist the programmer in achieving efficiency in a large proportion of these tasks. Some of the design issues arising from this latter aim are the topic of this paper. The alternative, allowing the programmer to design the representation, would surely also return a good part of the labour to the programmer.

The Syntax module is concerned with the manipulation of syntactic expressions in the language of the Gödel object program. It exports the types Name, Term and Formula, which are the types of terms representing object symbols, terms and formulas respectively.

The Programs module is concerned with the representation of complete Gödel object programs as terms in the meta-program. It exports the type Program for this purpose. We do not deal with the detailed structure of these program terms here, the interested reader is referred to Bowers [Bow92]. It is sufficient to note that in the Bristol implementation the module structure, language declarations, control declarations and statements of the object program are stored in balanced binary trees, permitting them to be accessed in logarithmic time.

We now take a brief look at some of the constants and functions used to form the ground representation in the Syntax module, as this will help to make the following discussion concrete. In Syntax, we are concerned with the representation of the syntactic structures of the object program. Let us begin with the symbols of the object programs's alphabet. Every symbol in a Gödel program has two names. Firstly, it has a *declared name*, which is the name given by the programmer when the symbol is declared. This is the name used wherever the symbol appears in the program, and for simplicity a symbol is usually identified with its declared name. Secondly, it has a *flat name*, which is an internal name used to identify symbols uniquely. Gödel allows generous overloading, and consequently the declared name of a symbol is not sufficient to identify it unambiguously. A flat name has four components: the name of the module in which the symbol is declared, its declared name, its category (base, constructor, constant, function, proposition or predicate) and its arity. Language conditions ensure that this quadruple uniquely identifies the symbol. It is the flat name that is represented by in the Syntax module.

The abstract data type Name is provided to represent symbol names. The Name type is the basic unit from which the representation of syntax is constructed. For representing Gödel programs it is convenient to have the four components of Gödel flat names explicit in the Name term so that they can be easily extracted and separated. Gödel flat names are therefore represented by the following function:

```
FUNCTION Name :
    String              % Module where symbol is declared.
  * String              % Declared name of symbol.
  * Category            % Symbol's category.
  * Integer             % Symbol's arity.
  -> Name.
```

The name Name is overloaded here: it can stand for both a type symbol and a function symbol. This is a convenient programming style, since it avoids the proliferation of symbols with names that are artificially differentiated simply because they are in different categories. Overloading like this is common in natural language, and the readability of the program is not compromised because it is always obvious from the context whether a type symbol or a function symbol is intended.

The base type Category gives the category of the symbol and is defined by six constants:

```
CONSTANT
    Base, Constructor, Constant,
    Function, Proposition, Predicate : Category.
```

For example, let there be a constant with declared name January in a module called Calendar. The 4-tuple <Calendar, January, *constant*, 0> is the flat name of this constant. At the meta-level, we can represent the declared name by the string "January" and the module name by the string "Calendar".[1] The representation of the symbol January is then the ground term Name("Calendar", "January", Constant, 0).

Terms in the object language are represented by terms of type Term in the meta-language:

---

[1]Strings are constants of type String exported by the module Strings. There are infinitely many, one for every possible sequence of characters.

```
FUNCTION CTerm :
    Name                        % Name of constant symbol.
    -> Term.

FUNCTION Term :
    Name                        % Name of function symbol.
    * List(Term)                % List of argument terms.
    -> Term.
```

The `Term` function is used to represent a compound term, and includes a list of representations of its arguments, and the `CTerm` function is used to represent a constant. Object atoms by are similarly represented by terms of type `Formula`:

```
FUNCTION PAtom :
    Name                        % Name of proposition symbol.
    -> Formula.

FUNCTION Atom :
    Name                        % Name of predicate symbol.
    * List(Term)                % List argument terms.
    -> Formula.
```

Of course, since this is a *ground* representation, object level variables are represented by ground terms at the meta-level. The particular variable represented is conveniently identified by a `String` constant.

```
FUNCTION Var :
    String                      % Variable name.
    -> Term.
```

Representations of formulas, which are all terms of type `Formula`, are built from terms representing atoms, the constant `Empty` (representing the empty formula), and functions representing connectives. Among these are the unary function ~' and binary functions &', \/', ->', <-' and <->'.

As an illustration, consider the representation under this scheme of the clause

        Append([], x, x) <-

in the module `Lists`. It is:

```
Atom(Name("Append", "Lists", Predicate, 3),
    [CTerm(Name("Nil", "Lists", Constant, 0)),
        Var("x"), Var("x")])
<-' Empty
```

Notice that this representation is very much larger than the formula it represents, both syntactically and in terms of it's internal form within a typical WAM-style implementation. If term with n arguments requires $n + 1$ machine words plus the size of its arguments for its object level representation, its meta-level representation would require $2n + 9$ machine words plus the size of the representation of its arguments.

Notice also that this representation is fully general in terms of permitting the representation of partially known structures. That is, meta-variables can appear in place of the Name terms that represent function or predicate symbols, or in place of some terms representing arguments, or in place of some or all of the argument list. For example, it is straightforward to use the facilities of the Syntax module to create a partial representation such as Term(y, [Var("w")|z]), where y and z are meta-variables. This represents some compound object term which has the variable w as its first argument, but about which nothing else is known. Although control declarations could prevent the creation of partial representations, this restriction has been avoided to give the programmer maximum flexibility. Partial representations are occasionally useful, for example in program synthesis, where they can act as constraints upon parts of a program under construction as described by Christiansen [Chr94]. Programmers working with the traditional Prolog style of meta-programming cannot normally represent terms with variable function symbols or arities. Some Prolog systems may permit terms with variable function symbols, but of course such a feature is outside first-order logic.

For compactness and readability, in what follows we shall usually omit the bulky Name terms that Gödel uses to represent symbols, and in their place write the declared name of the represented symbol with a prime. For example, F' is to be read as

   Name(module, "F", category, arity)

where the module, category and arity components will either be obvious or unimportant in the context.

6.2.2 Unification and substitutions

Of course, the essential difference between programs that use the ground representation and those that use the traditional, non-ground Prolog style of meta-programming is that, in simulating object level operations such as unification, the former must explicitly handle the representation of variables, while the latter can rely on the underlying system to perform this task on the meta-variables that stand in place of object variables.

An example of common Prolog meta-programming practice is given by programs that perform some variant of the default unification process, perhaps by adding the occur check. They usually work by using the built in primitives functor/3 and arg/3 to compare the functors and arguments of the input terms recursively. This type of program is often used for teaching; several examples can be found in Sterling and Shapiro [SS86]. These programs still rely on, and obtain their efficiency from, the underlying system's treatment of the meta-variables that stand for object variables. It goes without saying that these techniques are not declarative, and one consequence of this is that the programs change their input arguments (by binding the meta-variables they contain) in a way that seems awkward once a declarative style of meta-programming has become familiar.

Consider the problem of a meta-program that is to unify the representations of the object level terms F(A,y) and F(x,x). In the non-ground representation, we simply unify the representations by solving an equation such as F'(A',u) = F'(v,v), where F' and A' are the symbols of the meta-language representing F and A respectively. In the process, the meta-variable v is bound to the constant A', and because both occurrences of v are represented internally by the same physical memory location, the binding is automatically propagated to the second occurrence, and thus to u when u and v are bound together. All in all, each term is scanned once, and two memory locations are changed (ignoring any trailing that may be necessary) to obtain the term F'(A',A') representing F(A,A).

Contrast this with the ground approach. Here we want to perform a meta-level simulation of unification on the representations Term(F', [CTerm(A'),Var("x")]) and Term(F', [Var("v"),(Var("v")]). Beginning by comparing the terms from left to right, we soon need to record the binding of variable v to constant A. In non-ground, this was done by simply instantiating the meta-variable standing in place of the representation of v to the constant A' representing A, and of course all the other occurrences of this same meta-variable representing v were also instantly affected. If we try, naively, to proceed using an analogous algorithm in the ground case, we need to replace all occurrences of Var("v") by CTerm(A') in the representations of both terms. However, since the representation of v is a ground term, rather than a meta-variable, we cannot do this destructively as in non-ground. Instead, the process of searching both representations for occurrences of Var("v") and replacing each one with CTerm(A') will construct a complete new copy of the representation of each term. Now notice that this has been done simply to record a single variable binding. Doing this every time a variable is bound clearly involves an unacceptable overhead in space and time.

Recording bindings by constructing a representation of a substitution looks a more promising approach. Gödel's Syntax module exports an abstract type TermSubst for representations of substitutions, and predicates UnifyTerms and UnifyAtoms that take a pair of terms or atoms respectively and return the representation of their mgu. Syntax also provides predicates for other standard operations on substitutions, such as application, composition, and adding and deleting individual bindings. The intended interpretations of these predicates conform to the definitions in Lloyd [Llo87].

The direct and obvious way to represent substitutions is as lists of bindings.

CONSTRUCTOR Binding/1.

FUNCTION
/ : xFx(10) : Term * Term -> Binding(Term);
Subst : List(Binding(Term)) -> TermSubst.

So the term Subst([Var(v1)/t1,...,Var(vn)/tn]) represents the object level substitution $\{v_1/t_1, \ldots, v_n/t_n\}$, where Var(v1),...,Var(tn) represent the variables $v_1, \ldots, v_n$ and t1,...,tn are the representations of the terms $t_1, \ldots, t_n$ respectively.

During the unification process, it is necessary both to look up bindings in the substitution whenever a variable is encountered, and to add new bindings to the substitution as they are made. Looking up a binding means conducting a linear search of the list. Adding the binding of variable v_m to term t_m to the substitution θ already constructed means forming the composition $\theta\{v_i/t_i\}$. If θ is $\{v_1/t_1, \ldots, v_n/t_n\}$, then all occurrences of v_m in $t_1, \ldots, t_n$ must be replaced by t_m. Performing this manipulation on the representation [Var(v1)/t1,...,Var(vn)/tn] will necessarily involve constructing a copy of the list on current Prolog systems. We will return to this problem in the context of interpreters in the next section.

Two more possibilities suggest themselves for efficient simulation of unification in the ground representation: structure reuse and reflection. Structure reuse using liveness analysis techniques such as described by Mulkers [Mul93] might, in theory, permit fast unification by destructive assignment to variable representations in one or both of the structures to be unified, but there is a long way to go before practical analyses are accurate enough to make this work well.

Implementing unification by reflection, that is by actually performing the unification at the object level, is more promising. However, it suffers from the fact that in purely declarative meta-programming, the reflective implementation must be completely hidden. One possibility involves constructing the object level structures denoted by the ground representation, unifying them using the efficient unification routine in the underlying system, and then mapping the result back into the ground representation. However, the translation step is quite expensive, and since the operation is internally non-logical, it's logic cannot be made available to general optimisation processes such as partial evaluation and static analysis. These are important disadvantages. It's also possible to imagine an alternative reflective implementation, in which object level structures are maintained in parallel with their representations at the meta level. The difficulty here is that the object level unification essentially destroys these parallel structures, so copies of the structures to be unified must again be made before unifying them. Performing compile time analysis to determine when this is necessary leads back to the structure reuse idea.

6.3 Simple Interpreters

The greatest computational expense introduced by the ground representation occurs when representations of substitutions are manipulated, as in unification or resolution. In this section we look at interpreters that do no more than implement SLD-resolution, in order to demonstrate how meta-programs which use the ground representation may be made efficient. Such interpreters have the advantage of being both simple and familiar in principle, while at the same time being fully illustrative of the potential inefficiencies of the ground representation.

Even simple SLD-interpreters for the ground representation add value over the non-ground Vanilla interpreter, because the result is returned in the form of the ground representation of a computed answer substitution, rather than by instantiating meta-variables, and this is potentially more useful. In addition, it is straightforward to extend our SLD-interpreters, for example to return a proof term, or to return all the input clauses used in a derivation as part of a knowledge assimilation procedure like those described by Guessoum and Lloyd [GL90], [GL91]. The latter task requires the ground representation if it is to be done declaratively. We have not included enhanced interpreters in the discussion, because they are so easily imagined by the reader, and because the additional complexity might obscure the argument.

While we use only simple interpreters to illustrate the efficiency issues for declarative programming in this paper, this in no way implies that the optimisation techniques we propose are applicable only to such programs. The techniques described in this paper have been used in the construction of a range of sophisticated Gödel meta-programs, including coroutining interpreters, theorem provers, abstract analysis tools, declarative debuggers and program specialisers such as partial evaluators.

6.3.1 Instance Demo

In passing, it is worth discussing the interpreter proposed by Hill and Gallagher [HG94], called *Instance Demo*. Instance Demo is an interpreter for programs in the ground representation, yet it uses meta-variables for unification and relies on the underlying system to manage variable bindings and backtracking in exactly the same way as the Vanilla interpreter does in the non-ground representation. Figure 6.1 shows part of this interpreter, written using Gödel's meta-programming facilities. The predicate InstanceOf is intended to be true when its second argument represents an instance of the formula

```
PREDICATE Demo : Program * Formula * Formula.

Demo(program, goal, goal_instance) <-
  InstanceOf(goal, goal_instance) &
  IDemo(goal_instance, program).

PREDICATE IDemo : Formula * Program.

IDemo(empty, _) <-
  EmptyFormula(empty).

IDemo(conjunction, program) <-
  And(left, right, conjunction) &
  IDemo(left, program) &
  IDemo(right, program).

IDemo(atom, program) <-
  IClause(clause, program) &
  IsImpliedBy(atom, body, clause) &
  IDemo(body, program).
```

Figure 6.1
Instance Demo

represented by its first argument. However, when called with the second argument unknown, its effect is to return the formula in the first argument with all the representations of variables replaced by meta-variables. The predicate `IClause` uses `InstanceOf` to find instances of program clauses.

Instance Demo has the nice property that it is possible to remove the calls of `IClause` by partial evaluation, thus specialising the interpreter for a particular object program. The residual code then looks and operates exactly like a specialised Vanilla interpreter, and is equally efficient.

At first sight this looks a very promising way to obtain efficient interpreters for the ground representation. However, because Instance Demo is so similar to non-ground based interpreters, it suffers similar limitations. It effectively captures the declarative semantics of the object program but not the procedural semantics: it cannot be used to vary the search strategy much from that of the underlying system, and it cannot give information about variable bindings. The answer it provides to the object program and

```
...
Select(goal, left, sel, right) &
StatementMatchAtom(prog, _, sel, stat) &
RenameFormulas(avoid, [stat], [stat1]) &
IsImpliedBy(head, body, stat1) &
UnifyFormulas(head, sel, mgu) &
ComposeTermSubsts(so_far, mgu, subst) &
And(left, body, goal1) &
And(goal1, right, goal2) &
ApplySubstToFormula(goal2, mgu, new_goal) &
...
```

Figure 6.2
Part of the main loop of an interpreter using composition

goal is some instance of the original goal, but that instance may be incompletely specified (it may contain meta-variables), which makes it much less useful than an answer in the true ground representation.

6.3.2 SLD interpreter with composition

If we must build interpreters that deal directly with the undiluted ground representation of the object program, how should such interpreters work? Although there are relatively few examples in the literature, most of those that exist, for example in Bowen and Kowalski [BK82] or Hill and Lloyd [HL89], are based on explicitly mimicking the standard mechanism of SLD-resolution (as it appears for example in Lloyd [Llo87]).

Thus, the main loop of a typical interpreter first selects an atom in the current goal, then finds a clause with a matching predicate in the program, renames the variables in the clause, computes an mgu for the selected atom and the head of the renamed clause, composes this mgu with the sequence of mgus previously computed (which result will eventually form the computed answer we want), constructs a new goal from the remains of the current goal and the body of the renamed clause, and finally applies the mgu to the new goal.

Figure 6.2 shows part of an interpreter using composition, constructed from some of Gödel's meta-predicates. This interpreter has to do an enormous amount of work in each resolution step. As we have seen, the unification operation involves linear search within the representation of the partially constructed mgu, and making copies of it, but the mgu is usually quite small. The explicit Compose operation, however, must copy the entire substitution computed up to this point, and this contains bindings for many variables that have been introduced, used and subsequently dropped from the computation. This substitution can become very large indeed, and copying its representation is a major overhead. Renaming the variables of the input clause necessitates making a new copy of the clause, and similarly, the application of the mgu to the new goal necessitates copying the goal. All these copying operations are expensive in time and space, and as might be expected, the performance of this interpreter is a disaster.

We will now look at some ways to reduce this extra work.

6.3.3 Improvements: Unify Demo

The first priority is to avoid composing substitutions at all. We achieve this by altering the representation of substitutions, so that they are represented in an effectively *uncomposed* form. When a new binding is added to a substitution in this representation, it is simply added to the set of existing bindings and the rest of the substitution is unchanged. In this way, reference chains are built up, similar to the reference chains that form on the stack in a standard WAM implementation. To determine the result of applying a substitution in this form to some term, it is necessary to repeatedly follow all reference chains until a result is obtained that contains only variables for which there are no bindings in the substitution. For example, the the classical substitution $\{w/z, x/z, y/z\}$ might be represented by [Var("w")/Var("x"), Var("x")/Var("y"), Var("y")/Var("z")].

This less literal representation of substitutions differs from the classical representation, in that adding an arbitrary binding could lead to a loop in a reference chain. It is also not clear how to compose two substitutions in this representation; one way would be to translate them into classical substitutions first, by collapsing all the reference chains, and then use the classical composition algorithm. Naturally, this would be an expensive operation, but the point of this optimisation is to avoid composition entirely. If we limit the available operations on substitutions to creating an empty one, unification (with the occur check), and application, this representation is well behaved. This is easily done because the representation is an abstract type.

The representation of substitutions in uncomposed form is alone sufficient to avoid composition during unification, that is within a call of UnifyTerms, but during interpretation we must also avoid the explicit composition of the new mgu with the accumulated answer substitution so far computed. Passing the accumulated substitution to UnifyTerms as an input argument solves this problem; as UnifyTerms generates new bindings during the unification process, it can simply add them to the list of bindings already created by previous unifications.

Determining the binding of a variable in a given substitution now potentially requires accessing the substitution many times in order to follow the reference chains, whereas in the classical representation it only ever required one access. Each access involves making a linear search of the binding list for the variable concerned. Clearly, much can be gained by improving the efficiency of these accesses.

We introduce a new set of variable representations, in which the variables are uniquely identified by integer indexes. These are in addition to the Var("x") form, which is still required to represent all other variables, and these are all taken to have index 0. The new representations use the function

    FUNCTION V : Integer -> Term

and the meta-terms V(1),V(2),V(3)... are taken to represent the variables v_1, v_2, v_3... The root v is simply an arbitrary choice to construct syntactically correct Gödel variables. While it would not be practical to expect the authors of object programs to confine their choice of variable names to this limited set, we can ensure that all new variables introduced by a Gödel interpreter (chiefly when renaming clauses) are of this form.

```
PREDICATE Demo :
  Program * Formula * Integer * Integer * TermSubst * TermSubst.

Demo(_, empty_goal, sp, sp, subst, subst) <-
  EmptyFormula(empty_goal).

Demo(program, goal_atom, sp, new_sp, subst, new_subst) <-
  Atom(goal_atom) &
  StatementMatchAtom(program, goal_atom, statement) &
  StandardiseFormula(statement, sp, sp1, statement1) &
  IsImpliedBy(head, new_goal, statement1) &
  UnifyAtoms(goal_atom, head, subst, subst1).
  Demo(program, new_goal, sp1, new_sp, subst1, new_subst).

Demo(program, goal, sp, new_sp, subst, new_subst) <-
  And(left, right, goal) &
  Demo(program, left, sp, sp1, subst, subst1) &
  Demo(program, right, sp1, new_sp, subst1, new_subst).
```

Figure 6.3
Unify Demo

Now we have the possibility of using the integer index to obtain faster look up in the substitution, by using a declarative array implementation to represent the binding list. In practice we use a carefully designed and optimised binary tree, giving logarithmic access and update times.

We can also use the variable indexes to simplify the interpreter considerably, by reducing the information needed for safely renaming the input clause. Instead of carrying around formulas containing the variables to be avoided, (for example, in the heads of resultants), we merely have to remember the smallest variable index that has not previously been used. These considerations lead to the type of interpreter shown in Figure 6.3. The structure of this interpreter is analogous to that of the Vanilla interpreter, and it shares the selection rule and search strategy of the underlying system. It still performs explicitly the standard steps of input statement selection, standardisation apart, and unification but it is much simpler and more efficient than the composition interpreter. However, there is one more major optimisation that we can perform, and that is described in the next section.

6.3.4 SLD Demo and Resolve

Gödel supplies a system predicate, Resolve, which provides an optimised implementation of resolution. When resolving an atom with respect to some statement in an object program we must first rename the variables in the statement, replacing them with new variables which do not occur elsewhere in the current computation. The implementation of Resolve may take advantage of this knowledge to optimise the resolution process. The fact that these new variables are unique to the renamed statement allows Resolve to simplify many of the necessary unification operations and reduce the amount of occur-checking.

Figure 6.4 shows SLD Demo, a very simple Gödel meta-interpreter for definite programs which uses the Gödel predicate Resolve to resolve an atom in the current goal with respect to a statement selected from the object program.

The atom Resolve(goal_atom,statement,v_in,v1,subst,subst1,new_goal) performs the resolution of the atom goal_atom with the statement statement. The integers v_in and v1 are used to rename the statement with v_in being the integer value used in renaming before the resolution step is performed and v1 being the corresponding value after the resolution step has been performed. The representations of term substitutions subst and subst1 represent respectively the answer substitution before and after the resolution step. The last argument, new_goal, is the representation of the body of the renamed statement.

Figure 6.5 shows the performance of Unify Demo and SLD Demo interpreting (as object program) the well known Naive Reverse program on a list of 50 elements. The ground interpreters are compared with the object program executed directly on the Gödel system, and with the non-ground Vanilla interpreter interpreting the same program. Approximate performance data is adequate for the purposes of this discussion, so we need only consider the results from a single benchmark.

```
PREDICATE Demo :
  Program * Formula * Integer * Integer * TermSubst * TermSubst.

Demo(_, empty_goal, v, v, subst, subst) <-
  EmptyFormula(empty_goal).

Demo(program, goal_atom, v_in, v_out, subst, new_subst) <-
  Atom(goal_atom) &
  StatementMatchAtom(program, goal_atom, statement) &
  Resolve(goal_atom, statement, v_in, v1, subst, subst1, new_goal) &
  Demo(program, new_goal, v1, v_out, subst1, new_subst).

Demo(program, goal, v_in, v_out, subst, new_subst) <-
  And(left, right, goal) &
  Demo(program, left, v_in, v1, subst, subst1) &
  Demo(program, right, v1, v_out, subst1, new_subst).
```

Figure 6.4
SLD Demo

Program	Relative time for Naive Reverse
Object	1
Vanilla	10
Unify Demo	280000
SLD Demo	31000

Figure 6.5
Performance of Interpreters

The table indicates that SLD Demo is faster than Unify Demo by a factor of 10, demonstrating the effectiveness of combining renaming and unification in the Resolve predicate and optimising them. However, the interpreter is still 3 orders of magnitude slower than its non-ground equivalent, and obviously an overhead of this magnitude renders such interpreters almost useless for most practical applications. Note, however, that the interpreter and the system predicates it depends upon are written almost entirely in Gödel (a very small fraction is written directly in Prolog), and the Gödel code is translated to Prolog by the Gödel compiler. The Prolog system has no built-in support for the ground representation. In the remainder of this paper we consider two strategies for improving the performance of ground interpreters: compiling away some redundant computation by specialising the interpreter to its object program by partial evaluation; and providing in-built support for the ground representation at a low-level in the implementation.

6.4 Speeding Up SLD Demo

6.4.1 Partial evaluation

Partial evaluation is a program specialisation technique. A partial evaluator is a tool which takes a program and some partial input data to it and produces a specialised version of that program. A recent overview of partial evaluation and its application to several classes of programming language is given by Jones *et al* [JGS93]. The partial evaluation of logic programs by unfolding was put on a firm theoretical footing by Lloyd and Shepherdson [LS91] and we refer to that paper for a formal definition of partial evaluation for logic programming.

SAGE (Self-Applicable Gödel partial Evaluator), written by Gurr [Gur94], is a partial evaluator based mainly on finite unfolding. SAGE implements techniques developed with the aim of producing a declarative, effectively self-applicable partial evaluator for a full logic programming language (as opposed to some restricted subset of the language). A self-applicable partial evaluator is one which is capable of specialising itself and an *effectively* self-applicable partial evaluator is one which produces significantly improved, efficient residual code upon self-application.

SAGE performs the partial evaluation of a program w.r.t. some goal in four main phases:

1. static (termination) analysis;
2. partial evaluation;
3. optimisation of residual code;
4. replacement of original code by specialised code.

The primary motivation for the static analysis performed by *SAGE* is as a termination analysis. We produce an abstraction of the partial tree which is used to compute the subsequent partial evaluation and by analysis of this tree we partition the predicates into two sets. In the first set, which we refer to as the *safe* predicates, we place those predicates for which all atoms with this predicate may be unfolded without the risk of leading to an infinite unfolding. The complement of this set, the *unsafe* predicates, contains those predicates for which unrestricted unfolding of atoms with this predicate could not be guaranteed to terminate. This prior identification of the unsafe predicates permits SAGE to apply a more cautious strategy when unfolding them during the partial evaluation phase.

The main post-partial evaluation optimisation performed by *SAGE* is the removal of redundant terms. In general we expect to have constructed the partial evaluation of a set of atoms that define the specialisation of predicates in the original program to particular calls. Generally, certain arguments of these atoms are instantiated to non-variable terms before they are specialised. For example in a meta-interpreter with top level predicate Demo we may have specialised this interpreter to a particular object program by partially evaluating the atom Demo(<program>,query,answer), where <program> is the term representing the object program. The representation of an object program will generally be a very large term and is likely to be redundant (not utilised) in the residual code. We may therefore delete this term. To delete such a term we would replace the ternary predicate Demo with a new binary predicate, Demo_1 say, where any computed answer for Demo_1(query,answer) would be equivalent to that computed for Demo(<program>,query,answer).

6.4.2 Implementation of Resolve

The implementation of Resolve must handle the following operations:

- Renaming the statement to ensure that the variables in the renamed statement are different from all other variables in the current goal.
- Applying the current answer substitution to the atom to ensure that any variables it binds are correctly instantiated.
- Unifying the atom with the head of the renamed statement.
- Composing the mgu of the atom and the head of the statement with the current answer substitution to return the new answer substitution.

Each of these four operations is potentially very expensive when we are dealing with the explicit representation of substitutions and so the implementation of Resolve is optimised to make them highly efficient.

After the statement is renamed, any variables in it are guaranteed not to appear in either the current goal or the current substitution. This means that any bindings for variables in the head of the statement may be applied to the body of the statement and then discarded. Consequently only that part of the mgu of the atom and the renamed head of the statement that records the bindings of variables in the atom will need to be composed with the current substitution in order to produce the new substitution.

Having performed the unification of an atom with the head of a statement we must (in theory) combine the mgu of this unification with the current substitution. In practice it is more efficient for any bindings made to variables in the atom to be composed with the current substitution immediately. In order to achieve these compositions Bristol Gödel

Towards Fast and Declarative Meta-Programming   155

has a set of highly specialised predicates, each of which performs one specific unification operation.

Implementing **Resolve** based upon a set of highly specialised unification predicates has the added advantage that when partially evaluating a call to **Resolve** w.r.t. a known statement we may unfold the call until the residual code consists of a (possibly empty) conjunction of atoms with these predicates. When we specialise a meta-program such as the interpreter in Figure 6.4 to a known object program, the statements in the object program will be known. Therefore we may specialise **Resolve** with respect to each statement in the object program. Such a specialisation will remove the vast majority of the overhead associated with explicit unification in the ground representation.

We refer to the specialised unification predicates which **Resolve** is based upon as the *WAM-like predicates* as they are analogous to emulators for instructions in the WAM-engine. As such, these operations may be implemented at a very low level, leading to a computation time for the specialised form of a meta-program, such as that in Figure 6.4, comparable to that of the object program itself. In the next section we give an example of such a specialisation.

6.4.3 Applying SAGE to SLD Demo

Figure 6.6 illustrates the result of specialising the **Demo** interpreter of Figure 6.4 with respect to the standard **Append** program. In this specialised version of the interpreter we have made three optimisations:

1. the calls to **Resolve** have been specialised w.r.t. the two statements in the object program to produce the third and fourth statements respectively in the new predicate **Demo_1**

2. symbols (such as **Empty'** and **&'**), which are ordinarily hidden by Gödel's implementation of the ground representation as an abstract data type, have been promoted into the specialised program

3. the representation of the object program **Append**, which is now redundant, has been removed by replacing the predicate **Demo/6** by the new predicate **Demo_1/5**.

Specialising SLD Demo with respect to a given object program yields very considerable performance improvements. The specialised programs typically run 40 or more times faster than the original code. There are several sources of this improvement. Unfolding the predicates in **Syntax** that manipulate the abstract data type is an important one; this reduces the number of procedure calls and enables direct pattern matching on the constants and functions making up the abstract type. It is also significant that many of these predicates have control declarations which are expensive to evaluate at run-time. These control declarations are eliminated when the predicates they guard are unfolded.

Object program:
Append([],x,x).
Append([a|x],y,[a|z]) <- Append(x,y,z).

Specialised interpreter:
Demo_1(Empty',v,v,subst,subst).
Demo_1(left &' right,v_in,v_out,subst_in,subst_out) <-
  Demo_1(left,v_in,new_v,subst_in,new_subst) &
  Demo_1(right,new_v,v_out,new_subst,subst_out).
Demo_1(Atom(Append',[arg1,arg2,arg3]),v_in,v_out,subst_in,subst_out) <-
  GetConstant(arg1,Nil',subst_in,s1) &
  GetValue(arg2,arg3,s1,new_subst)
  Demo_1(Empty',v_in,v_out,new_subst,subst_out).
Demo_1(Atom(Append',[arg1,arg2,arg3]),v_in,v_out,subst_in,subst_out) <-
  GetFunction(arg1,Term(Cons',[sub11,sub12]),mode,subst_in,s1) &
  UnifyVariable(mode,sub11,v_in,v1) &
  UnifyVariable(mode,sub12,v1,v2) &
  GetFunction(arg3,Term(Cons',[sub21,sub22]),mode1,s1,s2) &
  UnifyValue(mode1,sub11,sub21,s2,new_subst) &
  UnifyVariable(mode1,sub22,v2,new_v) &
  Demo_1(Atom(Append',[sub12,arg2,sub22]),new_v,v_out,
      new_subst,subst_out).

Figure 6.6
Specialisation of Demo w.r.t. Append

There is a considerable redundant computation involved in repeatedly extracting the input statements from the Program term, where they are stored in a binary tree. For a recursive program such as Naive Reverse, the same statement is used many times over; specialisation completely removes this overhead by promoting all the statements of the object program into the clauses of the specialised interpreter, and eliminating the program term entirely. Finally, a specialised unification procedure is generated for the head of each clause in the object program, expressed in terms of the WAM-like predicates that remain from the unfolding of Resolve. Thus it can be seen that the partial evaluation of interpreters is analogous to the compilation of logic programs into WAM instructions. In the next section we give a brief description of these WAM-like predicates.

6.4.4 The WAM-like predicates

In Bristol Gödel there are seven WAM-like predicates. We illustrate each of these in Figure 6.7 by specialising `Resolve` to the statement `P(x,x,A,F(y,F(x,A))) <- Q(y)`. The body of this specialised statement contains one atom for each of the WAM-like predicates described below.

Statement: `P(x, x, A, F(y, F(x, A))) <- Q(y)`.

Specialised call to `Resolve`:

```
Resolve(
  Atom(P', [arg1, arg2, arg3, arg4]),
  statement,
  v, v1,
  subst_in, subst_out,
  Atom(Q', [sub1])
     ) <-
  UnifyTerms(arg1, arg2, subst_in, s1) &
  GetConstant(arg3, CTerm(A'), s1, s2) &
  GetFunction(arg4, Term(F', [sub1, sub2]), mode, s2, s3) &
  UnifyVariable(mode, sub1, v, v1) &
  UnifyFunction(mode, sub2, Term(F', [sub21, sub22]), mode1,
                s3, s4) &
  UnifyValue(mode1, arg1, sub21, s4, s5) &
  UnifyConstant(mode1, sub22, CTerm(A'), s5, subst_out).
```

Figure 6.7
Illustration of WAM-like predicates

The first three WAM-like predicates unify arguments of the head of the statement with the matching arguments of the atom as follows:

UnifyTerms(term1,term2,subst,subst1) attempts to unify the atom's two terms term1 and term2. UnifyTerms is the only one of these specific argument unification operations which enforces occur-checking and is used to unify repeated variables in the head of the statement. In this and the two subsequent atoms, subst is the current substitution and subst1 is this substitution after the relevant unification step.

GetConstant(term,constant,subst,subst1) attempts to unify the atom's term term with the constant constant.

The third of the WAM-like predicates, `GetFunction`, is more complex than the previous two and so we first give an overview of its operation. If an argument in the head of the statement is a term with a function at the top level, then there are two cases in which a call to `GetFunction` will succeed. In the first case the atom's matching argument is a variable. We must bind this variable to a renamed version of the term in the head of the statement (the necessary renaming is performed by subsequent calls and not at this point). In the second case the atom's matching argument is a term with a matching function at the top level. In this case the subsequent calls will unify the arguments of the atom's term with the corresponding arguments in the statement's term.

In precise terms, `GetFunction` behaves as follows:

GetFunction(term,function,mode,subst,subst1) attempts to unify the atom's term term with a term function, which has a function at the top level. If term is bound in the current substitution to a variable, then this variable is bound to function and mode is set to Write. The term function will be instantiated by subsequent calls to a renamed version of the term in the statement to which term is now bound. If, when GetFunction is called, term is bound in the current substitution to a term with a function at the top level which matches that of function, then function is instantiated to term and mode is set to Read.

The following predicates perform the unification operations necessary for processing the arguments of function terms in the head of the statement. This involves either renaming variables when in Write mode or unifying these arguments with the arguments of the matching function term in the atom when in Read mode. When the following calls are made in Write mode the terms term will be uninstantiated variables. When these calls are made in Read mode the terms term will have been set to terms in the atom being resolved.

UnifyVariable(mode,term,ind,ind1) in Write mode will instantiate term to the new variable Var("v",ind) and ind1 = ind+1. In Read mode, term will already be instantiated and ind1 is set to ind.

UnifyFunction(mode,term,function,mode1,subst,subst1) in Write mode will instantiate term to the term function and mode1 is set to Write. In Read mode this call behaves as for GetFunction, attempting to unify the atom's term term with a term function with a function at the top level. In this and the two subsequent atoms, subst is the current substitution and subst1 is this substitution after the relevant unification step.

UnifyValue(mode,term1,term,subst,subst1) in Write mode will instantiate term to term1. A check is made at this point to ensure that this does not introduce a loop into the substitution (an occur-check, in effect). In Read mode this call will unify (with occur-checking) the atom's two terms term1 and term.

UnifyConstant(mode,term,constant,subst,subst1) in Write mode will instantiate term to the constant constant. In Read mode this call attempts to unify the atom's term term with the constant constant.

The above seven predicates are analogous to emulators for the WAM instructions GetValue (in the case of UnifyTerms), GetConstant, GetFunction, UnifyValue, UnifyVariable and UnifyConstant, after which they are named. Note that a subtle difference in the manner in which the WAM implements the unification of nested function terms and the manner in which Resolve implements it means that the WAM (as originally defined by Warren [War83]) does not have an equivalent to the UnifyFunction instruction.

6.4.5 Further optimisations

Recall that substitutions are represented as uncomposed sets of bindings, and that, since looking up bindings in the substitution is a frequent occurrence during unification, the speed with which the binding for a particular variable can be located is an important factor in the efficiency of unification. For this reason, we added integer indexes to the representation of variables, and stored the bindings in a binary tree. Suppose we try to use a genuine array to store the bindings, to obtain constant rather than logarithmic access times? The difficulty is, of course, that arrays can only be updated destructively, and this destroys the soundness of the implementation if any references supposed to be to the original array remain after the update. However the update time is also constant, and fast compared with the typically logarithmic update time for a binary tree (because a copy must be made of the branch of the tree from the root to the updated leaf).

Consider the sequence of WAM-like predicates in Figure 6.7 that typically results from specialising Resolve, and the way these operate on substitutions. Figure 6.8 shows the sequence, with all arguments other than those of type TermSubst removed for clarity. As it is relatively simple, take the call of GetConstant(arg3,CTerm(A'),s1,s2). If arg3 is instantiated to the representation of a variable, this variable is dereferenced by following its binding chain in the input substitution s1. If the result of dereferencing arg3 is another variable, the binding of this variable to constant A is recorded by updating substitution s1 to give the output substitution s2. Given that the sequence of WAM-like predicates is executed from left to right, as it must be (DELAY declarations can be added to make sure this is so), if there are no other references to subst_in at the time Resolve is called, it will be safe to perform this update destructively, and similarly for updates made by all the other instructions in the sequence. We assume that all destructive updates are trailed where necessary, and therefore correctly undone on backtracking.

```
Resolve(..., subst_in, subst_out, ...) <-
  UnifyTerms(..., subst_in, s1) &
  GetConstant(..., s1, s2) &
  GetFunction(..., s2, s3) &
  UnifyVariable(...) &
  UnifyFunction(..., s3, s4) &
  UnifyValue(..., s4, s5) &
  UnifyConstant(..., s5, subst_out).
```

Figure 6.8
The pattern in which substitutions are passed

Program	Relative time for Naive Reverse
Unify Demo	280000
Unify Demo, specialised	500000
SLD Demo	31000
SLD Demo, no control	1980
SLD Demo, specialised	612
SLD Demo, specialised, arrays	216

Figure 6.9
Effect of optimising interpreters

Performing this analysis by hand, we can determine that it is safe to experiment with the use of arrays and destructive assignment for the representation of substitutions in the SLD Demo interpreter and its specialised form.

Figure 6.9 shows the results of some experiments with specialising Demo interpreters and using arrays to represent substitutions. We use the same Naive Reverse benchmark as before. The SICStus Prolog built-in predicate setarg/3 was used to provide the destructive array update.

It is interesting to note that specialising Unify Demo actually makes it worse by a significant amount. Because SAGE will not unfold the system predicate StandardiseFormula without knowing the initial free index, the residual code contains clauses such as that shown in Figure 6.10, generated from the second clause of Append. Building the representation of the clause on the heap before calling StandardiseFormula turns out to be more expensive than retrieving it from the Program term was in the unspecialised version. This example is included to demonstrate the sensitivity of partial evaluation to the way the source program is written.

In Figure 6.9 the line marked "no control" indicates the speed up obtained by compil-

```
Demo_1(Atom(Append',v_23),v_17,v_2,v_20,v_4) <-
  StandardiseFormula(
    Atom(Append',[Term(Cons',[Var("x"),Var("xs")]),Var("ys"),
              Term(Cons',[Var("x"),Var("zs")])])<-'
    Atom(Append',[Var("xs"),Var("ys"),Var("zs")]),
    v_17,v_18,v_21<-'v_9) &
  UnifyAtoms(Atom(Append',v_23),
    v_21,v_20,v_22) &
  Demo_1(v_9,v_18,v_2,v_22,v_4).
```

Figure 6.10
Part of Unify Demo specialised to Append

ing the unspecialised SLD Demo with a special Gödel compiler that ignores all control declarations and does not enforce safe negation, so the resulting program simply executes from left to right. This line shows that the run-time control in the Gödel system by itself costs a factor of 10 at least, and although this is one of the overheads removed by SAGE it is not an overhead created by the use of the ground representation, but rather a limitation of the present implementation. It is possible to remove almost all of the control overhead by performing a static analysis of the program at compile time, and an analyser to do this is under development. With respect to the speed obtained without checking control information, the partial evaluation gains a factor of 3 and the use of arrays another factor of 3, bringing the overall performance within a factor of 20 of the speed of the non-ground Vanilla interpreter.

During the execution of SLD Demo, new variable representations are constantly being created, each new variable having an index which points to a location in the substitution array, where the variable may eventually be bound. The free index value increases as new variables are needed. The substitution array is passed as a parameter to the WAM-like predicates and updated during their execution. The analogy between substitutions in the ground representation and the stack in the underlying system, and between the free variable index and stack pointer, is inescapable. They both implement the answer substitution accumulated during an SLD computation, and since the stack in the underlying system is designed for maximum efficiency, a promising way to obtain efficiency in interpreters for the ground representation must be to have substitutions emulate the system stack as closely as possible. When viewed in this way, the use of arrays to implement representations of substitutions seems a very natural solution to the problem of simulating unification efficiently in the ground representation. It simply involves introducing an extra level of indirection (relative to the object level) between an atom or term and the values to which its variables have been bound during a computation. The programmer specifies which indirection table (substitution) is to be used, by explicitly mentioning it as a parameter.

If the use of arrays and destructive assignment is to be a practical tool, some way must be found to ensure that programs remain declarative. For the SLD interpreter and similar programs in which there is just one substitution that is updated sequentially there is no difficulty, but we need to be able to make sure all programs are declarative, and do so automatically. One possibility would be to use an efficient but declarative array implementation, such as the "hairy structures" of Barklund and Millroth [BM87], which provide mutable data structures with a constant time overhead, and constant time access to the most recent version of the data structure. Another possibility would be to use a sophisticated abstract analysis aimed at deriving liveness information for possible structure reuse, or compile-time garbage collection, such as that of Mulkers [Mul93]. However, such analyses are complex to implement, slow to execute, and a little too general for our present needs.

A scheme that is close to the informal argument given above for the safety of SLD Demo would be ideal. In each clause of SLD Demo, and of the specialised version, there are no more than two occurrences of each variable of type TermSubst. This suggests something like the two-occurrence restriction of Janus (Saraswat *et al* [SKL90]), but making use of the Gödel type system to limit the restriction to variables of type TermSubst so that the impracticalities of Janus are avoided. Should a programmer write a clause with more than two occurrences of a TermSubst variable, the Gödel compiler would generate code to make a duplicate of the entire array at run-time, thus avoiding aliasing problems. While it may be a little heavy-handed, this approach is adequate for interpreters like SLD Demo and provides a starting point for developing a more discriminatory analysis.

6.4.6 Future work

By specialising it with SAGE, and using special-purpose low-level support in the form of arrays, we have brought SLD Demo from being 3000 times slower than a Vanilla interpreter with the same object program down to 20 times slower. Is there any prospect of narrowing the remaining gap? Most certainly there is. Even with arrays, our implementation of the WAM-like predicates is rather crude, and could be much improved by efficient low-level implementation. The SICStus setarg/3 predicate is more general than necessary, and the amount of trailing it performs could be reduced considerably in a special-purpose implementation.

Another possibility for for improving the efficiency of all manipulations of the ground representation is to try to decrease the size of the representation. It was noted in section 6.2.1 that the ground representation of a given term is very much larger than the term itself. As demonstrated convincingly in O'Keefe [O'K90], the size of the terms chosen to represent a given data set has an important effect on efficiency. Our choice of representation resulted from the requirement to implement the meta-programming facilities specified by Gödel, and to do so without customised low-level support from the implementation. With low-level support, it should be possible to reduce the size of the terms used for the ground representation to little larger than the object terms they represent, and without losing the generality over non-ground noted in section 6.2.1. Unfortunately, we have not yet been able to experiment with this idea in the existing Gödel system because it cannot be done without a major rewrite and access to the low-level implementation. It is hoped that a future version of Gödel will incorporate a compact representation as a fundamental design decision.

Partial evaluation has a drawback as a general optimisation technique: it is sometimes fragile. For example, choosing the wrong building block, such as `UnifyTerms` rather than `Resolve` as in Unify Demo, not only leads to a slower interpreter, but also one that is not amenable to partial evaluation by SAGE. On the other hand, an advantage of Gödel's module system is that some of the difficulty of effective program specialisation can be ameliorated by providing the programmer with a kit of procedures that specialise nicely. It is likely however that the speed up from specialising meta-programs will become less important as the language implementation improves.

6.5 Conclusions

If declarative programming is an important idea, then the problem of practical declarative meta-programming must be solved. The non-ground representation is not powerful enough for many important applications, so the central issue is how to make programs that use the ground representation run fast enough to be practical. Our initial experience with Gödel interpreters supported the common observation that programs using the ground representation are unacceptably slow. However, we believe that it is possible to improve the performance of these interpreters until they approach the speed of equivalent programs using the non-ground representation.

We have described a progression from the naive *unify-compose-apply* style of interpreter that explicitly mimics a standard formulation of SLD-resolution to a style that is much more easily optimised. Indeed, a significant part of the process of attaining practicality involves the creation of an idiom in which the use of these initially unfamiliar programming techniques will become natural.

A good part of the overhead that Gödel interpreters suffer from is due to extraneous language factors and limitations of the present implementation, rather than the specific use of the ground representation. In particular, evaluating control declarations, and the loss of indexing caused by the use of abstract data types increase the computation time significantly. While these overheads can be removed by partial evaluation, they could equally well be removed by improved Gödel compilation technology. Another overhead that is not entirely inevitable in the ground representation is due to the increased size of represented terms over the terms themselves. This overhead may be avoided by appropriate low-level support for the representation.

The only truly *inherent* overhead suffered by interpreters of the ground representation, compared with interpreters of the non-ground representation, lies in the management of variables. Without careful design, the additional computation from this source can cripple the meta-program. In a sense, writing an efficient ground interpreter from scratch is like building an efficient WAM implementation from scratch, because the programmer is entirely responsible for the management of variables. So interpreters and other meta-programs need to be designed with careful attention to the unification process and the representation of substitutions. However, we do not want to pass this design burden on to the programmer, but would rather relieve it; so Gödel provides, via the system modules Syntax and Programs, carefully chosen primitives such as Resolve, to be the building blocks from which efficient meta-programs can be constructed.

Once the interpreter has been constructed, most of the redundant computation it performs can be removed by specialising it with respect to a particular object program by partial evaluation. This process can change calls to Resolve into specialised unification procedures for all the statements of the object program, and so is analogous to compilation. Further speed up can be obtained by optimising the implementation of substitutions, using arrays to store the binding lists. There is a strong suggestion that this is the natural way to obtain efficient management of ground variables; substitutions can be viewed as operating in parallel to the stack in the underlying system. Alternatively, they provide an extra level of indirection between a structure and the values to which its variables are bound. This leads to a programming style in which application and composition of substitutions is discouraged, because these are expensive operations. Instead terms are viewed in the context of specified substitutions.

Taken literally, the performance figures we obtained still seem discouraging. Our simple SLD interpreter is at best an order of magnitude slower than the equivalent Vanilla interpreter. However, we think that the prospect of reducing this gap significantly is an excellent one, because the measurements were made using a crude experimental implementation, and there are important optimisations that remain to be tried, such as reducing the storage requirements of the representation. It must also be emphasised that something is gained in exchange for the extra computation: the ground representation is *declarative*, with all the advantages that declarative programming brings, such as clarity, increased opportunities for parallelism, and ease of processing by other meta-programs. One way to look at the remaining overhead is as an indication that Prolog is not powerful enough to build an efficient implementation of Gödel, and this is only to be expected since Gödel extends the functionality of Prolog considerably.

We claim that the combination of the techniques outlined here, if effectively implemented, can lead to declarative meta-programs that are easy to write, and come very close in execution speed to programs written using the traditional non-ground (and usually non-logical) Prolog approach. Such programs will have all the added functionality of the ground representation, being fully declarative and able to inspect variables. We *can* have efficient, declarative and powerful meta-programs.

References

[BK82]   K.A. Bowen and R.A. Kowalski. Amalgamating language and metalanguage in logic programming. In K.L. Clark and S.-A. Tarnlund, editors, *Logic Programming*, pages 153–172. Academic Press, 1982.

[BM87]   Jonas Barklund and Håkan Millroth. Integrating complex data structures in Prolog. In *1987 Symposium on Logic Programming*, pages 415–425, 1987.

[Bow92] A.F. Bowers. Representing Gödel object programs in Gödel. Technical Report CSTR-92-31, Department of Computer Science, University of Bristol, 1992.

[Chr94] H. Christiansen. Efficient and complete demo predicates for definite clause languages. Technical Report 51, Computer Science Department, Roskilde University, 1994.

[GL90]   A. Guessoum and J.W. Lloyd. Updating knowledge bases. *New Generation Computing*, 8(1):71–89, 1990.

[GL91]   A. Guessoum and J.W. Lloyd. Updating knowledge bases II. *New Generation Computing*, 10(1):73–101, 1991.

[Gur94] C.A. Gurr. *A Self-Applicable Partial Evaluator for the Logic Programming Language Gödel*. PhD thesis, Department of Computer Science, University of Bristol, January 1994.

[HG94]   P.M. Hill and J. Gallagher. Meta-programming in logic programming. Technical Report 94.22, School of Computer Studies, University of Leeds, 1994.

[HL89]   P.M. Hill and J.W. Lloyd. Analysis of meta-programs. In H.D. Abramson and M.H. Rogers, editors, *Meta-Programming in Logic Programming*, pages 23–52. MIT Press, 1989. Proceedings of the Meta88 Workshop, June 1988.

[HL94]   P.M. Hill and J.W. Lloyd. *The Gödel Programming Language*. MIT Press, 1994.

[JGS93] N. Jones, C.K. Gomard, and P. Sestoft. *Partial Evaluation and Automatic Software Generation*. Prentice Hall, 1993.

[Llo87] J.W. Lloyd. *Foundations of Logic Programming*. Springer-Verlag, second edition, 1987.

[LS91]   J.W. Lloyd and J.C. Shepherdson. Partial evaluation in logic programming. *The Journal of Logic Programming*, 11(3&4):217–242, 1991.

[Mul93] Anne Mulkers. *Live Data Structures in Logic Programs*. Lecture Notes in Computer Science 675. Springer-Verlag, 1993.

[O'K90] Richard A. O'Keefe. *The Craft of Prolog*. MIT Press, 1990.

[SKL90]   Vijay A. Saraswat, Ken Kahn, and Jacob Levy. Janus: A step towards distributed constraint programming. In Saumya Debray and Manuel Hermenegildo, editors, *Logic Programming: Proceedings of the 1990 North American Conference*, pages 421–446, 1990.

[SS86]   L. Sterling and E. Shapiro. *The Art of Prolog*. MIT Press, 1986.

[TZ88]   J. A. Thom and J. Zobel. Nu-Prolog reference manual, version 1.3. Technical report, Machine Intelligence Project, Department of Computer Science, University of Melbourne, 1988.

[vH92]   F.V. van Harmelen. Definable naming relations in meta-level systems. In A. Petorossi, editor, *Meta-Programming in Logic, Proceedings of the 3rd International Workshop, META-92*. Springer-Verlag, 1992.

[War83]   D.H.D. Warren. An abstract PROLOG instruction set. Technical Note 309, SRI International, 1983.

7 Composing Logic Programs by Meta-Programming in Gödel

Antonio Brogi and Simone Contiero

Abstract

Increasing attention is being paid to Gödel, a new declarative programming language aimed at diminishing the gap between theory and practice of programming with logic. We investigate the possibility of employing Gödel as a meta-language for re-using and composing existing definite programs. Two alternative implementations of a set of meta-level operations for composing definite programs are presented. The first implementation consists of extending the vanilla meta-interpreter using the non-ground representation of object level programs. The second implementation exploits the meta-programming facilities offered by Gödel, which support the construction of meta-interpreters using the ground representation of object level programs. The two implementations are then compared and the merits of Gödel as a meta-language are discussed.

7.1 Introduction

The idea of programming with logic proposed by Kowalski in [Kow74] found its first realisation in the Prolog language, which has been widely considered as *the* logic programming language for the last twenty years. The development of logic programming, however, has shown the deficiencies of Prolog as a declarative language and the consequent gap between theory and practice of programming with logic, as illustrated for instance by Apt [Apt93]. Several efforts have been devoted to design new programming languages aimed at being the declarative successors of Prolog (e.g., Clark, Mc Cabe and Gregory [CMG82] and Naish [Nai87]). The Gödel language, proposed by Hill and Lloyd [HL94], is starting to emerge as a new declarative general-purpose programming language in the family of logic programming languages. Gödel is a strongly typed language, has a module system and places considerable emphasis on meta-programming. These features of Gödel, in particular the emphasis on meta-programming, suggest the intriguing question whether or not existing logic programs can be suitably re-used in Gödel.

The ultimate objective of this work is to investigate the adequacy of Gödel as a meta-language for re-using and composing existing definite programs. From this perspective, we consider a simple extension of logic programming which introduces a set of meta-level operations for composing definite programs. These operations, originally presented by Mancarella and Pedreschi [MP88] and by Brogi [Bro93], form an algebra of logic programs with interesting properties for reasoning about programs and program compositions. From a programming perspective, the operations enhance the expressive power of logic programming by supporting a wealth of programming techniques, ranging from software engineering to artificial intelligence applications (e.g., Brogi et al. [BMPT90, BMPT94, BT90]). In this paper the implementation in Gödel of such a set of program composition operations is discussed.

Meta-level operations over object level programs can be naturally implemented by means of meta-programming techniques. More precisely, as shown by Brogi et al. [Bro93, BMPT94], several composition operations over logic programs can be implemented by extending the well known *vanilla* meta-interpreter. The actual realisation of extended vanilla meta-interpreters in Gödel presents various implementation choices that lead to different solutions. One of these choices is the representation of object level constructs, which is one of the crucial issues in meta-programming. In logic programming, object level expressions are usually represented by terms at the meta-level, and the critical issue is how to represent object level variables at the meta-level. Two basic alternative representations of object level variables are employed, as ground terms and as variables (or, more generally, non-ground terms). The first is called *ground* representation and the second *non-ground* representation by Hill and Lloyd in [HL89]. The ground representation is very versatile and adequate for many applications of meta-programming, such as program transformation and compiler writing. Though the non-ground representation is less versatile, it has been widely adopted in the practice of meta-programming with logic, for instance for the construction of several expert systems (e.g., Sterling and Shapiro [SS86]). Indeed, in the absence of suitable support for managing the complexity of meta-programming with the ground representation, programmers have been attracted by the simplicity and efficiency of the non-ground representation, which are due to the fact that the non-ground representation directly exploits the basic unification mechanism of the meta-language. The two representations have been thoroughly discussed for instance by Hill and Lloyd [HL89], Kowalski [Kow90], Levi and Ramundo [LR93], and in this volume by Martens and De Schreye [MD95] and by Brogi and Turini [BT95].

Two alternative implementations in Gödel of program composition operations are presented here. Both implementations are based on an extended vanilla meta-interpreter, and they differ in the choice of the representation of object level programs.

The first implementation adopts the non-ground representation of object level programs. This choice offers a simple and concise way of extending the vanilla meta-interpreter to deal with program composition operations. The implementation is equipped with suitable support for the non-ground representation, which is not directly supported by the Gödel system. This support frees the user from the need of explicitly providing the non-ground representation of object programs.

The second implementation adopts the ground representation of object level programs, for which Gödel provides considerable support. The Gödel approach to meta-programming is based on abstract data types. For instance, the system module Programs offers a large number of operations on an abstract data type that is the type of terms representing object level programs. The abstract data type view supports a declarative high-level style of meta-programming as the user has to be concerned neither with the internal representation of the data type nor with the implementation of the associated operations. The ground representation offered by Gödel, however, relies on a naming policy that does contrast with the naming policy employed in the context of logic program composition. We show how this problem can be tackled by suitably extending Gödel's support for generating the ground representation of object programs.

The two implementations are then extended in order to support the composition of a larger class of definite programs in which expressions denoting program compositions are allowed in the clause of object programs, as recently proposed by Brogi, Renso and Turini in [BRT94]. The various implementations are then compared in order to highlight the merits of the alternative representations and, most important, the adequacy of Gödel as a meta-language for composing logic programs. The performances of the different implementations are discussed on the basis of some experimental tests, including the application of partial evaluation techniques to the meta-interpreters. The analysis of the implementations also outlines some possible extensions of Gödel which may improve the flexibility of the language.

A preliminary report on this activity was described by Brogi and Contiero in [BC94]. The plan of the paper follows. A set of meta-level operations for composing logic programs is introduced in section 7.2. Two alternative implementations of these operations in Gödel are described in sections 7.3 and 7.4, respectively. Section 7.5 is devoted to present the extension of the two implementations to deal with a larger class of object programs. An analysis of the performances of the various implementations is made in section 7.6, while some concluding remarks are drawn in section 7.7.

7.2 Program Composition Operations

This section is devoted to briefly introduce a set of composition operations that form an algebra of logic programs, originally defined by Brogi et al. in [Bro93, MP88].

Four basic operations for composing definite logic programs are introduced: Union (denoted by $\cup$), intersection ($\cap$), encapsulation ($*$), and import ($\triangleleft$).

Program composition operations can be defined in different ways. The approach chosen here is based on the characterisation of the operational behaviour of program compositions. Such a characterisation can be expressed by directly extending the standard notion of SLD refutation (e.g., see Apt [Apt90]) to deal with program expressions. The standard SLD refutation relation may be defined by means of inference rules of the form

$$\frac{Premise}{Conclusion}$$

asserting that $Conclusion$ holds whenever $Premise$ holds. We write $P \vdash G$ if there exists a refutation for a goal G in a program P. For the sake of simplicity, we consider only ground programs and goals in that we are only interested here in characterising the (ground) success set of a program. Program clauses are therefore represented by means of the following rule:

$$\frac{P \text{ is a plain program} \ \wedge \ A \leftarrow G \in ground(P)}{P \vdash (A \leftarrow G)} \quad (1)$$

where $ground(P)$ denotes the set of ground instances of the clauses of program P.

We now introduce three rules defining SLD-resolution for ground programs and goals.

$$\frac{}{P \vdash Empty} \quad (2)$$

$$\frac{P \vdash G_1 \ \wedge \ P \vdash G_2}{P \vdash (G_1, G_2)} \quad (3)$$

$$\frac{P \vdash (A \leftarrow G) \ \wedge \ P \vdash G}{P \vdash A} \quad (4)$$

Rule (2) states that the empty goal, denoted by $Empty$, is solved in any program P. Rule (3) deals with conjunctive goals. It states that a conjunction (G_1, G_2) is solved in a program P if G_1 is solved in P and G_2 is solved in P. Finally, rule (4) deals with atomic goal reduction. To solve an atomic goal A, choose a clause from program P and recursively solve the body of the clause in P.

The derivation relation $\vdash$ can be generalised to the case of program compositions in a simple way. Namely, each composition operation is modelled by adding new inference rules to rules (1)—(4).

$$\frac{P \vdash (A \leftarrow G)}{P \cup Q \vdash (A \leftarrow G)} \quad (5)$$

$$\frac{Q \vdash (A \leftarrow G)}{P \cup Q \vdash (A \leftarrow G)} \quad (6)$$

$$\frac{P \vdash (A \leftarrow G_1) \;\wedge\; Q \vdash (A \leftarrow G_2)}{P \cap Q \vdash (A \leftarrow G_1, G_2)} \qquad (7)$$

$$\frac{P \vdash A}{P^* \vdash (A \leftarrow Empty)} \qquad (8)$$

$$\frac{P \vdash (A \leftarrow G_1, G_2) \;\wedge\; Q \vdash G_2}{P \triangleleft Q \vdash (A \leftarrow G_1)} \qquad (9)$$

Rules (5) and (6) state that a clause $A \leftarrow G$ belongs to the program expression $P \cup Q$ if it belongs either to P or to Q. Rule (7) states that a clause $A \leftarrow G$ belongs to $P \cap Q$ if there is a clause $A \leftarrow G_1$ in P and a clause $A \leftarrow G_2$ in Q such that $G = (G_1, G_2)$. Rule (8) states that the program expression P^* contains a unit clause $A \leftarrow Empty$ for each atom A that is provable in P. Finally, rule (9) deals with the import operation. It states that the clauses in $P \triangleleft Q$ are obtained from the clauses in P by dropping the calls to Q, provided that they are provable in Q. The extended derivation relation $\vdash$ defined by rules (1)—(9) characterises the operational behaviour of arbitrary composition of programs.

The use of the composition operations $\cup$, $\cap$, $*$ and $\triangleleft$ for programming finds natural application in several domains, ranging over expert systems, hypothetical and hierarchical reasoning, knowledge assimilation and modularisation. The description of such applications is outside the scope of this paper and is reported by Brogi et al. in [Bro93, BMPT90, BMPT94, BT90]. An example of the application of the operators to modular programming is also reported in this volume by Brogi and Turini [BT95].

7.3 Non-Ground Implementation

We now present an implementation in Gödel of the set of program composition operations introduced in the section 7.2. The implementation consists of extending the vanilla meta-interpreter using the non-ground representation of object level programs. The definition of the vanilla meta-interpreter in Gödel is illustrated in subsection 7.3.1. The extended meta-interpreter is presented in subsection 7.3.2, and the associated support for the non-ground representation of definite programs is described in subsection 7.3.3.

7.3.1 Vanilla meta-interpreter

The standard vanilla meta-interpreter using the non-ground representation of object programs can be written in Gödel as illustrated by Hill and Lloyd in Chapter 10 of [HL94]. We consider here a more general form of the vanilla meta-interpreter, where the `Solve` predicate has an extra argument to explicitly denote the name of the object program to be interpreted. The module `Vanilla` below contains the definition of this more general form of vanilla meta-interpreter in Gödel.

```
MODULE          Vanilla.
IMPORT          Object_Program.
PREDICATE       Solve : Program_Name * OFormula.
DELAY           Solve(x,y)   UNTIL GROUND(x) & NONVAR(y).

Solve(x, Empty).
Solve(x, y And z) <- Solve(x, y) & Solve(x, z).
Solve(x, y) <- Statement(x, y If z) & Solve(x, z).
```

In Gödel, the `PREDICATE` declaration declares `Solve` to be a binary predicate whose first argument has type `Program_Name` and whose second argument has type `OFormula`. The type `Program_Name` is used for meta-level terms representing the name (viz. a constant) of the object level program. The type `OFormula` is used for the type of meta-level terms representing object level formulae. The connectives `&` and `<-` are represented by the functions `And` and `If`, respectively. The `DELAY` declaration is a control declaration stating that calls to `Solve` will delay until first argument (i.e. the program name) is ground and the second argument is not a variable. Statements in the object program to be interpreted are represented in the imported module `Object_Program` using the predicate `Statement` and the constant `Empty`.

The module `Object_Program` (imported by `Vanilla`) contains the meta-level representation of the program to be interpreted. For instance, consider the program consisting of the module M below, which defines the relations `Arc` and `Path` over a graph.

```
MODULE     M.
BASE       Node.
CONSTANT   Bristol, London, Pisa : Node.
PREDICATE  Arc, Path : Node * Node.
Path(x, y) <- Arc(x, y).
Path(x, y) <- Arc(x, z) & Path(z, y).
Arc(Bristol, London).
Arc(London, Pisa).
```

Composing Logic Programs by Meta-Programming in Gödel

The meta-level representation of M is reported below. Object level symbols are represented by themselves, including object level variables which are represented by meta-level variables.

EXPORT	Object_Program.
BASE	Program_Name, OFormula, Node.
CONSTANT	Empty : OFormula;
	M : Program_Name;
	Bristol, London, Pisa : Node.
FUNCTION	And : xFy(110) : OFormula * OFormula -> OFormula;
	If : xFy(100) : OFormula * OFormula -> OFormula;
	Arc, Path : Node * Node -> OFormula.
PREDICATE	Statement : Program_Name * OFormula.
LOCAL	Object_Program.

Statement(M, Path(x,y) If Arc(x,y)).
Statement(M, Path(x,y) If Arc(x,z) And Path(z,y)).
Statement(M, Arc(Bristol,London) If Empty).
Statement(M, Arc(London,Pisa) If Empty).

The module Object_Program consists of an EXPORT and a LOCAL part. The EXPORT part contains the declarations of types, constants, functions and predicates that are exported by the module. The BASE declaration declares the types used in the object program (viz. Node), as well as the types Program_Name and OFormula, which are also used in the importing module Vanilla. The CONSTANT declaration declares the name of the object program to be interpreted, and the constants occurring in the object program. The FUNCTION declaration declares the function symbols used in both modules (If, And), as well as the object level predicates (e.g. Arc), which are represented by functions at the meta-level. Notice that the module conditions of Gödel require that types, constants and functions used in both modules (such as OFormula, Empty and If) must be declared in the imported module Object_Program. Finally, the LOCAL part of module Object_Program contains the clauses defining the predicate Statement, which is used to represent the object program to be interpreted by the vanilla meta-interpreter.

It is worth making a couple of remarks here. First, the lack of parametric modules in Gödel does not allow a flexible use of the Solve meta-interpreter since the name of the module containing the object program to be interpreted must be fixed in the module Vanilla. The availability of parametric modules would increase the flexibility of Gödel and, in particular, the possibility of parameterising a module w.r.t. the modules to be imported would allow a more flexible use of the meta-interpreter. Second, Gödel does not provide any special support for meta-programming with the non-ground representation. This means that, though it is easy to write a vanilla meta-interpreter in Gödel, the non-ground representation of the object programs must be given *explicitly* by the programmer.

7.3.2 Extended vanilla meta-interpreter

The Solve meta-interpreter can be extended in a simple and concise way in order to implement meta-level operations for composing logic programs. Following Brogi et al. [Bro93, BMPT94], each program composition operation is represented at the meta-level by a function symbol: ∪ by Union, ∩ by Intersection, ∗ by Encapsulate, and ◁ by Import. The idea is to use the first argument of Solve for representing arbitrary compositions of object programs, such as P Union Q, rather than just a single object program. The meaning of each function symbol denoting a program composition operation can be defined by extending the definition of the vanilla meta-interpreter. Intuitively, this corresponds to turning the inference rules given in section 7.2 into meta-level axioms.

The module Extended_Vanilla below contains the definition of the Solve meta-interpreter suitably extended to deal with programs composition operations.

```
EXPORT     Extended_Vanilla.
IMPORT     Object_Programs.
PREDICATE  Solve : Program_Expression * OFormula.
DELAY      Solve(x,y)  UNTIL GROUND(x) & NONVAR(y).

LOCAL      Extended_Vanilla.
PREDICATE  Clause : Program_Expression * OFormula.
Solve(x, Empty).
Solve(x, y And z) <-
     Solve(x, y) &
     Solve(x, z).
Solve(x, y) <-
     Clause(x, y If z) &
     Solve(x, z).
Clause(x Union y, z If w) <-
     Clause(x, z If w).
Clause(x Union y, z If w) <-
     Clause(y, z If w).
```

```
Clause(x Intersection y, z If (w1 And w2)) <-
    Clause(x, z If w1) &
    Clause(y, z If w2).
Clause(Encapsulate(x), y If Empty) <-
    Solve(x, y).
Clause(x Import y, w If z) <-
    Clause(x, w If u) &
    Partition(u, z, v) &
    Solve(y, v).
Clause(x, y If z) <-
    Statement(x, y If z).
...
```

The EXPORT part contains the declaration of the predicate Solve, whose first argument now has type Program_Expression. This is the type of a meta-level term representing a program expression, that is a term constructed via the functions Union, Intersection, Encapsulation and Import starting from a set of program names.

The definition of Solve in the LOCAL part of Extended_Vanilla extends the definition of Solve given in the module Vanilla. The only differences are the type of the first argument (which is now Program_Expression) and the substitution of the predicate Statement with a new predicate Clause. The latter is introduced for the meta-level representation of compositions of object programs. Intuitively speaking, the definition of Clause extends the definition of Statement by induction on the structure of program expressions. For instance, the definition of Clause in the case of Union states that a clause z <- w belongs to a program composition x ∪ y if it belongs either to x or to y. The definition of Clause for Intersection states that a clause z <- (u & v) belongs to the composition x ∩ y if z1 <- w1 belongs to x, z2 <- w2 belongs to y, z1 and z2 unify via a mgu ϑ, and $z = (z1)\vartheta$, and $(u \ \& \ v) = (w1 \ \& \ w2)\vartheta$. Notice that the adoption of the non-ground representation of object programs allows this statement to exploit the basic unification mechanism. The meta-level representation of an encapsulated program expression Encapsulate(x) consists of assertions of the form Clause(Encapsulate(x), y If Empty) for each y provable in x. The Import operation is defined as follows. The statements in a composition x Import y are obtained from the statements of x by possibly dropping part of their body if this is provable in the imported y and possibly instantiating the remaining part of the body. Finally the last definition of Clause resorts to the predicate Statement (used to represent the single object programs to be composed), which is defined in the imported module Object_Programs.

Basic modularity principles suggest that the Solve meta-interpreter and the representation of the object programs should be arranged into separate modules. Such a separation makes it easier to use the meta-interpreter with different collections of object programs. The structuring of the module Extended_Vanilla partly supports such a possibility in the sense that the meta-interpreter and the representation of the object programs are arranged into two separate modules:

Extended_Vanilla
↓
Object_Programs

Notice that such a separation requires the employment of two predicates (Clause and Statement) for the meta-level representation of object program compositions. This is due to the module conditions of Gödel that do not allow one to spread the definition of a predicate over different modules.

We also implemented a more modular solution that establishes a one-to-one correspondence between object level programs and Gödel modules containing their meta-level representation, as illustrated by the following figure:

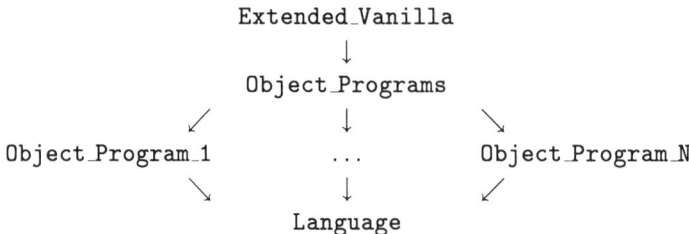

Roughly speaking, each object program P_i is represented by means of a predicate Statement_i defined in a Gödel module Object_Program_i. In this second implementation, Object_Programs simply plays the role of a bridge between the extended vanilla meta-interpreter and the representation of the object programs, and contains a clause

    Statement(x, y If z) <- Statement_i(x, y If z).

for each imported module Object_Program_i. Finally, types and symbols used in both modules are declared in the module Language at the bottom of the hierarchy, as required by Gödel's module conditions.

7.3.3 Support for the non-ground representation

As we pointed out, Gödel provides considerable support for meta-programming with the ground representation, while it does not provide any special support for the non-ground representation. This means that, though it is easy to write a vanilla meta-interpreter in Gödel, the non-ground representation of the object programs must be given *explicitly* by the programmer. When using Gödel as a meta-language for composing object definite programs, the problem is how to automatically generate a Gödel module containing the non-ground representation of a (collection of) given definite program(s).

In order to free programmers from the need of explicitly providing the non-ground representation of their definite programs, we developed suitable support for the generation of such a representation. The main issue to be faced is concerned with types, since Gödel is a strongly typed language while definite programs are not generally equipped with type declarations. One solution might be to try to infer as much type information as possible from the object program, by resorting to program analysis techniques. Rather than trying to infer (incomplete) type information from untyped logic programs, we simply employed a single type OTerm to represent *any* object level term.

The module Non_Ground_IO supports the generation of the non-ground representation of a (collection of) definite program(s) to be imported by the Extended_Vanilla module. Notice that we actually implemented two variants of the support corresponding to the two module structures for the extended vanilla presented in subsection 7.3.2. Since these supports have a similar structure we shall discuss only the support for the first solution.

EXPORT	Non_Ground_IO.
IMPORT	FlocksIO.
PREDICATE	Represent : List(String) * String.
DELAY	Represent(x,y) UNTIL GROUND(x) & GROUND(y).

Non_Ground_IO imports FlocksIO, containing the abstract data type Flock that has been revealed to be very convenient for parsing the files containing the object logic programs. The predicate Represent can be used to generate the non-ground representation of a collection of definite programs. The first argument of Represent is a list of strings that denotes the names of the files containing the definite programs to be represented at the meta-level. The second argument of Represent is a string denoting the name of the Gödel module in which the non-ground representation of the object programs will be written. Notice that if the object level programs are intended to be interpreted by the vanilla meta-program of module Extended_Vanilla then the second argument of Represent is necessarily the string "Object_Programs", since Gödel does not provide parametric modules at the moment. The possibility of specifying the name of the target Gödel module is however offered by the support also in the prospect of a support for parametric modules in the near future, as discussed by Hill [Hil93].

To illustrate the use of the support, suppose that the user wants to query program expressions obtained by composing a collection of given (object) definite programs. First, the non-ground representation of the object programs is generated by means of the module Non_Ground_IO that creates the Gödel module Object_Programs:

[Non_Ground_IO] <- Represent(["P","Q","R"], "Object_Programs").

Then, the user loads the module Extended_Vanilla (which imports Object_Programs) and queries it by means of meta-level goals such as:

[Extended_Vanilla] <- Solve(P Union (Q Import R), G(x)).

7.4 Ground Implementation

We now present a second implementation of the set of program composition operations introduced in section 7.2. This second implementation exploits the meta-programming facilities offered by Gödel for the ground representation of object level programs. We first illustrate a simple ground meta-interpreter which is analogous in structure and operation to the vanilla meta-interpreter introduced in subsection 7.3.1. Then we show the extended meta-interpreter and the associated support for the generation of the ground representation of object logic programs.

7.4.1 Demo meta-interpreter

Gödel provides considerable support for meta-programming with the ground representation, in which object level expressions are represented by ground terms at the meta-level. For instance, the correspondent of the vanilla meta-interpreter of subsection 7.3.1 (module Vanilla) can be written in Gödel using the ground representation of object level programs, as illustrated in the module Demo below by Gurr [Gur94a].

```
EXPORT      Demo.
IMPORT      Programs.
PREDICATE   Demo : Program * Formula * TermSubst.

LOCAL       Demo.
Demo(program, goal, answer) <-
      StandardiseFormula(goal, 0, var_index, new_goal) &
      EmptyTermSubst(empty_subst) &
      Demo1(program, new_goal, var_index, _, empty_subst, subst) &
      RestrictSubstToFormula(new_goal, subst, answer).

PREDICATE Demo1 : Program * Formula * Integer * Integer * TermSubst *
                  TermSubst.
```

```
Demo1(program, goal, v, v, subst, subst) <-
    EmptyFormula(goal).
Demo1(program, goal, v_in, v_out, subst_in, subst_out) <-
    And(left, right, goal) &
    Demo1(program, left, v_in, new_v, subst_in, new_subst) &
    Demo1(program, right, new_v, v_out, new_subst, subst_out).
Demo1(program, goal, v_in, v_out, subst_in, subst_out) <-
    Atom(goal) &
    MyStatementMatchAtom(program, goal, stment) &
    Resolve(goal, stment, v_in, new_v, subst_in, new_subst, new_goal) &
    Demo1(program, new_goal, new_v, v_out, new_subst, subst_out).
```

The module Demo imports the system module Programs, which contains a variety of predicates for handling the ground representation of Gödel programs. The PREDICATE declaration declares Demo to be a ternary predicate with arguments of type Program, Formula and TermSubst, which represent the ground representation of a program, of a formula and of a term substitution, respectively.

The LOCAL part of Demo contains the definition of the predicate Demo. The Demo predicate first calls the predicates StandardiseFormula and EmptyTermSubst to initialise the variables in the object goal and to get the representation of the empty substitution. Demo then calls Demo1 that performs the interpretation of the object program and returns the representation of the computed substitution. Finally such a substitution is restricted to the initial object goal.

The definition of Demo1 closely resembles the definition of Solve with the non-ground representation of object programs (module Vanilla in subsection 7.3.1). To illustrate the reading of this meta-interpreter, let us focus on the last statement in the definition of Demo1, namely the case of atomic goal reduction. Given an atomic goal, the predicate MyStatementMatchAtom selects a statement in the program whose predicate (proposition) in the head is the same as the predicate (proposition) in the goal. The predicate Resolve performs the clause reduction step: Given a goal, a selected statement and a substitution returns the new goal and a new substitution obtained by composing the initial one with the substitution computed during the reduction step.

One of the tedious aspects of meta-programming with the ground representation is that unification and substitutions must be handled explicitly. Gödel offers several system modules, such as Programs, which provide a large set of operations for working with the ground representation. It is worth noting that the abstract data type view of modules hides most of the complexity of these operations and supports a declarative style of meta-programming.

7.4.2 Extended Demo Meta-Interpreter

We now extend the Demo meta-interpreter using the ground representation in order to implement meta-level operations for composing definite programs. The idea is to follow the same approach presented in section 7.3. As in the non-ground case, program composition operations are represented by function symbols whose meaning is defined by meta-level axioms. The key difference is that in the ground case object programs are represented by ground terms rather than referred to by constant names as in the non-ground case. The ground representation of a single object program can be generated by means of the system predicate ProgramCompile, which given the name of a program returns the term containing its ground representation.

There is, however, a problem to face when extending the ground vanilla meta-interpreter with the ground representation to deal with program expressions. Indeed Gödel meta-programming facilities strictly rely on the internal ground representation of Gödel object programs, which coherently mirrors the naming policy of the language. Each symbol is internally represented by a *flat name* that is a quadruple containing the name, category and arity of the symbol, as well as the name of the module in which the symbol is declared. Therefore symbols with same name, category and arity that are declared in different modules are distinguished. This contrasts with the naming policy adopted in the logic program composition setting, where a predicate definition may be spread over different modules (e.g., see Brogi et al. [BMPT94] and Monteiro and Porto [MP89]).

A first solution to this obstacle is to exploit the meta-programming support offered by Gödel to generate the ground representation of object programs, and to meta-program a unification mechanism for flat names in order to identify symbols with same name, category and arity declared in different modules. The implementation of this solution presented two major drawbacks. First, the introduction of an an extra representation of substitutions demands the construction from scratch a corresponding support. Second, the introduction of a further interpretation layer for handling the unification mechanism heavily affects the performance of the system. A more efficient solution might be to provide a new version of predicates such as Resolve suitably modified in order to support the desired unification mechanism. This solution, however, would require an even greater deal of work to the programmer.

The alternative approach we chose consists of extending the support for the generation of the ground representation of object programs in order to avoid the undesired distinction among flat names. This way, the extended Demo meta-interpreter is let free from the problem of flat names, which is completely solved during the pre-processing phase. Before presenting the extended support for this second solution (subsection 7.4.3), let us show the corresponding Demo vanilla meta-interpreter.

```
EXPORT      Extended_Demo.
IMPORT      Program_Expressions.
PREDICATE   Demo: Program_Expression * Formula * TermSubst.

LOCAL       Extended_Demo.
Demo(pexp, goal, answer) <-
     StandardiseFormula(goal, 0, var_index, new_goal) &
     EmptyTermSubst(empty_subst) &
     Demo1(pexp, new_goal, var_index, _, empty_subst, subst) &
     RestrictSubstToFormula(new_goal, subst, answer).

PREDICATE Demo1 : Program_Expression * Formula * Integer * Integer *
                  TermSubst * TermSubst.
...
Demo1(pexp, goal, v_in, v_out, subst_in, subst_out) <-
     Atom(goal) &
     MyStatementsMatchAtom(pexp, goal, v_in, v1, stment) &
     Resolve(goal, stment, v1, new_v, subst_in, new_subst, new_goal) &
     Demo1(pexp, new_goal, new_v, v_out, new_subst, subst_out).

PREDICATE   MyStatementsMatchAtom : Program_Expression * Formula *
                                    Integer * Integer * Formula.
MyStatementsMatchAtom(p Union q, goal, v_in, v_out, statement) <-
     MyStatementsMatchAtom(p, goal, v_in, v_out, statement).
MyStatementsMatchAtom(p Union q, goal, v_in, v_out, statement) <-
     MyStatementsMatchAtom(q, goal, v_in, v_out, statement).
...
MyStatementsMatchAtom(Enc(p), goal, v_in, v_out, statement) <-
     EmptyTermSubst(empty_subst) &
     Demo1(p, goal, v_in, v_out, empty_subst, subst_out) &
     BuildStatement(goal, subst_out, statement).
MyStatementsMatchAtom(Prog(_,p), goal, v, v, statement) <-
     MyStatementMatchAtom(p, goal, statement).
```

Extended_Demo imports the module Program_Expressions, which contains the declaration of the type Program_Expression with the associated function symbols. The only change in the definition of the Demo predicate is the type of its first argument. The definition of Demo1 extends the definition of Demo1 given in the module Demo much in the same way as in the non-ground case of section 7.3. Intuitively, the predicate MyStatementsMatchAtom extends MyStatementMatchAtom in the same way as Clause extends Statement in the vanilla meta-interpreter of subsection 7.3.2. It is worth noting that in this extended ground meta-interpreter substitutions must be explicitly handled. For instance, in the case of an encapsulated program expression Enc(p), the representation of the empty substitution as well as of the answer computed by Demo1 must be explicitly handled in order to build a statement belonging to the ground representation of Enc(p).

7.4.3 Support for the ground representation

In order to employ the module Extended_Demo for querying an arbitrary composition of object logic programs, the meta-interpreter must be provided with the ground representation of such a program expression. We have therefore equipped the module Extended_Demo with a suitable support capable of generating the ground representation of a composition of definite programs.

Given a program expression E over definite logic programs, the support generates the Gödel term denoting the ground representation of E. Such a term is obtained by composing the ground representations of the single object programs forming the program expression. The ground representation of a definite program P is obtained by first transforming P into an equivalent Gödel program P_G, and then applying ProgramCompile to P_G. Similarly to subsection 7.3.3, each logic program is transformed into an untyped Gödel module by employing a unique type.

The support is in charge of solving the problem of flat names, so that symbols with same name, category and arity, though occurring in different programs, are identical in their ground representation. This is done by assigning the same name to all the modules and by arranging them in separate directories. The structure of this support is more complex than the non-ground one, since it must combine Gödel modules and utilities, and operating system commands. Notice, however, that the complexity of the ground support is transparent to the user, which only has to invoke a shell script called ProgramsCompile. Indeed, in order to compile a collection of object programs, the user simply calls ProgramsCompile with the names (of the files containing) the object programs. For instance, the compilation of the object programs P, Q and R is obtained by invoking:

ProgramsCompile P Q R

Finally, in order to allow the user to query the extended Demo meta-interpreter, we developed a user interface module Test_Demo reported below. The predicate Go allows the user to specify both the program expression and the goal of the intended query as strings. First the ground representation of the program expression is built via the predicate StringToProgramExpression. Then the string identifying the goal is converted into a formula w.r.t. the name of any of the object programs (which have been given the same name M during the pre-processing phase). Finally, the answer possibly computed by Demo is converted back from the internal Gödel representation of substitutions into a string.

```
MODULE      Test_Demo.
IMPORT      Extended_Demo, Program_ExpressionsIO, Answers.
PREDICATE   Go : String * String * String.

Go(pe_string, goal_string, answer_string) <-
        StringToProgramExpression(pe_string, program_expression) &
        ProgramInProgramExpression(program_expression, program) &
        StringToProgramFormula(program, "M", goal_string, [goal]) &
        Demo(program_expression, goal, answer) &
        RestrictSubstToFormula(goal, answer, computed_answer) &
        AnswerString(program, "M", computed_answer, answer_string).
```

7.5 Extending the Language of Program Expressions

In section 7.2 we have introduced a set of meta-level operators for composing definite programs. Such operations define a language of program expressions, where an expression is defined by the production:

$$Exp ::= P \mid Exp \cup Exp \mid Exp \cap Exp \mid (Exp)^* \mid Exp \triangleleft Exp$$

where P is (a name of) a definite program.

The language of program expressions has been recently extended by Brogi, Renso and Turini in [BRT94] in order to allow program expressions also in the clause bodies of the object programs rather than only in top-level queries. Simply stated, the syntax of object programs is extended so that meta-level formulae of the form $G \: In \: E$, where G is an object goal and E a program expression, can occur in the clause bodies of the object programs.

The meaning of the meta-annotation In is simply formalised by adding the following inference rule to the rules (1)—(9) of section 7.2.

$$\frac{Q \vdash G}{P \vdash (G \; In \; Q)} \qquad (10)$$

The rule states that solving a goal of the form $G \; In \; Q$ in a program expression P simply amounts to solving G in the new context Q.

As illustrated by Brogi, Renso and Turini [BRT94], the introduction of the meta-annotation In notably increases the expressive power of the language of program expressions. Indeed the dynamic forms of program compositions and context switching supported by In broaden the application area of the language of program expressions by supporting, for instance, forms of object-oriented programming. A thorough discussion of the In feature, including a bottom-up characterisation of the resulting language, can be found in [BRT94].

In this section we discuss how the two implementations presented in sections 7.3 and 7.4 have been suitably extended to deal with the In meta-feature. In both cases, we employ an ambivalent syntax for the object programs and represent a meta-level formula of the form $G \; In \; E$ by a formula $In(G, E)$, where In is a predicate name and G and E are terms.

7.5.1 Extending the non-ground implementation

The In meta-feature can be quite easily encorporated in the non-ground implementation. Following the somewhat standard approach of this kind of representation, In is simply represented at the meta-level by a function symbol In with the declaration:

`FUNCTION In : ObjectFormula * PRGEXP -> ObjectFormula.`

As far as the meta-interpreter is concerned, the implementation of this feature is really straightforward. Indeed it simply amounts to turning the inference rule (10) into the meta-level axiom:

`Solve(x,In(y,z)) <- Solve(z,y).`

to be added to the extended vanilla meta-interpreter of module Extended_Vanilla. Intuitively speaking, the clause states that a meta-level goal of the form In(y,z) is provable in x if y is provable in the new context z.

7.5.2 Extending the ground implementation

In spite of a certain lack of flexibility, it is still possible to implement the construct In using the ground representation. Recall that in this case the object programs must be program-compiled in order to generate their ground representation. This fact has some relevant consequences.

A first problem concerns types. The extended syntax of object programs allows formulae of the form In(P(..), PExp) to occur in clause bodies, where P is a predicate symbol. Intuitively speaking, the ambivalence of the syntax is obtained by declaring all predicate symbols also as function symbols. More precisely, suppose that the initial language of object programs (without In formulae) is determined by a set $\mathcal{P}$ of predicate symbols, a set $\mathcal{F}$ of function symbols and a set $\mathcal{C}$ of constant symbols. Then we consider a new language in which the three set are replaced by the following ones: $\mathcal{P}' = \mathcal{P} \cup \{In\}$, $\mathcal{F}' = \mathcal{F} \cup \mathcal{P}$, and $\mathcal{C}' = \mathcal{C} \cup \Pi$ where Π is the set of object program names. In order to perform such an extension always safely, In must not belong to $\mathcal{P}$ (viz. $\mathcal{P} \cap \{In\} = \emptyset$), the set of predicate and function symbols must be disjoint in the initial language (viz. $\mathcal{P} \cap \mathcal{F} = \emptyset$), and the names of programs must be disjoint by the other constants (viz. $\mathcal{C} \cap \Pi = \emptyset$). Notice that each symbol is uniquely identified by the pair of its name and arity. Notice also that these conditions are sufficient but not strictly necessary. For instance the name of a program might be even used as a generic constant provided that it does not occur in any In formula. The resulting Gödel program is trivially well-typed, since each construct other than program expressions has the same type.

The second argument of a In formula denotes a program expression. Therefore we can always infer the type of this argument and assign it a special type Pe. We then declare a constant of type Pe for each program name and a function symbol for each composition operation. Notice that at this level the program expressions are denoted by means of the *names* of the programs and by the function symbols they are built with. Therefore this representation differs from that used at the meta-level, which instead relies on the *ground representations* of the object programs.

The implementation of the In meta-feature does not dramatically affect the structure of the meta-interpreter. Actually the meta-interpreter has to be extended only in order to consider a further case when dealing with atomic goal reduction. Namely if the atomic goal to be reduced is a In goal then the meta-interpreter switches the control to the program expression specified by In. Otherwise the behaviour of the meta-interpreter is just the same as before. The part of the Gödel code corresponding to the new steps is reported below.

```
Demo1(pe, goal, v_in, v_out, subst_in, subst_out) <-
      Atom(goal) &
      ExtractGoalName(pe,goal,_,name_string,arity,_) &
      (name_string ~= "In" \/ arity ~= 2 ) &
      MyStatementsMatchAtom(pe, goal, v_in, v1, stment) &
      Resolve(goal, stment, v1, new_v, subst_in, new_subst, new_goal) &
      Demo1(pe, new_goal, new_v, v_out, new_subst, subst_out).
Demo1(pe, goal, v_in, v_out, subst_in, subst_out) <-
      Atom(goal) &
      ExtractGoalName(pe,goal,p,name_string,2,[arg1,arg2]) &
      name_string = "In" &
      NewContext(p,[arg1,arg2],subst_in,new_pe,last_goal) &
      Demo1(new_pe,last_goal, v_in, v_out, new_subst, subst_out).

PREDICATE ExtractGoalName : Program_Expression * Formula * Program *
                            String * Integer * List(Term).
ExtractGoalName(pe,goal,p,name_string,arity,args) <-
      ProgramInProgramExpression(pe,p) &
      PredicateAtom(goal,name,args) &
      ProgramPredicateName(p,_,name_string,arity,name).

PREDICATE NewContext : Program * List(Term) * TermSubst *
                       Program_Expression * Formula.
NewContext(p,[arg1,arg2],subst_in,new_pe,last_goal) <-
      ProgramTermToString(p,"M",arg1,st1) &
      ProgramTermToString(p,"M",arg2,st2) &
      InfixForm(st2,inf_st2) &
      Convert(inf_st2,new_pe) &
      StringToProgramFormula(p,"M",st1,[new_goal]) &
      PredicateAtom(new_goal,_,args) &
      FunctionTerm(arg1,_,args1) &
      UnifyList(args,args1,subst_in,new_subst) &
      ApplySubstToFormula(new_goal,new_subst,last_goal).
```

The first clause of Demo1 is obtained by the extending the corresponding clause in module Extended_Demo (subsection 7.4.2) with the control that the predicate of the selected object goal is not In. The second clause of Demo1 deals with *In* formulae. First the two arguments of In are transformed into strings, and then the ground representations of the new program expression and of the new goal are obtained. Finally the new goal is solved in the new context.

7.6 Performances

In this section we briefly discuss some results obtained by testing the meta-interpreters described so far on simple examples, built from the definite program *Factorial* reported below.

$fact(z, s(z))$.
$fact(s(X), W) : -fact(X, Y), mul(s(X), Y, W)$.

$mul(z, X, z)$.
$mul(s(X), Y, Z) : -mul(X, Y, W), plus(Y, W, Z)$.

$plus(z, X, X)$.
$plus(s(X), Y, s(Z)) : -plus(X, Y, Z)$.

We denote by *Fact*, *Mul* and *Plus* the programs consisting only of the clauses defining the predicates *fact*, *mul* and *plus* respectively. Moreover let *Fact_with_In* denote the following program:

$fact(z, s(z))$.
$fact(s(X), W) : -fact(X, Y), in(mul(s(X), Y, W), union(mul, plus))$.

which contains an occurrence of the *In* construct in its second clause.

The following table shows the average times for various queries w.r.t. different composition of the above programs. The experiments were carried on using Gödel release 1.4.

	Demo	Extended_Demo	Vanilla	Extended_Vanilla
$Factorial$	0.939 s	1.041 s	0.166 s	0.169 s
$Fact \cup (Mul \cup Plus)$	.	1.144 s		0.198 s
$Fact \cup (Mul \cup Plus)^*$		1.315 s		0.239 s
$Fact\_with\_In$		2.044 s		0.195 s

First of all we observe that the implementations based on the non-ground representation are 5-6 times faster than those based on the ground representation. This is true both for the single object program meta-interpreters and for the extended meta-interpreters. Such a result is due to the fact that the non-ground representation is simpler and does not have to pay the overhead needed for handling of the ground representation. So, far from being surprising, these tests only confirm a well known situation. It should be observed, however, that the gap between the non-ground and the ground representation is partly due to the fact that, in the present implementation of Gödel, the ground interpreter performs the occur check for object program unification while the non-ground does not perform it.

The extended meta-interpreters are slightly slower than the standard single object program meta-interpreters. This is quite reasonable because the latter meta-interpreters actually are a special case of the former meta-interpreters which need to perform some additional operations. It is also interesting to observe that modularity has a cost in terms of efficiency, even if limited. Indeed, in the non-ground approach, we have investigated two alternative solutions (see subsection 7.3.2): One in which all the object programs are represented by a single Gödel module at the meta-level, and another in which there is a one-to-one correspondence between object level programs and Gödel modules representing them at the meta-level. While the latter is more modular, the former is more efficient. (In the table only the results of the more efficient solution are reported.)

We finally observe that while the implementation of the In meta-feature is particularly efficient in the non-ground case, it notably increases the execution times in the ground case.

7.6.1 Application of the partial evaluator SAGE

Gödel meta-programs using the ground representation suffer from a computational expense which is in addition to the so-called interpretation overhead present in all meta-programs. As shown by Gurr [Gur94b], the partial evaluation of Gödel programs may almost entirely remove these additional overheads. Such an achievement could make the ground representation really usable by reducing the existing gap in performances w.r.t. the non-ground representation.

A partial evaluator for Gödel programs, called SAGE, has been recently developed by Gurr [Gur94b]. SAGE (Self Applicable Gödel partial Evaluator) is written in Gödeland is based on the technique of finite unfolding. Though SAGE can be used to specialise any Gödel program, it is mainly able to specialise programs using the ground representation.

Roughly speaking, the basic idea of using finite unfolding for partially evaluating a program P w.r.t. a certain goal G is the following. First (partial) search trees for P and G are constructed, and then the specialised program is extracted from the leaves of these trees. In the case of single object program meta-interpreters, given the meta-interpreter and an object program, the partial evaluator returns a new meta-interpreter, specialised w.r.t. the given object program. We do not discuss here the issues of partial evaluation in logic programming, which is thoroughly discussed by Lloyd and Shepherdson in [LS91]. Moreover a compete description of SAGE is given by Gurr in [Gur94b, Gur94a], and a discussion of the application of partial evaluation techniques to Gödel programs is reported in this volume by Bowers and Gurr [BG95].

We have done some experiments using SAGE (version dated April 1994) in order to try to specialise the meta-interpreter we have developed. Some results are reported in the following table and, though not excellent, are indeed encouraging. The execution times of the specialised versions are reported in bold under the execution times of the unspecialised programs. In some cases SAGE reported a failure during the optimisation process, and these cases are denoted by "-" in the table.

	Demo	Extended_Demo	Vanilla	Extended_Vanilla
$Factorial$	0.939 s **0.260** s	1.041 s **0.956** s	0.166 s **0.144** s	0.169 s **0.153** s
$Fact \cup (Mul \cup Plus)$		1.144 s **1.169** s		0.198 s **0.147** s
$Fact \cup (Mul \cup Plus)^*$		1.315 s **1.101** s		0.239 s —
$Fact\_with\_In$		2.044 s —		0.195 s —

The experiments show that in the case of the ground representation we obtain a time reduction of about 66 % for the single object program meta-interpreter (and this rate may grow with larger object programs). It should be noted, however, that the effectiveness of partial evaluation in improving the ground meta-interpreter is due to the fact they are very slow to start with.

As far as our extended meta-interpreter is concerned, the optimisation is limited to a rate of about 20 %. There are several reasons for this. First SAGE is not always able to specialise a program, even if practically possible, because it sometimes lacks some information which would ensure it not to get to a situation of infinite unfolding. Moreover SAGE can not specialise a predicate only in certain chosen parts of program. This is a serious problem in the case of our meta-interpreter since there are predicates, such as MyStatementsMatchAtom, which may be deeply specialised but only on part of their definitions. Intuitively speaking, this is due to the fact that the definition of some composition operations, such as Encapsulation, makes the definition of the predicates MyStatementsMatchAtom and Demo1 mutually recursive, and hence SAGE does not make any specialisation for the risk of infinite unfolding. Finally, in the current implementation of Gödel all system modules but Syntax are closed, and closed modules in some cases do not allow one to specialise predicates defined in these modules.

It is worth observing that SAGE requires, in some cases, a quite deep knowledge of its functioning by the user, and in general this is not desirable. In spite of these observations, the current release of SAGE, though still prototypical, may indeed produce good results in many situations and be of great help for the user. On the other hand, some improvements are needed to increase the applicability and usability of this tool.

7.7 Discussion

The ultimate objective of our work was to investigate the adequacy of Gödel as a metalanguage for re-using and composing existing definite programs. First of all, we would like to draw from our own experience a few considerations on Gödel as a programming language. The type discipline and the module system, with the associated abstract data type view, are features of Gödel essential for the incremental and disciplined development of software. Moreover some Gödel system modules, such as Units and Flocks, have been revealed to be very convenient for analysing and handling terms and programs.

Another relevant feature of Gödel is the emphasis placed on meta-programming. Gödel provides ample support for meta-programming with the ground representation. Several system modules offer operations for generating and handling the representation of programs. We found Gödel's abstract data type view of modules particularly powerful in the case of meta-programming. Indeed, system modules such as **Programs** and **Syntax** support a declarative style of meta-programming by hiding the complexity of the ground representation. In this respect, however, it is worth mentioning that one of the deficiencies of the current implementation of Gödel is the tracer. For instance terms with types used in the ground representation, such as **Formula**, are not visible to the user. As a consequence the tracer is often of little help when analysing the behaviour of meta-programs.

We developed two alternative implementations in Gödel of a suite of meta-level operations for composing definite programs. The two implementations are based on an extended vanilla meta-interpreter and they differ each other for the chosen representation of object level programs. The first implementation, described in section 7.3, employs the non-ground representation of object programs and is quite simple and efficient. The second implementation, described in section 7.4, heavily exploits Gödel's support for meta-programming. The choice of addressing the naming problem during the pre-processing phase notably simplifies the structure of the extended meta-interpreter with the ground representation, which closely resembles its non-ground correspondent.

As we have shown, the non-ground implementation offers much better performances than the ground one. However the latter presents some interesting aspects. One of them is the extensibility of the implementation to deal with the composition of object Gödel programs. Actually the current implementation of program composition operations already deals with the composition of object Gödel programs since the source definite programs are translated into corresponding Gödel programs. Such programs, however, do not contain control declarations or type information. An interesting research direction is to investigate how the implementation of program composition operations can be extended to deal with arbitrary Gödel programs. This may allow us to push the program composition approach to extend the Gödel language itself. Another interesting aspect of the ground implementation is that it seems to be more suitable than the non-ground one for implementing some program composition operations. For instance, certain forms of program compositions require a fine-grained manipulation of the object programs. An example is the overriding operator introduced by Brogi [Bro93] which is defined in terms of the predicate names occurring in a program rather than in terms of its clauses. The support offered by Gödel for accessing the ground representation of object programs simplifies the implementation of this kind of composition operations.

As far as the problems of efficiency are concerned, the ground implementation can greatly benefit from the application of tools that are currently under development for Gödel and that rely on the ground representation. For this reason, the extended **Demo** meta-interpreter using the ground representation was designed according to the requirements of the partial evaluator SAGE. We expect that the application of partial evaluation techniques will sensibly improve the performances of our implementation. In this direction our first attempts to exploit SAGE have given quite promising results. Unfortunately, the moving from a single program environment to a multi-program setting involves a number of consequences which prevent, at the moment, a whole exploitation of the partial evaluation techniques offered by SAGE.

Finally, we would like to mention some other possible extensions and improvements that might be encorporated in Gödel. The availability of parametric modules, for instance, would notably increase the flexibility of the language. Even more importantly, the possibility of parameterising the language of a Gödel module would make it possible to (at least partly) overcome one of the most severe limits of the language. Namely, Gödel meta-programming support does not easily allow one to meta-program extensions of the language itself. This is due to the fact that the `ProgramCompile` operation generating the ground representation of a program requires the latter to be a (pure) Gödel program.

Acknowledgements

We would like to thank J.W. Lloyd and C. Gurr for the many suggestions and encouragement. Thanks also to A. Bowers for his valuable comments. The very detailed suggestions given by the anonymous referee have greatly contributed to improve the structure of the paper.

References

[Apt90] K. R. Apt. Logic programming. In J. van Leeuwen, editor, *Handbook of Theoretical Computer Science*, pages 493–574. Elsevier, 1990. Vol. B.

[Apt93] K. R. Apt. Declarative programming in Prolog. In D. Miller, editor, *Proc. International Symposium on Logic Programming*, pages 11–35. MIT Press, 1993.

[BC94] A. Brogi and S. Contiero. Gödel as a meta language for composing logic programs. In F. Turini L. Fribourg, editor, *Proceedings of Fourth International Workshops on Logic Program Synthesis and Transformation (LOPSTR 94) and Meta-programming in Logic (META 94)*, number 883 in LNCS. Springer-Verlag, 1994.

[BG95] A. Bowers and C. Gurr. Towards Fast and Declarative Meta-programming. This volume.

[BMPT90] A. Brogi, P. Mancarella, D. Pedreschi, and F. Turini. Hierarchies through Basic Metalevel Operators. In M. Bruynooghe, editor, *Proceedings of the Second Workshop on Meta-programming in Logic*, pages 381–396, 1990.

[BMPT94] A. Brogi, P. Mancarella, D. Pedreschi, and F. Turini. Modular Logic Programming. *ACM Transactions on Programming Languages and Systems*, 16(4):1361–1398, 1994.

[Bro93] A. Brogi. *Program Construction in Computational Logic*. PhD thesis, University of Pisa, March 1993.

[BRT94] A. Brogi, C. Renso, and F. Turini. Amalgamating language and meta-language for composing logic programs. In M. Alpuente and R. Barbuti, editors, *Proceedings Joint Conference on Declarative Programming (GULP-PRODE 94)*, 1994.

[BT90]   A. Brogi and F. Turini. Metalogic for Knowledge Representation. In J.A. Allen, R. Fikes, and E. Sandewall, editors, *Principles of Knowledge Representation and Reasoning: Proceedings of the Second International Conference*, pages 100–106. Morgan Kaufmann, 1990.

[BT95]   A. Brogi and F. Turini. Meta-logic for Program Composition: Semantics issues. This volume.

[CMG82] K.L Clark, F.G. McCabe, and S. Gregory. IC-Prolog language features. In K.L. Clark and S.A. Tarnlund, editors, *Logic Programming*, pages 253–266. Academic Press, 1982.

[Gur94a] C.A. Gurr. A guide to specialising Gödel programs with the partial evaluator sage. Technical report, University of Edinburgh, 1994.

[Gur94b] C.A. Gurr. *A self-applicable partial evaluator for the logic programming language Gödel.* PhD thesis, University of Bristol, 1994.

[Hil93] P.M. Hill. A parameterised module system for constructing typed logic programs. In R. Bajcsy, editor, *Proceedings IJCAI'93*, pages 874–880. Morgan Kaufmann, 1993.

[HL89]   P.M. Hill and J.W. Lloyd. Analysis of metaprograms. In H.D. Abramson and M.H. Rogers, editors, *Metaprogramming in Logic Programming*, pages 23–52. The MIT Press, 1989.

[HL94]   P.M. Hill and J.W. Lloyd. *The Gödel Programming Language.* The MIT Press, 1994.

[Kow74] R.A. Kowalski. Predicate logic as a programming language. In *IFIP 74*, pages 569–574, 1974.

[Kow90] R.A. Kowalski. Problems and Promises of Computational Logic. In J.W. Lloyd, editor, *Computational Logic, Symposium Proceedings*, pages 1–36. Springer-Verlag, 1990.

[LR93]   G. Levi and D. Ramundo. A Formalization of Metaprogramming for Real. In D.S. Warren, editor, *Proceedings Tenth International Conference on Logic Programming*, pages 354–373. The MIT Press, 1993.

[LS91]   J.W. Lloyd and J.C. Shepherdson. Partial evaluation in logic programming. *Journal of Logic and Computation*, 11:217–242, 1991.

[MD95]   B. Martens and D. De Schreye. Two semantics for definite meta-programs, using the non-ground representation. This volume.

[MP88]   P. Mancarella and D. Pedreschi. An algebra of logic programs. In R. A. Kowalski and K. A. Bowen, editors, *Proceedings Fifth International Conference on Logic Programming*, pages 1006–1023. The MIT Press, 1988.

[MP89]   L. Monteiro and A. Porto. Contextual logic programming. In G. Levi and M. Martelli, editors, *Proceedings Sixth International Conference on Logic Programming*, pages 284–302. The MIT Press, 1989.

[Nai87] L. Naish. Negation and quantifiers in NU-Prolog. In E. Shapiro, editor, *Proceedings Third International Conference on Logic Programming*, pages 624–634. Springer-Verlag, 1987.

[SS86]   L. Sterling and E. Shapiro. *The Art of Prolog.* The MIT Press, 1986.

8 Meta-Programming with Theory Systems

Jonas Barklund, Katrin Boberg, Pierangelo Dell'Acqua and Margus Veanes

Abstract
A theory system is a collection of interdependent theories, some if which stand in a meta/object relationship, forming an arbitrary number of meta-levels. The main thesis of this chapter is that theory systems constitute a suitable formalism for constructing advanced applications in reasoning and software engineering. The Alloy language for defining theory systems is introduced, its syntax is defined and a collection of inference rules is presented. A number of problems suitable for theory systems are discussed, with program examples given in Alloy. Some current implementation issues and future extensions are discussed.

8.1 Outline

A conventional logic program can be seen as the nonlogical axioms of a single theory. This chapter presents a thesis that we obtain a more powerful tool for applications in artificial intelligence and software engineering if we consider systems of theories, where pairs of theories may stand in an object/meta relationship, rather than single theories.

We proceed in 7 steps:

1. Arguing that multi-level programming should be a powerful tool for many advanced applications, in particular artificial intelligence and software engineering (Sect. 8.2).

2. Introducing theory systems as an approach to multi-level programming (Sect. 8.3).

3. Defining the formal syntax and one possible inference system of a language, *Alloy*, in which theory systems can be programmed (Sect. 8.4).

4. Defining the models of Alloy programs (Sect. 8.5).

5. Presenting examples of problem solving using theory systems expressed in Alloy (Sect. 8.6).

6. Discussing self-reference and how to program it in Alloy (Sect. 8.7).

7. Proposing some future extensions, supporting technologies and some current implementation issues (Sects. 8.8–8.9).

We end with some notes and conclusions.

For a general introduction to meta-programming in logic programming, the reader is referred to the overviews by Barklund [Bar94] and Hill & Gallagher [HG94].

8.2 Artificial Intelligence and Software Engineering

The studies of artificial intelligence in general and expert systems in particular make it clear that truly useful problem solvers must be constructed in a quite different way than has been tried in the past. Among the problems with current approaches are:

1. Lack of robustness with respect to domains.
2. Low adaptability of problem solving methods.
3. Failure to capture "common sense" reasoning.

These problems are indeed very difficult but we believe that the marginal success so far is largely because the attempts at addressing them have been carried out mostly using single-level architectures (cf. Sterling [Ste84]). By single-level architectures we mean systems without provisions for reasoning about any part of their own beliefs or procedures and for adapting themselves according to these observations. The three problems mentioned above could be approached as follows:

1. Given a program that solves problems in some domain, the system might transform this program to adapt it to another domain. Also, given a program that represents a piece of knowledge, together with some suitably represented new knowledge, the system might create a new program that incorporates both the knowledge present in the old program and the new knowledge, after resolving any discrepancies between them.

2. Given a subprogram that carries out a particular form of reasoning, the system might transform it to a similar program that carries out a somewhat different form of reasoning, better adapted to some circumstances.

3. This is the most difficult problem of these three. McCarthy defined a program having common sense as one that "automatically deduces for itself a sufficiently wide class of immediate consequences of anything it is told and what it already knows" [McC68]. The heuristics for exploring the interesting consequences of new information or finding the information necessary for solving a problem are naturally expressed as meta-knowledge. These heuristics might need to be revised over time, as they turn out to be more or less successful. This can be seen as a metameta-level problem, indicating that one should not be restricted to only two levels.

All three of these potential solutions involve writing programs that are capable (i) of observing parts of other programs, (ii) of examining those programs' conclusions and perhaps also the reasoning behind these conclusions, and (iii) of creating new programs, presumably starting from existing programs.

The reader should note that the preceding sentence could just as well have been a statement about advanced software engineering; the same basic operations seem to be useful in both application areas. Our thesis is that a useful methodology for building correct software is one where a program is constructed "implicitly" by writing a meta-program that takes a number of "standard programs", transforming and combining them to produce a program that performs the desired task. The "standard programs" would be of various kinds, some of them simple program pieces that perform various kinds of recursion, for example, but some of them might be sophisticated and complex programs that carry out a computation for the same domain as the program to be produced.

The meta-programs may in some cases be very simple, merely composing and transforming the given programs in certain ways. However, if the produced programs must satisfy particular criteria, for example, real-time constraints, then the meta-programs may have to do a much more detailed analysis or perhaps even run the generated programs as a step in their construction.

The advantage with the outlined approach is that if the standard programs are completely understood, the produced programs will be as well. Moreover, all future modifications to the produced programs are done by changing the program that generated them, which is likely to lead to fewer mistakes than manual work. This programming paradigm could truly be called "high-level programming".

Although the main body of work on artificial intelligence, reasoning and expert systems has been spent on single-level formalisms, we are certainly not alone in observing that a multilevel formalism should provide a better tool for attacking the fundamental problems. We mention some related formalisms at the end of this chapter; further references can be found in the remainder of this book.

8.3 Logic Programming with Multiple Theories

Formally, a *theory* is a set of sentences in some language, including the logical axioms of the language, that is closed under the inference rules of the language. Once the language is fixed, any set of sentences in that language defines a theory, obtained by adding the logical axioms of the language and closing it under inference. In logic programming, the language might be that of definite clauses with SLD-resolution and the logical axioms those concerning (Herbrand) equality. A program is then a set of definite clauses defining a single theory.

In applications that involve reasoning it is often appropriate to compute with more than one theory. For example, we could write a program that simulates the reasoning of a collection of agents, representing the beliefs of each agent as a theory (if we employ the "sentential" view of beliefs, perhaps first used explicitly by McCarthy [McC79]). If the language prevents us from having more than one theory in our program, then these "internal" theories have to be represented in some other way, perhaps as data structures with the programmer writing an ad hoc interpreter to simulate inference. There is a large class of applications in "reasoning" and software engineering, perhaps also in other areas, that are naturally written using multiple theories; therefore multiple theories ought to be supported directly in the language.

Theory systems constitute a useful formalism for writing these kinds of programs, because the theories in a theory system are suitable for representing reasoning agents or parts of them, programs to be manipulated, programs that manipulate them, etc. The meta/object relationship between theories provides the inspection and control facilities needed in both kinds of applications.

8.3.1 Theory systems

We propose now a simple structure for theory systems that appears to be adequate for our purposes. A *theory system* is a mapping from (ground) theory terms to theories. Any theory τ contains theorems about theories named $\tau \diamond \cdots$ ('$\diamond$' is a distinguished function symbol that we write using infix notation). In fact, the restriction of a theory system to theory terms of the form $\tau \diamond \cdots$, for some τ, is a theory system in itself. Such a theory system can be thought of being defined by τ.

It is convenient to say that a theory t_1 is a *meta-theory* of any theory identified as $t_1 \diamond t_2 \diamond \cdots \diamond t_k$, where $k > 1$. Conversely we say that those theories are *object theories* with respect to t_1.

We use the symbol '⊢' for relating theory terms and sentences. A *theoremhood statement* $t_1 \vdash \ulcorner u_1 \vdash \Psi \urcorner$ says that $\ulcorner u_1 \vdash \Psi \urcorner$ is a theorem of t_1. We mentioned that theories may contain theorems about other theories, and $\ulcorner u_1 \vdash \Psi \urcorner$ in t_1 expresses that Ψ is a theorem in the theory $t_1 \diamond u_1$ (cf. Fig. 8.1). Note that a subset of the theorems of t_1, namely those on the form $\ulcorner u_1 \vdash \cdots \urcorner$ (the left shaded area in the figure), have a one-to-one correspondence with the theorems of $t_1 \diamond u_1$, and similarly for another subset of t_1 (the right shaded area) and $t_1 \diamond u_j$.

The other kind of statement that we use for defining theory systems is called a *coincidence statement*. If the program defining the theory system in Fig. 8.1 contains a coincidence statement $t_1 \diamond u_j \equiv t_i$, then the theories $t_1 \diamond u_j$ and t_i have exactly the same theorems. (The relation denoted by '$\equiv$' is an equivalence relation, i.e., it is reflexive, symmetric and transitive.) More importantly, that statement ensures a one-to-one correspondence between a subset of t_1 (the right shaded area) and t_i. In absence of such a coincidence statement, there is no connection whatsoever between theories, unless one is a meta-theory of the other. In particular, proving $\ulcorner t_i \vdash \Phi \urcorner$ in t_1 in order to determine

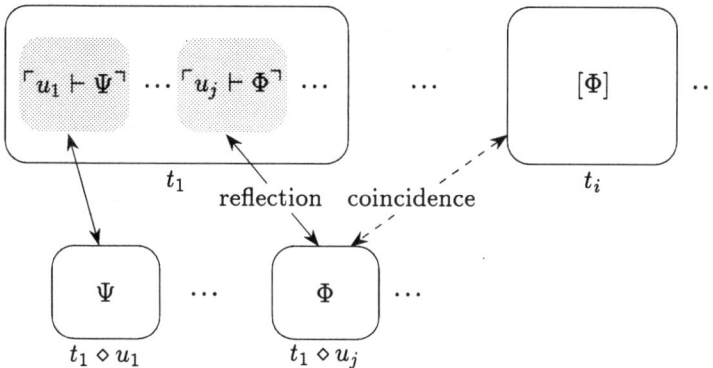

Figure 8.1
A generic theory system.

in t_1 whether Φ is a theorem of t_i requires that $t_1 \diamond t_i \equiv t_i$.

8.3.2 Representation

We will assume that all theories use the same definite clause language but that the set of terms of this language is rich enough that for any variable, function or predicate symbol σ, there is some unique constant σ' which *represents*, or *names*, σ. Similarly, for each well-formed expression α, there must be some unique ground term α' that represents α.

Our final requirement on the definite clause language is that for any theoremhood statement and coincidence statement there is some unique ground atom representing it.

We can now define precisely the relationship between a meta-theory and an object theory. Consider a theory system and a pair of theories identified by some theory terms τ_1 and $\tau_1 \diamond \tau_2$; the first is thus a meta-theory of the second. Our *theoremhood reflection principle* states that

$$\tau_1 \vdash \ulcorner \tau_2 \vdash \kappa \urcorner \Leftrightarrow \tau_1 \diamond \tau_2 \vdash \kappa$$

and can be seen as a correctness statement for interaction between a meta-theory and an object theory.

Our *coincidence reflection principle* states that

$$\tau_1 \vdash \ulcorner \tau_2 \equiv \tau_3 \urcorner \Leftrightarrow \tau_1 \diamond \tau_2 \equiv \tau_1 \diamond \tau_3$$

and can be seen as a correctness statement for coincidence of internal theories.

Both these principles are valid for every theory system.

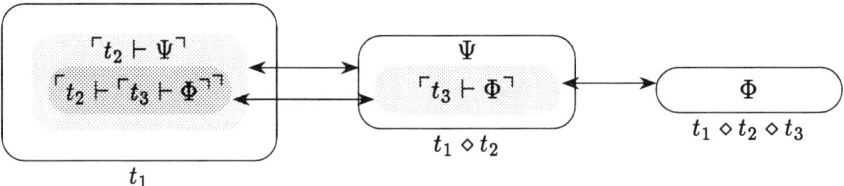

Figure 8.2
Three meta-levels of theories.

The traditional *local reflection principle* for a single theory T in mathematical logic [Smo77] reads

$$Pr_T(\ulcorner\phi\urcorner) \Rightarrow \phi$$

and states the correspondence between a provability statement and what is to be proved, namely that if the provability predicate holds for an encoding of a formula ϕ, then ϕ holds as well. We call our statements reflection principles by analogy, as they state correspondences between names of statements and what these statements are about.

These implications and equivalences should not be confused with the inference rules sometimes referred to as "reflection principles" but for which a better name is "reflection rules" or "linking rules" (cf. the discussions by Giunchiglia, Serafini & Simpson [GSS92] and Costantini, Dell'Acqua & Lanzarone [CDL94]). However, in Sect. 8.4.3 we will present two reflection rules corresponding to the two implications of the theoremhood reflection principle.

Fig. 8.2 depicts part of a theory system in which a theory t contains a theorem $\ulcorner u \vdash \ulcorner v \vdash \Phi \urcorner \urcorner$. The reflection principle requires that the theory $t \diamond u$ contains the theorem $\ulcorner v \vdash \Phi \urcorner$ and thus that $t \diamond u \diamond v$ contains Φ. The figure thus illustrates that theory systems may be arbitrarily deep and that the theoremhood reflection principle applies at any level.

Finally, the following is an example of a program defining a simple theory system.

$$Tim \vdash \ulcorner \dot{x} \vdash Tasty(\dot{x}) \urcorner \leftarrow Cannibal(y) \wedge Names(x,y) \quad (1)$$

$$Tim \vdash Cannibal(Tom) \quad (2)$$

$$Tim \diamond Tom \equiv Tom \quad (3)$$

The theoremhood statements 1 and 2 specify two axioms of the theory *Tim*. According to the theoremhood reflection principle, statement 1 also says something about theories named $Tim \diamond \cdots$. One such theory is $Tim \diamond Tom$, which coincides with *Tom*, according to statement 3. When the theories are thought of as representing the beliefs of agents, we can read the statements as saying "Tim believes that all cannibals find themselves tasty",

"Tim believes that Tom is a cannibal" and "Tim's view of Tom's beliefs is correct", respectively. From this reading we can deduce that Tom finds himself tasty, and in Sect. 8.4.3 we will show how to derive this conclusion using an inference system.

8.4 Syntax

We will now define the syntax, inference rules and informal semantics of Alloy, a language for computing with theory systems. What we define in this section can be seen as the "core" syntax of Alloy: the language of definite clauses extended with name terms, name atoms, theoremhood statements and coincidence statements.

The language, at this stage, does not contain negation, except that denials are introduced as part of proving goals (as usual in SLD-resolution). In Sect. 8.6.2 we use negation in some examples, which are therefore not meaningful until Alloy is extended with negation.

8.4.1 Formal syntax

The Alloy language has two components: the system component for defining theory systems and the theory component for defining individual theories.

Alphabet. Besides punctuation symbols, the part of the alphabet that is common to both components of the language consists of a class of *variables* and, for each $n \geq 0$, a class of *function symbols* and a class of *predicate symbols* of arity n. Collectively, function and predicate symbols are referred to as *functors*. As usual, function and predicate symbols of arity 0 are referred to as *constants* and *propositional constants*, respectively. The class of predicate symbols include the binary symbols '=' and '*Name*', two binary symbols that will be denoted by '⊢'' and '≡'', and the propositional constants '*True*' and '*False*'. The function symbols include the binary symbol '⋄'.

The alphabet has also a collection of *connectives* '←', '∧' and '?', and *naming symbols* '⌈', '⌉', '·' ("dot") and '|'. A dot is used in combination with variables only. If x is a variable, then $\dot{x}$ is called a variable *with a dot*, $\ddot{x}$ a variable *with two dots*, etc., in general a variable with one or more dots is called a *dotted* variable.

In addition, the system component of the language has the binary operators '⊢' and '≡'.

We will use letters P, Q and R to stand for predicate symbols, F and G to stand for function symbols, x, y and z to stand for variables.[1]

---

[1] The letters may also be subscripted.

Theory component. In the following we define the expressions of the theory component. We will do so by a simultaneous inductive definition of terms, atoms, queries and sentences as separate subclasses and refer to them collectively as *theory expressions.*

In the definition of terms and atoms we will make use of the notion of an expression being a schema of another. Intuitively, E is a schema of an expression e whenever some (or none) subexpression occurrences of e have been replaced by dotted variables ("holes") in E. In general, we say that E is a k-*level schema* of e if one of the following conditions holds.

1. $e = E$,
2. e is not a connective and E is a variable with k dots,
3. $e = e_0(e_1, \ldots, e_n)$ and
    - $E = E_0(E_1, \ldots, E_n)$, or
    - $E = E_0(E_1, \ldots, E_j | X)$ for some j, $0 \leq j < n$,

    where each E_i is a k-level schema of e_i, $0 \leq i \leq n$, and X a variable with k dots.
4. $e = \ulcorner d \urcorner$ and $E = \ulcorner D \urcorner$, where D is a $(k+1)$-level schema of d.

By simply *schema* we mean a 1-level schema. It is an immediate consequence of the definition that if E is a schema of e and a subexpression of e has been replaced by a variable with k dots in E, then this dotted variable occurs nested within $k-1$ pairs of '$\ulcorner$' and '$\urcorner$'. For example, if F is a binary functor, then $\dot{x}(y, \ulcorner G(\ddot{z}) \urcorner)$ is a schema of $F(y, \ulcorner G(\dot{z}) \urcorner)$.

Terms. The class of *terms* is the least class satisfying the following conditions.

1. Each variable and constant is a term.
2. If F is a function symbol of arity n and $t_1, t_2, \ldots, t_n$ are terms, then $F(t_1, t_2, \ldots, t_n)$ is a term.
3. If X is a schema of a functor or a theory expression, then $\ulcorner X \urcorner$ is a term, called a *name term.*

Letters t and u will be used for terms.

Atoms. The class of *atoms* is the least class satisfying the following conditions.

1. If P is a predicate symbol of arity n and $t_1, \ldots, t_n$ are terms, then $P(t_1, \ldots, t_n)$ is an atom, called a *predication.*
2. If T is a schema of a term t and S a schema of a sentence s, then $\vdash'(\ulcorner T \urcorner, \ulcorner S \urcorner)$ is an atom, called a *name atom.*

3. If T is a schema of a term t and U a schema of a term u, then $\equiv'(\ulcorner T \urcorner, \ulcorner U \urcorner)$ is an atom, called a *name atom*.

Letters A and B will be used for atoms. We will use the shorthand $\ulcorner T \vdash S \urcorner$ for $\vdash'(\ulcorner T \urcorner, \ulcorner S \urcorner)$ and $\ulcorner T \equiv U \urcorner$ for $\equiv'(\ulcorner T \urcorner, \ulcorner U \urcorner)$.[2]

It follows easily from the definitions of terms and atoms that a variable with n dots is embedded within k, $k \geq n$, nested levels of naming. For a variable with n dots, the lowest dot makes a "hole" in the innermost pair of '$\ulcorner$' and '$\urcorner$', the next dot in the next pair and so on. If $k = n$, then the variable is called free in the corresponding term or atom. For example z is the only free variable of the name term

$$\ulcorner \ulcorner F(x, \dot{y}, \ddot{z}) \urcorner \urcorner.$$

A name term or name atom is said to be *proper* if it is ground. Consider for example

$$\ulcorner F(\dot{x}_1) \vdash \dot{x}_2(\dot{x}_1, y, A) \leftarrow \dot{z} \urcorner.$$

This name atom is not proper because it contains four free variable occurrences of three different variables. Only for proper names can we tell which expression they name.

Queries. The class of *queries* is the least class satisfying the following conditions.

1. An atom is a query; *True* is called the *empty query*.
2. If C and D are queries, then $C \wedge D$ is a query.

Letters C and D will be used for queries.

Sentences. The class of *sentences* is the least class satisfying the following conditions.

1. If A is an atom and C is a query, then $A \leftarrow C$ is a sentence, called a *program clause*.
2. If C is a query, then $C?$ is a sentence, called a *goal*.
3. If C is a query, then $\leftarrow C$ (shorthand for *False* $\leftarrow C$) is a sentence, called a *denial*.

The variables of a program clause and a denial are universally quantified. A goal, on the other hand, is the negation of a denial and thus existentially quantified:

$$\neg(\leftarrow C) \Leftrightarrow \neg(\forall(\mathit{False} \leftarrow C)) \Leftrightarrow \neg(\mathit{False} \leftarrow \exists C) \Leftrightarrow \exists C \Leftrightarrow C?.$$

---

[2] We could let $\vdash'$ be *Demo*, in which case $\ulcorner T \vdash S \urcorner$ would be shorthand for the familiar $\mathit{Demo}(\ulcorner T \urcorner, \ulcorner S \urcorner)$.

System component. The language of the system component has two kinds of expressions: theoremhood statements and coincidence statements.

- If t is a term (called a *theory term* in this context) and s is a sentence, then $t \vdash s$ is a *theoremhood statement*.
- If t_1 and t_2 are (theory) terms, then $t_1 \equiv t_2$ is a *coincidence statement*.

Collectively they are referred to as *system expressions*.

8.4.2 Normalized language

In order to be able to handle terms conventionally, we want each term to have a *normal form*, where the naming symbols '⌜', '⌝', '·' and '|' have been eliminated. We call the elimination process *normalization* and the result a *normalized term*. In this context $\ulcorner - \urcorner$ is a function mapping expressions to expressions; $\ulcorner - \urcorner$ is required to be *compositional* in order to enhance the expressive power of the language. This means that if e is a compound expression $e_0(e_1, \ldots, e_n)$, then $\ulcorner e \urcorner$ can be expressed as a composition of all $\ulcorner e_i \urcorner$, $0 \leq i \leq n$. In addition, $\ulcorner \dot{v} \urcorner = v$.

Clearly there exist several different normalizations. Probably the most general approach is to have a binary function symbol 'o', denoting a composition function that produces the name of a compound expression from the name of a functor or a connective and a list of names of expressions. Using this approach, the notion of lists is needed; this can be accomplished by using a binary function symbol '•' and a constant 'Λ' to represent the empty list. (We will use the less cumbersome notation $[e_1, e_2, \ldots, e_n | x]$ for •(e_1, •(e_2, ··· •(e_n, x) ···)).)

The alphabet is assumed to have a unique name e' for each symbol e, in such a way that the mapping $e \mapsto e'$ is injective.[3] If these names are all terms, then the normalization can be described by the following transformations.

$$\ulcorner \dot{e} \urcorner \longrightarrow e$$
$$\ulcorner e \urcorner \longrightarrow e' \quad \text{if } e \text{ is a symbol}$$
$$\ulcorner e_0(e_1, \ldots, e_n | \dot{x}) \urcorner \longrightarrow \mathrm{o}(\ulcorner e_0 \urcorner, [\ulcorner e_1 \urcorner, \ldots, \ulcorner e_n \urcorner | x])$$
$$\ulcorner e_0(e_1, \ldots, e_n) \urcorner \longrightarrow \mathrm{o}(\ulcorner e_0 \urcorner, [\ulcorner e_1 \urcorner, \ldots, \ulcorner e_n \urcorner])$$

We can however take advantage of the restriction that we imposed on the definition of schemas, namely disallowing holes for connectives, and make the following modifications to the above transformations. For each connective c, the alphabet has a corresponding function symbol c' of the same arity as c. If e_0 in the last case above is for example '∧',

---
[3] If e' itself is a symbol it has a name e'', etc.

then
$$\ulcorner e_1 \wedge e_2 \urcorner \longrightarrow \wedge'(\ulcorner e_1 \urcorner, \ulcorner e_2 \urcorner).$$

We get similar transformations for the other connectives. For example, using this normalization, the normal form of $\ulcorner\ulcorner F(x,\dot{y},\ddot{z})\urcorner\urcorner$ is obtained as follows.

$$\begin{aligned}
\ulcorner\ulcorner F(x,\dot{y},\ddot{z})\urcorner\urcorner &\xrightarrow{*} \ulcorner \circ(F', [x', y, \dot{z}])\urcorner \\
&\xrightarrow{*} \circ(\circ', [F'', \ulcorner[x', y, \dot{z}]\urcorner]) \\
&\xrightarrow{*} \circ(\circ', [F'', \circ(\bullet', [x'', \ulcorner[y, \dot{z}]\urcorner])]) \\
&\xrightarrow{*} \circ(\circ', [F'', \circ(\bullet', [x'', \circ(\bullet', [y', \ulcorner[\dot{z}]\urcorner])])]) \\
&\xrightarrow{*} \circ(\circ', [F'', \circ(\bullet', [x'', \circ(\bullet', [y', \circ(\bullet', [z, \Lambda'])])])])
\end{aligned}$$

Unnecessary naming of 'o', '•' and 'Λ' can be avoided by defining the transformation so that 'o', '•' and 'Λ' become "transparent" with respect to naming, i.e., $\ulcorner \Lambda \urcorner \longrightarrow \Lambda$, $\ulcorner \circ(e_1, e_2)\urcorner \longrightarrow \circ(\ulcorner e_1\urcorner, \ulcorner e_2\urcorner)$ and $\ulcorner \bullet(e_1, e_2)\urcorner \longrightarrow \bullet(\ulcorner e_1\urcorner, \ulcorner e_2\urcorner)$. Note that this does not violate the injectivity of the naming function. Assuming this modification then, for example,

$$\ulcorner\ulcorner F(x,\dot{y},\ddot{z})\urcorner\urcorner \xrightarrow{*} \circ(F'', [x'', y', z]).$$

8.4.3 Inference system

Equality, naming and unification. Before normalizing the language of a program, we extend each of its theories with every axiom on the form $Names(\ulcorner t\urcorner, t)$, where t is a term.

After normalizing, as described in Sect. 8.4.2, the usual Herbrand equality theory, as axiomatized by Clark [Cla78], can be used. (However, computation of the naming relation ought to be integrated with unification in order to delay computation of names of non-ground terms.)

As all correct normalizations will behave in the same way, with respect to equality of the normalized expressions, it would alternatively be conceivable to extend Herbrand equality to name expressions without normalization.

Inference rules. The inference system that we are to explain here is by no means the only possible inference system for Alloy, in fact, it is not even complete. We choose this inference system for presentation because it is simple and because it is complete for propositional programs. For an actual implementation we are presently developing a more goal-oriented inference system, outlined in Sect. 8.9.

The main purpose of the inference system is to be able to prove statements of the form $\tau \vdash C?$, i.e., that a goal $C?$ is a theorem of some theory τ. This can either be accomplished by a refutation, i.e., by assuming $\tau \vdash \leftarrow C$ and proving $\tau \vdash \leftarrow$ (inconsistency in τ), or by a proof that may include refutations as subproofs. A successful refutation of a denial $\tau \vdash \leftarrow C$ is always ended by cancelling $\tau \vdash \leftarrow C$ and concluding $\tau \vdash C?$ (through the application of the RR rule described below).

We shall present seven inference rules. The first rule is ordinary SLD-resolution within a theory. Let $\triangle C$ denote the atom selected from a query C and $\triangledown C$ the rest of the query. It is assumed that the predicate symbol of the selected atom is not *Names*.

$$\text{RS} \frac{\tau \vdash \leftarrow C \quad \tau \vdash A \leftarrow D}{\tau \vdash (\leftarrow \triangledown C \wedge D)\theta} \quad \theta = mgu(A, \triangle C)$$

The second rule is a "Relativized RAA" rule, allowing us to make subproofs that are refutations.

$$\frac{\begin{array}{c}[\tau \vdash \leftarrow C]^i \\ \vdots \\ \tau \vdash \leftarrow\end{array}}{\tau \vdash C?} \text{ RR cancel } i$$

The third and fourth rules are the reflection rules, justifiable from the theoremhood reflection principle (Sect. 8.3.2). They make use of the meta/object relationship between a pair of theories in both directions: If a theory τ_M reasons that its internal theory τ_O contains some sentence κ, then $\tau_M \diamond \tau_O$ indeed contains κ, and vice versa.

$$\text{TD} \frac{\tau_M \vdash \ulcorner \tau_O \vdash \kappa \urcorner}{\tau_M \diamond \tau_O \vdash \kappa} \qquad \frac{\tau_M \diamond \tau_O \vdash \kappa}{\tau_M \vdash \ulcorner \tau_O \vdash \kappa \urcorner} \text{TU}$$

The fifth and sixth rules are similar to the third and fourth, but are instead justifiable from the coincidence reflection principle. They express that if a theory has as a theorem stating that two of its internal theories coincide, then we may infer that these theories do coincide, and vice versa.

$$\text{CD} \frac{\tau \vdash \ulcorner \tau_1 \equiv \tau_2 \urcorner}{\tau \diamond \tau_1 \equiv \tau \diamond \tau_2} \qquad \frac{\tau \diamond \tau_1 \equiv \tau \diamond \tau_2}{\tau \vdash \ulcorner \tau_1 \equiv \tau_2 \urcorner} \text{CU}$$

The seventh rule uses a coincidence between two theories to transfer a theorem of one of them to the other.

$$\frac{\tau_1 \equiv \tau_2 \quad \tau_1 \vdash \kappa}{\tau_2 \vdash \kappa} \text{CE}$$

From these inference rules one could derive others, for example, an indirect SLD-resolution inference.

$$\frac{\tau_M \vdash \ulcorner \tau_O \vdash \leftarrow C \urcorner \quad \tau_M \vdash \ulcorner \tau_O \vdash A \leftarrow D \urcorner}{\tau_M \vdash \ulcorner \tau_O \vdash (\leftarrow \nabla C \wedge D)\theta \urcorner} \quad \theta = mgu(A, \triangle C)$$

This derived inference rule can be justified:

$$\text{TD} \frac{\tau_M \vdash \ulcorner \tau_O \vdash \leftarrow C \urcorner}{\tau_M \diamond \tau_O \vdash \leftarrow C} \quad \frac{\tau_M \vdash \ulcorner \tau_O \vdash A \leftarrow D \urcorner}{\tau_M \diamond \tau_O \vdash A \leftarrow D} \text{TD}$$
$$\frac{\tau_M \diamond \tau_O \vdash (\leftarrow \nabla C \wedge D)\theta}{\tau_M \vdash \ulcorner \tau_O \vdash (\leftarrow \nabla C \wedge D)\theta \urcorner} \text{TU} \quad \text{RS}$$

Another useful derived rule is for indirect reasoning with coinciding theories,

$$\frac{\tau \vdash \ulcorner \tau_1 \equiv \tau_2 \urcorner \quad \tau \vdash \ulcorner \tau_1 \vdash \kappa \urcorner}{\tau \vdash \ulcorner \tau_2 \vdash \kappa \urcorner},$$

justified as follows.

$$\text{CD} \frac{\tau \vdash \ulcorner \tau_1 \equiv \tau_2 \urcorner}{\tau \diamond \tau_1 \equiv \tau \diamond \tau_2} \quad \frac{\tau \vdash \ulcorner \tau_1 \vdash \kappa \urcorner}{\tau \diamond \tau_1 \vdash \kappa} \text{TD}$$
$$\frac{\tau \diamond \tau_2 \vdash \kappa}{\tau \vdash \ulcorner \tau_2 \vdash \kappa \urcorner} \text{TU} \quad \text{CE}$$

As an example, consider again the cannibal example of Sect. 8.3.1. Here is how to prove the statement $Tom \vdash Tasty(Tom)?$.

$$\begin{array}{r}
(1) \quad [Tim \vdash \leftarrow \ulcorner Tom \vdash Tasty(Tom) \urcorner]^1 \\
\hline
(2) \quad Tim \vdash \leftarrow Cannibal(y) \wedge Names(\ulcorner Tom \urcorner, y) \\
\hline
Tim \vdash \leftarrow Names(\ulcorner Tom \urcorner, Tom) \\
\hline
Tim \vdash \leftarrow True \\
\hline
Tim \vdash \ulcorner Tom \vdash Tasty(Tom) \urcorner? \\
\hline
(3) \quad Tim \diamond Tom \vdash Tasty(Tom)? \\
\hline
Tom \vdash Tasty(Tom)?
\end{array}$$

(with RS, RS, NM, RR cancel 1, TD, CE)

We mentioned above that this inference system is incomplete. What must be done in order to increase the number of provable statements is taking care of proofs that involve improper names. See Sect. 8.9 for a further discussion of how this can be done.

8.5 Semantics

Let $\mathcal{I}$ be the set of theory terms and $\mathcal{M} = \{\mathfrak{M}_\tau\}_{\tau \in \mathcal{I}}$ a family of (arbitrary first order[4]) structures for the language of theory expressions under a given normalization. The elements of $\mathcal{M}$ are called *theory structures*. A *system structure* is a pair $\langle \mathcal{M}, \equiv \rangle$, where $\equiv$ is an elementary equivalence relation on $\mathcal{M}$. (Two first order structures are said to be elementarily equivalent whenever they have the same set of logical consequences.)

Let P be an Alloy program, i.e., a set of system expressions, and let $\langle \mathcal{M}, \equiv \rangle$ be a system structure. If the theory structures are Herbrand interpretations, we can assume without loss of generality that P is ground; P could then be a Herbrand instantiation (possibly infinite) of an underlying non-ground program. We say that $\langle \mathcal{M}, \equiv \rangle$ is a *model* of P if the following hold:

$$\tau \vdash \varphi \in P \Rightarrow \mathfrak{M}_\tau \models \varphi; \qquad (4)$$

$$\tau_1 \equiv \tau_2 \in P \Rightarrow \mathfrak{M}_{\tau_1} \equiv \mathfrak{M}_{\tau_2}; \qquad (5)$$

$$\mathfrak{M}_{\tau_1} \models \vdash'(\ulcorner \tau_2 \urcorner, \ulcorner \varphi \urcorner) \Leftrightarrow \mathfrak{M}_{\tau_1 \diamond \tau_2} \models \varphi; \qquad (6)$$

$$\mathfrak{M}_{\tau_1} \models \equiv'(\ulcorner \tau_2 \urcorner, \ulcorner \tau_3 \urcorner) \Leftrightarrow \mathfrak{M}_{\tau_1 \diamond \tau_2} \equiv \mathfrak{M}_{\tau_1 \diamond \tau_3}; \qquad (7)$$

$$\mathfrak{M}_{\tau_1 \diamond (\tau_2 \diamond \tau_3)} \equiv \mathfrak{M}_{(\tau_1 \diamond \tau_2) \diamond \tau_3} \qquad (8)$$

$$t \text{ is a ground term} \Rightarrow \mathfrak{M}_\tau \models Names(\ulcorner t \urcorner, t). \qquad (9)$$

Conceptually, the set of theoremhood statements of P is partitioned by the theory terms. Each part is identified by a theory term, the denotation of which is a model for that part (4). A coincidence statement between any two theory terms enforces the structures they denote to be elementarily equivalent (5). The theoremhood and coincidence reflection principles must be satisfied by the theory structures (6, 7). Furthermore, $\diamond$ must be associative with respect to elementary equivalence between the denoted structures (8). From formulas 6 and 8 we can easily deduce that

$$\mathfrak{M}_{\tau_1} \models \ulcorner \tau_2 \diamond \tau_3 \vdash \varphi \urcorner \Leftrightarrow \mathfrak{M}_{\tau_1} \models \ulcorner \tau_2 \vdash \ulcorner \tau_3 \vdash \varphi \urcorner \urcorner.$$

Finally, the *Names* predicate symbol must denote a naming relation (restricted to terms), i.e., one that relates any ground term with its name (9). It is also clear that the set of logical consequences of any theory structure is closed under SLD-resolution, as the set of theorems of any first order structure is complete.

Considering the special case when P is just a Horn clause program, i.e., when all the sentences of P are of the the form $\tau \vdash \varphi$ where φ is a Horn clause and τ is the only theory

---

[4] Here we need not restrict ourselves to Herbrand interpretations only.

term, then the notion of system structure collapses to that of a first order structure. In that case conditions 5–7 are trivially satisfied. The only extra requirement, not part of a standard definition of a model of P, would be (9).

In our approach we have not altered the notion of logical consequence, as was done for example by Jiang [Jia94], in order to handle meta-reasoning. Instead we introduce the notion of system structure, following closely the informal semantics, giving us a notion of semantics which is a modest extension of a first order semantics in the sense that the basic building blocks, theory structures, are still first order structures. A more thorough investigation of the semantics of Alloy will be the subject of a future publication.

8.6 Applications Using Theory Systems

In this section we shall present a number of useful applications of meta-programming with theory systems, some of them commonly known, some of them new. We shall show how fragments of these applications can be programmed elegantly in Alloy. Our ambition is twofold. Firstly, we wish to convince the reader of the strength and versatility of meta-programming with theory systems, continuing and extending the work by Bowen & Kowalski [BK82], Sterling [Ste84], Bowen [Bow85], Brogi & Turini [BT91] and others. Secondly, we hope to illustrate programming in Alloy and how many problems can be programmed in a much more straightforward and concise way than in single-level programming or single-theory meta-programming.

8.6.1 Reasoning agents

Many forms of reasoning for artificial agents have been proposed, such as abductive reasoning, inductive reasoning, non-monotonic reasoning, case based reasoning, temporal reasoning and so on. A favourite approach of many philosophers and other researchers in artificial intelligence is to invent a new specialized logic for each one of these forms of reasoning. There are many problems with this approach. One is that it is not clear at all that these logics can be combined to build artificial agents capable of more than one form of reasoning. Another is that there are often no efficient implementation techniques known for these new logics.

A more sensible method is to employ a single logic, with known properties, which can be implemented; such as some subset of classical logic. However, many of the forms of reasoning mentioned above cannot be mapped straightforwardly to classical logic. (This has even been used as an argument against using logic at all for reasoning agents.)

Fortunately, there is a partial solution. If we go from using single-level logic languages to meta-logic languages for theory systems, we obtain a modest extension of classical

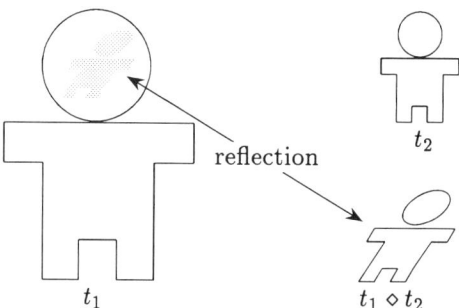

Figure 8.3
An agent t_1, which has a (distorted) view of the beliefs of another agent t_2, constituting a theory $t_1 \diamond t_2$.

logic in terms of semantics but we get a substantial extension in terms of reasoning capabilities, because we can express various forms of reasoning in the logic itself. This approach becomes even more sensible when one recognizes that many forms of reasoning actually contain a substantial element of meta-level reasoning. For example, default reasoning involves observing that some question cannot be decided and making a hypothesis (although it is not always recognized as such) about the answer.

In Alloy we can represent an agent's beliefs by a theory, which internally defines a system of theories. Some of these theories might represent (correctly or incorrectly) the agent's view of other agents' beliefs, ambitions and motives; cf. Fig. 8.3. Other theories might represent the agent's beliefs about the surroundings and about various domains. Presumably, there are also theories that encode various problem solving strategies and tactics.

This approach has several advantages.

- Modularity. An agent's mind is internally structured.

- Multiple levels. It is possible to represent beliefs and procedures at various meta-levels, e.g., theories synthesizing problem solving procedures to be used in specific domains represented by "lower" theories.

- No parapsychology. As the theory representing the beliefs of an agent is clearly separated from the theory representing another agent's beliefs about the first agent's beliefs, our formalism does not create "mind-reading" and confusion (unless explicitly programmed).

- Generality. Various properties of knowledge and beliefs (see the following section) can be programmed into the system but they are not automatically present.

As an example of programming multiple agents that reason about each other, consider the traffic problem illustrated in Fig. 8.4.

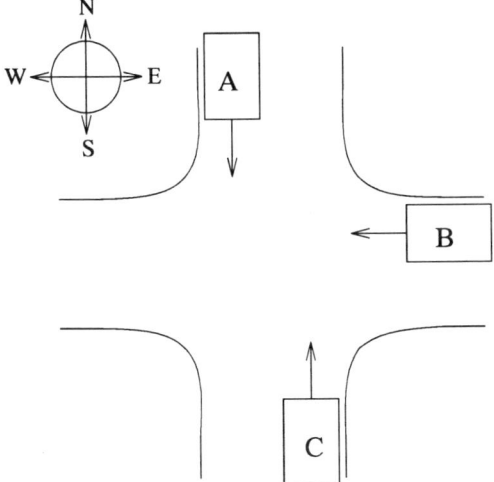

Figure 8.4
"Can the driver in car C, coming from south, pass the crossing?"

Three cars are simultaneously approaching a four-way crossing. There are no other signs or traffic lights, so the rule is that drivers should give way to cars coming on their right side. Using a simple application of this rule, we obtain that car A can pass, while cars B and C must wait, because they give way to some car on their right side. However, the driver of car C could instead reason that car B must wait, because the driver of car B will see car A on her right entering the crossing and give way to it. Hence, the driver of car C might conclude that he can safely pass.

Our purpose here is not to argue whether it would be legal or not for the driver of car C to pass, based on the argument above[5], but to show that such multiagent reasoning can be programmed straightforwardly in Alloy.

The following statement encodes the problem of the driver of car C.

$$\textit{Traffic} \vdash \ulcorner D(C, South) \vdash \textit{Pass}([D(A, North), D(B, East), D(C, South)])? \urcorner \qquad (10)$$

The theory *Traffic* is where our reasoning about the drivers will take place. Each theory *Traffic* $\diamond\, D(x, y)$ represents (our view of) the beliefs of the driver of car x, coming from direction y. Each theory *Traffic* $\diamond\, D(x_1, y_1) \diamond D(x_2, y_2)$ represents (our view of) the beliefs that the driver of car x_1, coming from direction y_1 has about the beliefs of the driver of car x_2, coming from direction y_2 has, etc.

---

[5]It is easy to observe that many real drivers seem to reason exactly this way.

A theorem $Pass(z)$ in a theory $\cdots \diamond D(x, y)$ would mean that the driver in question would believe that she can pass a crossing in which she sees the cars listed in z. Similarly, a theorem $Wait(x)$ would mean that she would believe she has to stop.

The first two clauses are interesting, because they help us to encode a form of *group belief*.

$$Traffic \vdash Driver(D(x,y), D(x,y))$$
$$Traffic \vdash Driver(D(x,y) \diamond p, d) \leftarrow Driver(p, d)$$

Every atom on the form $Driver(D(x_1, y_1) \diamond \cdots \diamond D(x_n, y_n), D(x_n, y_n))$ is a theorem in *Traffic*. For example, we can derive $Traffic \vdash Driver(D(C, South), D(C, South))$ and $Traffic \vdash Driver(D(C, South) \diamond D(B, East) \diamond D(A, North), D(A, North))$. Note that each such theorem is about a theory term encoding some driver's view of some driver's view of ... some driver's beliefs, and the ultimate driver in such a chain. We can use the predicate *Driver* in *Traffic* for expressing that something should be believed by every driver and that every driver should believe that other drivers believe so, etc., arbitrarily deep.

The following three clauses define the actual reasoning.

$$Traffic \vdash \ulcorner \dot{t}_1 \vdash Pass(c) \leftarrow Not\text{-}in\text{-}crossing(\dot{x}_1, c) \urcorner \leftarrow$$
$$Driver(t, d) \land Names(t_1, t) \land Names(x_1, x) \land Gives\text{-}way\text{-}to(d, x)$$
$$Traffic \vdash \ulcorner \dot{t}_1 \vdash Wait(c) \leftarrow In\text{-}crossing(\dot{x}_1, c) \land$$
$$Names(x_2, x_1) \land Names(c_1, c) \land$$
$$\ulcorner \dot{x}_2 \vdash Pass(\dot{c}_1)? \urcorner \urcorner \leftarrow$$
$$Driver(t, d) \land Names(t_1, t) \land Names(x_1, x) \land Gives\text{-}way\text{-}to(d, x)$$
$$Traffic \vdash \ulcorner \dot{t}_1 \vdash Pass(c) \leftarrow In\text{-}crossing(\dot{x}_1, c) \land$$
$$Names(x_2, x_1) \land Names(c_1, c) \land$$
$$\ulcorner \dot{x}_2 \vdash Wait(\dot{c}_1)? \urcorner \urcorner \leftarrow$$
$$Driver(t, d) \land Names(t_1, t) \land Names(x_1, x) \land Gives\text{-}way\text{-}to(d, x)$$

The first clause says that any driver will reason: if there is no car approaching from such a direction that I must give way to it, then I may pass.

The second clause says that any driver will reason: if there is a car approaching from such a direction that I must give way to it, and I believe that driver will reason that he can pass, then I must wait.

The third clause says that any driver will reason: if there is a car approaching from such a direction that I must give way to it, but I believe that driver will reason that he must wait, then I can pass anyway.

The next four clauses of *Traffic* simply determine who must yield to whom.

$$Traffic \vdash Gives\text{-}way\text{-}to(D(\_, North), D(\_, West))$$
$$Traffic \vdash Gives\text{-}way\text{-}to(D(\_, West), D(\_, South))$$
$$Traffic \vdash Gives\text{-}way\text{-}to(D(\_, South), D(\_, East))$$
$$Traffic \vdash Gives\text{-}way\text{-}to(D(\_, East), D(\_, North))$$

The predicates *In-crossing* and *Not-in-crossing* are list membership/nonmembership predicates; these predicates are here part of the group belief of drivers (alternatively we could have placed them in every theory *Traffic* ⋄ · · ·).

$Traffic \vdash \ulcorner t_1 \vdash In\text{-}crossing(x, [x|\_]) \urcorner \leftarrow$
    $Driver(t,d) \land Names(t_1, t)$
$Traffic \vdash \ulcorner t_1 \vdash In\text{-}crossing(x, [\_|c]) \leftarrow In\text{-}crossing(x, c) \urcorner \leftarrow$
    $Driver(t,d) \land Names(t_1, t)$
$Traffic \vdash \ulcorner t_1 \vdash Not\text{-}in\text{-}crossing(x, []) \urcorner \leftarrow$
    $Driver(t,d) \land Names(t_1, t)$
$Traffic \vdash \ulcorner t_1 \vdash Not\text{-}in\text{-}crossing(x, [y|c]) \leftarrow x \neq y \land Not\text{-}in\text{-}crossing(x, c) \urcorner \leftarrow$
    $Driver(t,d) \land Names(t_1, t)$

A full proof of the original statement 10 is rather long, but involves proving the following statements, among others.

$Traffic \diamond D(C, South) \diamond D(B, East) \diamond D(A, North) \vdash$
    $Pass([D(A, North), D(B, East), D(C, South)])?$
$Traffic \diamond D(C, South) \diamond D(B, East) \vdash$
    $Wait([D(A, North), D(B, East), D(C, South)])?$
$Traffic \diamond D(C, South) \vdash Pass([D(A, North), D(B, East), D(C, South)])?$
$Traffic \vdash \ulcorner D(C, South) \vdash Pass([D(A, North), D(B, East), D(C, South)])? \urcorner$

8.6.2 Properties of knowledge

Some formalisms intended for knowledge representation, reasoning and meta-reasoning (such as Konolige's modal logic of knowledge [Kon86]) build various properties of knowledge or belief into the formalism. Five well-known properties of this kind are (using the notation of Konolige, where $bel(S)$ is the set of beliefs of an agent S, while $[S]\phi$ is the proposition that agent S believes ϕ):

Saturation (K). Reasoners are closed under inference, so $bel(S)$ is saturated.

Knowledge (T). For knowledge, beliefs must be true, so $\phi \in bel(S) \Rightarrow \phi$ is true.

Consistency (D). Reasoners are supposed to be consistent in their knowledge, so $\phi \in bel(S) \Rightarrow \neg\phi \notin bel(S)$.

Positive introspection (4). If reasoners believe something, they also believe that they believe it, so $\phi \in bel(S) \Rightarrow [S]\phi \in bel(S)$.

Negative introspection (5). If reasoners do not believe something, they also believe that they do not believe it, so $\phi \notin bel(S) \Rightarrow \neg[S]\phi \in bel(S)$.

Alloy is intended, among other things, for applications of this kind, but only the first property has been built into the language. Instead, we might express these properties as part of our meta-programs. This makes it possible to model also reasoning agents that do not have these properties, or who have quite different properties. Let us show how these properties could be represented in a suitably extended version of Alloy, one by one. We will assume that there is a theory A which defines an internal theory system, in which the beliefs of some agent is represented by a theory identified as B in A (and thus as $A \diamond B$ outside A).

Saturation (K) This property is built in, as Alloy theories are closed under inference. This means that Alloy can only represent directly agents whose beliefs are closed under inference.

Knowledge (T) This postulate can be expressed for some particular binary predicate P as a theoremhood statement

$$A \vdash P(x, y) \leftarrow \ulcorner B \vdash P(\dot{u}, \dot{v})?\urcorner \wedge Names(u, x) \wedge Names(v, y).$$

If we would like to express the **T** postulate for *any* predicate symbol we should do it in a meta-theory of A.

A variant of the **T** postulate can be expressed as

$$A \vdash \ulcorner W \vdash P(\dot{u}, \dot{v})\urcorner \leftarrow \ulcorner B \vdash P(\dot{u}, \dot{v})?\urcorner,$$

in which A has an internal theory W which contains A's view of the world. This statement says that if A believes that B believes some P atom, then that atom is also contained in A's beliefs about the world.

Consistency (D) Consistency of the reasoner B could be expressed as an integrity constraint

$$A \vdash \leftarrow \ulcorner B \vdash \dot{p}\urcorner \wedge \ulcorner B \vdash not\, \dot{p}\urcorner$$

(However, Alloy currently has no inference rules that take integrity constraints or negation into account.)

Positive introspection (4) This is straightforward:

$$A \vdash \ulcorner B \vdash \ulcorner B \vdash \dot{p} \urcorner \urcorner \leftarrow \ulcorner B \vdash \dot{q} \urcorner \wedge \mathit{Names}(p, q)$$

Negative introspection (5) If Alloy were to be extended with negation, then negative introspection is also easy:

$$A \vdash \ulcorner B \vdash \mathit{not} \ulcorner B \vdash \dot{p} \urcorner \urcorner \leftarrow \mathit{not} \ulcorner B \vdash \dot{q} \urcorner \wedge \mathit{Names}(p, q)$$

8.6.3 Program composition operators

Brogi, Mancarella, Pedreschi and Turini have proposed an algebra of operators for composing logic programs [BMPT90]. The operators are $P \cup Q$, $P \cap Q$, P^* and $P \triangleleft Q$, for union, intersection, encapsulation and import of programs, respectively. Their meta-interpretive definition can be coded elegantly in Alloy, provided that we choose one unary and three binary function symbols for constructing theory terms that stand for the theories resulting from these operations.

We let all the theories of a logic program with theory operators constitute a theory system internal to a theory M. The definition of M contains five theoremhood statements that define the theorems of theories named by operator expressions. Here we represent the operators by the function symbols U, I, E and T, respectively.

$$M \vdash \ulcorner U(\dot{p}, \dot{q}) \vdash \dot{a} \leftarrow \dot{c} \urcorner \leftarrow \ulcorner \dot{p} \vdash \dot{a} \leftarrow \dot{c} \urcorner$$
$$M \vdash \ulcorner U(\dot{p}, \dot{q}) \vdash \dot{a} \leftarrow \dot{c} \urcorner \leftarrow \ulcorner \dot{q} \vdash \dot{a} \leftarrow \dot{c} \urcorner$$
$$M \vdash \ulcorner I(\dot{p}, \dot{q}) \vdash \dot{a} \leftarrow \dot{c} \urcorner \leftarrow \mathit{Partition}(c, c_1, c_2) \wedge \ulcorner \dot{p} \vdash \dot{a} \leftarrow \dot{c_1} \urcorner \wedge \ulcorner \dot{q} \vdash \dot{a} \leftarrow \dot{c_2} \urcorner$$
$$M \vdash \ulcorner E(\dot{p}) \vdash \dot{a} \leftarrow \mathit{True} \urcorner \leftarrow \ulcorner \dot{p} \vdash \dot{a}? \urcorner$$
$$M \vdash \ulcorner T(\dot{p}, \dot{q}) \vdash \dot{a} \leftarrow \dot{c_1} \urcorner \leftarrow \ulcorner \dot{p} \vdash \dot{a} \leftarrow \dot{c} \urcorner \wedge \mathit{Partition}(c, c_1, c_2) \wedge \ulcorner \dot{q} \vdash \dot{c_2}? \urcorner$$

This straightforward program, which uses a ground representation, is no less elegant than the program by Brogi & Contiero [BC94] that uses a non-ground representation. (We assume that the ternary predicate *Partition* has been defined to compute the partition of a conjunction into a pair of (possibly empty or unitary) conjunctions.

For example, consider a program in the algebra with three "basic" theories *Rules*, *Public* and *Private* [Bro93]. In the Alloy program, the clauses of these theories should appear as theoremhood statements $M \diamond \mathit{Rules} \vdash \cdots$, $M \diamond \mathit{Public} \vdash \cdots$, and $M \diamond \mathit{Private} \vdash \cdots$, respectively. We can then add a coincidence statement such as $\mathit{GiveCredits} \equiv M \diamond U(T(\mathit{Rules}, \mathit{Private}), \mathit{Public})$ in order to define a theory *GiveCredits* which can subsequently be queried. Any query to *GiveCredits* will then be computed in the composed theory $(\mathit{Rules} \triangleleft \mathit{Private}) \cup \mathit{Public}$.

8.6.4 Implicit programming

Essentially all programs today are written manually by programmers. The programmers build on past experience and sometimes even directly on programs written in the past. (Indeed, this happens every time an existing program needs modification; we may see it as writing a program that is to perform almost the same computation as an existing program.) This might happen in many ways. Sometimes a program piece can be reused as is, when the abstraction it provides is exactly the one sought for. Typically pieces of the existing program need to be systematically rewritten in some way, for example, an extra argument might need to be added to a procedure or a base case replaced. If the existing program needs extensive rewriting, perhaps only its basic structure remains, such as the recursion pattern.

When really done systematically, this is a useful methodology. If the existing program does what is expected from it and each small change transforms it in a known way, then we may have confidence that the program resulting from a sequence of such changes computes what we expect. It is a serious problem today that modifications of the kind outlined above can rarely be carried out flawlessly. The resulting program then does not do what is expected and expensive corrective work is required. We may never know when the program becomes error-free.

Suppose we could partly automate this process, so the programmer could instead take a program or a program fragment and specify exactly which modifications must be done. The requested transformation would then be applied and the process continued until the desired program had been created. Given a collection of generally useful program fragments, the programmer might even build an entirely new program by incorporating and transforming these components. An alternative, often discussed in the realm of functional programming, is to provide very powerful abstractions so every problem can be coded in terms of these high-level abstractions. This approach is mathematically very appealing but has not yet turned out to be a practical approach to programming. The process outlined above is closer to an approach taken by actual programmers and also seems to be useful for reasoning exactly about the resulting programs. A more detailed comparison between these approaches seems necessary in the future.

As an example, let us show a simple program that adds an extra argument to a predicate. The transformation program is in a theory called T.

$$T \vdash \ulcorner Extend(\dot{m}, \dot{p}) \vdash \dot{b} \leftarrow \dot{d} \urcorner \leftarrow$$
$$\ulcorner \dot{m} \vdash \dot{a} \leftarrow \dot{c} \urcorner \wedge$$
$$NonoccurringVariable(\ulcorner \dot{a} \leftarrow \dot{c} \urcorner, v)$$
$$TransAtom(a, b, p, v) \wedge$$
$$TransQuery(c, d, p, v) \wedge$$

$$T \vdash TransAtom(\ulcorner \dot{p}(|\dot{x})\urcorner, \ulcorner \dot{p}(\dot{v}|\dot{x})\urcorner, p, v) \leftarrow True$$
$$T \vdash TransAtom(\ulcorner \dot{q}(|\dot{x})\urcorner, \ulcorner \dot{q}(|\dot{x})\urcorner, p, v) \leftarrow p \neq q$$
$$T \vdash TransQuery(\ulcorner True\urcorner, \ulcorner True\urcorner, \_, \_) \leftarrow True$$
$$T \vdash TransQuery(\ulcorner \dot{a} \wedge \dot{c}\urcorner, \ulcorner \dot{b} \wedge \dot{d}\urcorner, p, v) \leftarrow$$
$$TransAtom(a, b, p, v) \wedge$$
$$TransQuery(c, d, p, v)$$

(We assume that the predicate *NonoccurringVariable* has been defined to compute (in its second argument) some variable name that does not occur in the name given as the first argument.)

In order to use this program, we must make T's view of the inspected and the defined theories coincide with the actual theories that we wish to inspect and define:

$$T \diamond JohnsBrain \equiv JohnsOldBrain$$
$$T \diamond Extend(JohnsBrain, Likes) \equiv JohnsNewBrain$$

Henceforth, the theory *JohnsNewBrain* will be exactly like the theory *JohnsOldBrain*, except that any clause which contains a predication $Likes(\cdots)$ has been replaced by a clause in which these predications have all been replaced by $Likes(\cdots, v)$, where v is some variable that did not occur in the original clause.

8.7 Self-reference

The reader may have noted that we have avoided using any circular theories. There is no automatic mechanism which gives a theory access to information about its own provability.

There are several advantages with systems that do not contain self-referring theories, i.e., theories that do not really reflect upon themselves but at most upon "views" of themselves. For example, there will be no paradoxes and implementation becomes simpler and more efficient.

The disadvantage with prohibiting or avoiding self-reference is, of course, a reduced expressivity. It is not possible to define agents that truly introspect. It is an open question at this time how serious a restriction it would be to prohibit self-reference completely, but it is clear that one can often make do with a sufficiently high tower of theories, each being a meta-theory for the theories below it. A very close approximation to a single theory which is a meta-theory for itself is obtained through an infinite tower of identical theories, each being a meta-theory for the theories below it. Such a tower can be expressed in Alloy.

If we wished to make an Alloy theory $T \diamond U$ truly self-referential, e.g., through a theory I in $T \diamond U$, we could add one of the two equivalent statements $T \diamond U \diamond I \equiv T \diamond U$ and $T \vdash \ulcorner U \diamond I \equiv U \urcorner$ to the program. It is easy to show that with either statement, in any model $\langle \mathcal{M}, \equiv \rangle$ of the program, we will have that $\mathfrak{M}_{T \diamond U \diamond I} \equiv \mathfrak{M}_{T \diamond U}$. That is, whatever $T \diamond U$ "observes" in the theory it calls I, is really also in $T \diamond U$ itself. This is a "two-way" self-reference: $T \diamond U$ may query itself by querying the theory it calls I, or it may compute clauses and add them to itself if it contains clauses such as $\ulcorner I \vdash \dot{p} \urcorner \leftarrow \cdots p \cdots$.

One could allow $T \diamond U$ to query itself but not add clauses to itself by instead adding here are three simple (and equivalent) ways: a theoremhood statement

$$T \vdash \ulcorner U \diamond I \vdash \dot{p} \urcorner \leftarrow \ulcorner U \vdash \dot{p} \urcorner,$$

to the program. It is easy to show that $\mathfrak{M}_{T \diamond U} \subseteq \mathfrak{M}_{T \diamond U \diamond I}$, i.e., that whatever is satisfied by $\mathfrak{M}_{T \diamond U}$ is also satisfied by $\mathfrak{M}_{T \diamond U \diamond I}$, so $T \diamond U \diamond I$ includes an "image" of $T \diamond U$.

However, note that there is no clause that could be added to $T \diamond U$ in order to achieve this effect. The rationale is simply that self-reference must be "sanctioned" from outside a theory.

8.8 Abduction

Abduction is a form of reasoning with a purpose to determine hypotheses that explain an observation, typically in the context of knowledge assimilation [Kow79, Kow90b]. Abductive reasoning seems particularly interesting in combination with meta-reasoning. Suppose the beliefs of John are represented by a theory $Beliefs(John)$, which internally defines a theory system in which there is a theory $Beliefs(Mary)$, representing John's beliefs about Mary's beliefs. Suppose further that

$Beliefs(John) \vdash$
$\quad SmilesAt(a,b) \leftarrow \ulcorner Beliefs(\dot{u}) \vdash Likes(\dot{u},\dot{v}) \urcorner \wedge Names(u,a) \wedge Names(v,b),$

i.e., a statement that those who believe they like him smile at him. If John notices Mary smiling at him, we can assume that a belief $SmilesAt(Mary, John)$ appears among John's beliefs, calling for an explanation. By performing abductive reasoning, the hypothesis $\ulcorner Beliefs(Mary) \vdash Likes(Mary, John) \urcorner$ appears as a good candidate for inclusion in $Beliefs(John)$ because it would imply the observation. John therefore might assume that Mary believes she likes him.

This is of course merely a simple example but the area of agents performing meta-reasoning about each other's actions, beliefs, motives and ambitions is clearly one where abductive reasoning needs to be carried out as part of the meta-reasoning.

Abductive reasoning can be carried out in many ways. One way is to add inference rules for abductive reasoning, obtaining new abductive proof procedures [KM90]. However, it is also possible to realize abductive reasoning through meta-level deduction, as suggested by Bowen & Kowalski [BK82]. Such achievement of abductive reasoning through meta-reasoning is a topic that ought to be explored further using theory systems.

8.9 Implementation and Language Extensions

In our implementation efforts, we are extending Luther [Bev92], an instance of Warren's abstract Prolog machine [War83]. The idea is that the generalized SLD-resolution rule should be essentially as efficient as in Prolog, regardless of the number of "indirection" levels. This can be made possible by representing the clauses of all theories, also those that only exist as a "view" in some other theory, by ordinary abstract machine code. An interesting difficulty is when a program clause is not an explicit axiom in a theory but is obtained through some computation in a meta-theory of the current theory. This we intend to solve by never actually creating the program clause but rather use directly the parts of the program clause that are explicit in the meta-theory and then carry out a computation in the meta-theory. The following example should illustrate the technique. Consider the following program fragment.

$$T_M \diamond T_O \equiv T_O$$
$$T_M \vdash \ulcorner T_O \vdash P(\dot{x}, F(\dot{y})) \leftarrow Q(\dot{y}) \wedge \dot{z} \urcorner \leftarrow R(\dot{x}, \dot{y}, \dot{z})$$

If we were to prove a P atom in T_O (or in $T_M \diamond T_O$), we could first carry out a computation in T_M of the complete name of some clause $P(\cdots, F(\cdots)) \leftarrow Q(\cdots) \wedge \cdots$ where the dotted parts were filled in by the R atom in T_M. However, computing the whole clause could well be a waste of resources, as is easy exemplified: Suppose that the goal atom is actually $P(42, G(54))$. Unification of the goal atom with the head of the generated program clause will always fail immediately and the computation in T_M of the program clause would be worthless. What we do instead is to compile as part of the code reachable from T_O a clause

$$P(x_1, F(y_1)) \leftarrow Names(x, x_1) \wedge Names(y, y_1) \wedge R(x, y, z) \wedge Q(y_1) \wedge \alpha[z].$$

We see that all parts of the clause that were explicitly given in the meta-level clause are present in this clause. The two *Names* atoms constrain the variables x and y so that any value they obtain must be a name of something that can be unified with x_1 and x_2, respectively. The expression $\alpha[z]$ can best be described as a call to whatever becomes the value of z. In the worst case, this might require using an interpreter but it seems to us that in this situation, the value of z is usually taken from some context where there is machine code available for the named query. In this case, that code can be used (with some care). If we consider again the goal atom $P(42, G(54))$ we see that this clause will fail before computing any part of the body.

As mentioned before, the style of computation described above realizes a different inference system from the one described in Sect. 8.4.3. In this system, computations in various theories can be interleaved, as shown by the example. The idea is to be as goal-directed as possible.

It is clear that negation of some kind must be added to the language, either explicit negation, negation as failure or both. If we incorporate negation as failure in Alloy, we will investigate the merits of a monotonic version of negation as failure, where the theory in which a finitely failed proof is obtained is given explicitly.

It would also be very interesting to incorporate some form of abductive procedure in Alloy, because of the natural links between meta-reasoning and abduction pointed out in Sect. 8.8. Denials are already formally present in the language and would then function as integrity constraints when given as part of a program [Kow90b].

8.10 Notes and Related Work

There have been a few changes in the definition of Alloy since our previous publication [BBD94].

1. Theory terms now include expressions on the form $\cdots \diamond \cdots$.

2. In addition to program clauses, Alloy now has goals and denials.

3. What used to be called a tagged program clause is now called a theoremhood statement and may contain any sentence.

4. Representation statements have been generalized to coincidence statements (a representation statement $t \triangleright u$ can be written as $t \diamond u \equiv u$). This allowed us to generalize the reflection rules and simplify the inference system considerably.

5. There is an SLD-resolution style inference rule instead of an inference rule for program clauses.

It should be obvious for the knowledgeable reader that the development of Alloy is very much inspired by work of Kowalski [Kow90a, Kow90b], and by Reflective Prolog of Costantini & Lanzarone [CL89].

There have recently appeared some proposals for systems for meta-reasoning with a similar philosophy as ours. Attardi & Simi [AS94] use what they call "relativized truth" but obtain a system quite similar to ours. One significant difference is that they choose to duplicate their inference system (a natural deduction system): the rules are present once for the object level and again for the meta-level. Moreover, among their basic axioms for the meta-level, there is one which ensures positive introspection. We have preferred to have no such epistemic bias, except for saturation. Giunchiglia et al. [GSS92] have defined a multilevel deduction system with distinct levels, called MK. There is only one theory per meta-level but the communication between meta-levels is similar to that in Alloy. This seems to be the basis for the reasoning part of GETFOL, a system that is also capable of code introspection and revision [GC94].

Our proposal for a meta-programming based software engineering methodology is related to the proposal by Kowalski about using meta-language for assembling programs [Kow82] and the work by Brogi et al. about using theory operators for building programs, which is discussed in more detail in Sect. 8.6.3.

Bowen & Weinberg [BW85] and Bacha [Bac87] have investigated compilation of partially known clauses in a context similar to ours.

Sato [Sat92] proposes an approach to meta-programming through a complete truth predicate tr in three valued logic. Sato's definition of tr is self referential, and gives in the general case an inconsistent definition of tr in two valued logic by being paradoxical. As a slight modification of the definition of tr he introduces a three valued complete $demo$ predicate.

The language is fully amalgamated, like the theory part of Alloy to which it corresponds. (Note, however, that the system part and the theory part of Alloy are clearly separated both syntactically and semantically.)

The main similarity with our approach to meta-programming is the ability to reason with several levels, which is made possible by tr being self-referential; thus making it possible to express $tr(\ulcorner \ldots tr(\ldots, \ldots) \ldots \urcorner, \ldots)$ (the nesting can be of arbitrary depth). Furthermore, like naming in Alloy, the structural coding makes it possible to decompound terms and formulas to their least parts and look, for example, at codes of functors.

Jiang [Jia94] proposes an ambivalent approach to meta-reasoning, by introducing a language called AL where syntactically no distinction is made between terms, formulas or functors. Jiang takes a radically different approach from ours by defining what he calls a "Herbrand-based" semantics, which does not build upon the standard notion of logical consequence in first order model theory. It is hard to form a definitive opinion of the proposed semantics because as it is presented, it is not well-defined and thus cannot be understood without having to guess the intentions. Neither does he present an inference system, nor hint at any possible implementation of the proposed ideas. (It should be noted, however, that AL to some extent captures meta-programming as it is often done in Prolog, which has an operational semantics.)

Syntactically, the main similarity with our approach is the possibility to express reasoning across several meta-levels. The main distinction is that there is no naming or coding involved, formulas can occur directly as subexpressions in other formulas. The program clause 1 could for example be expressed as

$$Bel(Tim, \forall x(Cannibal(x) \rightarrow Bel(x, Tasty(x))).$$

The idea is that whether an expression is to be interpreted as a function or a relation is determined by the context where it appears.

Christiansen has proposed an amalgamated language in which there are two levels of reasoning [Chr94]. The operational semantics of the language is based on instance predicates, relating names of formulas such that one is an instance of the other. As was shown by Kowalski [Kow90a, Kow90b] and further developed by Hill & Gallagher [HG94], such instance predicates can be used with meta-variables replacing names of subexpressions in a way which turns out to be operationally similar to the way in which variables are represented using non-ground representations.

8.11 Conclusion

As can be seen from this article, Alloy is a language still under development. We can already conclude, however, that it allows a direct way of expressing multilevel knowledge, in particular recursive beliefs.

The main difference between Alloy and the mainstream of meta-logic programming lies in the support for arbitrary many meta-levels and in that self-reference is the exception rather than the rule.

One may certainly doubt that a language claimed to be so powerful is efficiently implementable and this can only be proved by an actual implementation, which is under way. One reason for hope is the belief that much of the computation will still be deduction within a single theory (which may be someone's view of someone's view of ... a theory) and this should be possible to support with essentially the efficiency of an ordinary Prolog system. The difficulties seem to lie in the meta-programming specific parts and in the fact that there are so many ways to use a piece of information in a meta-programming setting. For example, a program clause may be actually used for deduction, a name for it may be used as data, so may a name for a name for it, etc. Program clauses computed from names with "holes" is likely to be another (manageable) obstacle to efficient computation.

Acknowledgements

This research has been influenced by valuable discussions with our colleagues, particularly Stefania Costantini, Gaetano Lanzarone, and Andreas Hamfelt, and our partners in the Compulog 2 project, particularly Antonio Brogi, Pat Hill, Bob Kowalski and John Lloyd.

The research reported herein was supported financially by the Swedish National Board for Technical and Industrial Development (NUTEK) under contract No. 92-10452 (ESPRIT BRP 6810: *Computational Logic 2*).

J. B. thanks his family for their continuing support.

References

[AS94] G. Attardi and M. Simi. Building proofs in context. In F. Turini, editor, *Proc. META 94*, LNCS 883, Berlin, 1994. Springer-Verlag.

[Bac87] H. Bacha. Meta-level programming: a compiled approach. In J.-L. Lassez, editor, *Proc. 4th Intl. Conf. on Logic Programming*, pages 394–410, Cambridge, Mass., 1987. MIT Press.

[Bar94] J. Barklund. Metaprogramming in logic. UPMAIL Technical Report 80, Uppsala Univ., Computing Science Dept., 1994. To be published in *Encyclopedia of Computer Science and Technology*, Marcel Dekker, New York.

[BBD94] J. Barklund, K. Boberg, and P. Dell'Acqua. A basis for a multilevel metalogic programming language. In F. Turini, editor, *Proc. META 94*, LNCS 883, Berlin, 1994. Springer-Verlag.

[BC94] A. Brogi and S. Contiero. Gödel as a meta-language for composing logic programs. In F. Turini, editor, *Proc. META 94*, LNCS 883, Berlin, 1994. Springer-Verlag.

[Bev92] J. Bevemyr. The Luther WAM emulator. UPMAIL Tech. Rep. 72, Comp. Sci. Dept., Uppsala Univ., Uppsala, 1992.

[BK82] K. A. Bowen and R. A. Kowalski. Amalgamating language and metalanguage in logic programming. In K. L. Clark and S.-Å. Tärnlund, editors, *Logic Programming*, pages 153–72. Academic Press, London, 1982.

[BMPT90] A. Brogi, P. Mancarella, D. Pedreschi, and F. Turini. Composition operators for logic theories. In J. W. Lloyd, editor, *Computational Logic*, pages 117–34. Springer-Verlag, Berlin, 1990.

[Bow85] K. A. Bowen. Meta-level programming and knowledge representation. *New Generation Computing*, 3:359–383, 1985.

[Bro93] A. Brogi. *Program Construction in Computational Logic*. PhD thesis, Dipartimento di Informatica, Università di Pisa, 1993.

[BT91] A. Brogi and F. Turini. Metalogic for knowledge representation. In J. A. Allen, R. Fikes, and E. Sandewall, editors, *Principles of Knowledge Representation and Reasoning: Proc. 2nd Intl. Conf.*, pages 61–69, Los Altos, Calif., 1991. Morgan Kaufmann.

[BW85] K. A. Bowen and T. Weinberg. A meta-level extension of Prolog. In J. Cohen and J. Conery, editors, *Proc. 1985 Symp. on Logic Programming*, pages 78–86, Washington, D.C., 1985. IEEE Comp. Soc. Press.

[CDL94] S. Costantini, P. Dell'Acqua, and G. A. Lanzarone. Extending Horn clause theories by reflection principles. In C. MacNish, D. Pearce, and L. M. Pereira, editors, *Logics in Artificial Intelligence*, LNAI 838, pages 400–413, Berlin, 1994. Springer-Verlag.

[Chr94] H. Christiansen. Efficient and complete Demo predicates for definite clause languages. Technical Report 51, Dept. of Computer Science, Roskilde University, 1994.

[CL89] S. Costantini and G. A. Lanzarone. A metalogic programming language. In G. Levi and M. Martelli, editors, *Proc. 6th Intl. Conf. on Logic Programming*, pages 218–33, Cambridge, Mass., 1989. MIT Press.

[Cla78] K. L. Clark. Negation as failure. In H. Gallaire and J. Minker, editors, *Logic and Data Bases*, pages 293–322. Plenum Press, New York, 1978.

[GC94] F. Giunchiglia and A. Cimatti. Introspective metatheoretic reasoning. In F. Turini, editor, *Proc. META 94*, LNCS 883, Berlin, 1994. Springer-Verlag.

[GSS92] F. Giunchiglia, L. Serafini, and A. Simpson. Hierarchical meta-logics: Intuitions, proof theory and semantics. In A. Pettorossi, editor, *Meta-Programming in Logic*, LNCS 649, pages 235–249, Berlin, 1992. Springer-Verlag.

[HG94] P. M. Hill and J. Gallagher. Meta-programming in logic programming. Technical Report 94.22, School of Computer Studies, Univ. of Leeds, 1994. To be published in *Handbook of Logic in Artificial Intelligence and Logic Programming*, Vol. 5, Oxford Science Publ., Oxford Univ. Press.

[Jia94] Y. Jiang. Ambivalent logic as the semantic basis of metalogic programming. In P. Van Hentenryck, editor, *Logic Programming, Proc. 11th Intl. Conf*, pages 387–401, Cambridge, Mass., 1994. MIT Press.

[KM90] A. C. Kakas and P. Mancarella. Abductive logic programming. In *Proc. NACLP90 Workshop on Non-Monotonic Reasoning and Logic Programming*, Austin, Texas, 1990. MCC.

[Kon86] K. Konolige. *A Deduction Model of Belief.* Pitman, London, 1986.

[Kow79] R. A. Kowalski. *Logic for Problem Solving.* North Holland, New York, 1979.

[Kow82] R. A. Kowalski. The use of metalanguage to assemble object level programs and abstract programs. Report, Imperial College, London, 1982.

[Kow90a] R. A. Kowalski. Meta matters. Invited presentation at Second Workshop on Meta-Programming in Logic, 1990.

[Kow90b] R. A. Kowalski. Problems and promises of computational logic. In J. W. Lloyd, editor, *Computational Logic*, pages 1–36. Springer-Verlag, Berlin, 1990.

[McC68] J. McCarthy. Programs with common sense. In M. Minsky, editor, *Semantic Information Processing*, pages 403–418. MIT Press, Cambridge, Mass., 1968.

[McC79] J. McCarthy. First order theories of individual concepts and propositions. In B. Meltzer and D. Michie, editors, *Machine Intelligence 9*, pages 120–147. Edinburgh University Press, Edinburgh, 1979.

[Sat92] T. Sato. Meta-programming through a truth predicate. In K. Apt, editor, *Proc. Joint Intl. Conf. Symp. on Logic Programming 1992*, pages 526–540. MIT Press, Cambridge, Mass., 1992.

[Smo77] C. Smorynski. The incompleteness theorems. In J. Barwise, editor, *Handbook of Mathematical Logic*, pages 821–865. North-Holland, Amsterdam, 1977.

[Ste84] L. S. Sterling. Logical levels of problem solving. *J. Logic Programming*, 1:138–45, 1984.

[War83] D. H. D. Warren. An abstract Prolog instruction set. SRI Tech. Note 309, SRI Intl., Menlo Park, Calif., 1983.

III META-LOGICS FOR KNOWLEDGE MANAGEMENT

9 Using Meta-Logic to Reconcile Reactive with Rational Agents

Robert A. Kowalski

Abstract
In this paper I outline an attempt to reconcile the traditional Artificial Intelligence notion of a logic-based rational agent with the contrary notion of a reactive agent that acts "instinctively" in response to conditions that arise in its environment. For this purpose, I will use the tools of meta-logic programming to define the observation-thought-action cycle of an agent that combines the ability to perform resource-bounded reasoning, which can be interrupted and resumed any time, with the ability to act when it is necessary.

9.1 Introduction

The traditional notion of an intelligent agent in Artificial Intelligence is that of a rational agent that has explicit representations of its own goals and of its beliefs about the world. These beliefs typically include beliefs about the actions that the agent can perform and about the effects of those actions on the state of the world.

This traditional notion of intelligent agent has been challenged in recent years by the contrary notion of an agent that reacts "instinctively" to conditions in its immediate environment. A reactive agent need possess neither an explicit representation of its own goals nor any "world model".

In this paper, I shall outline an attempt to reconcile these two conflicting views. For this purpose, I will extend the notion of knowledge assimilation, Kowalski [Kow79], using the tools of meta-logic programming, to combine assimilation of inputs, reduction of goals to subgoals and execution of appropriate subgoals as actions.

Expressed in informal, simplified, procedural, meta-logic programming style, the definition of the top-most level of a logic-based agent might take the form:

> to "cycle" at time T,
>   observe one input at time T,
>   assimilate the input from time T to $T+m$,
>   reduce current goals to subgoals from time $T+m$ to $T+m+n$,
>   perform any requisite atomic action from time $T+m+n$ to $T+m+n+1$,
>   "cycle" at time $T+m+n+1$.

The parameters m and n adjust the resources allocated for rational processing of inputs and goals respectively, in relation to those allocated to making observations and performing actions. If the parameters m and n are relatively small, the agent will have little time to "think" before acting. The behaviour will be similar to that of a reactive agent. If m and n are sufficiently large, the agent will be able to generate a complete plan before beginning to act. The behaviour will be similar to that of a traditional, rational agent. For intermediate values of m and n the agent will be able to plan several steps ahead before committing to a particular course of action.

The definition of the *cycle* predicate is similar to the procedural characterisation of a deliberate agent given by Genesereth and Nilsson [GN87]. The most important difference between our procedure and theirs is that we aim to give the definition both a procedural and a declarative interpretation in the spirit of logic programming. As we will see later, a key factor contributing to the declarative interpretation of the procedure is the organisation of inputs and outputs into input and output streams. Both the agent and the agent's environment are interpreted as logically defined processes which communicate with each other by means of these input and output streams. Input and output are symmetric in the sense that the input stream of the agent is the output stream of the environment and vice versa.

Before discussing the *cycle* predicate in greater detail, I will discuss the changes necessary to the more conventional demo and assimilate predicates.

9.2 The Proof Predicate, Demo

The familiar, conventional demo predicate for a Horn clause object language is defined non-deterministically:

$$demo(KB, P) \leftarrow axiom(KB, P \leftarrow Q) \wedge demo(KB, Q)$$
$$demo(KB, P \wedge Q) \leftarrow demo(KB, P) \wedge demo(KB, Q)$$
$$demo(KB, true)$$

Here $demo(KB, P)$ expresses that conclusion P can be demonstrated from "knowledge base" KB; $axiom(KB, P \leftarrow Q)$ expresses that $P \leftarrow Q$ is a clause represented explicitly as an axiom in KB. For simplicity, I use the ambivalent syntax of Kowalski and Kim [KK91] and Jiang [Jia94]. Moreover, I have considered only the propositional case. The general case can be reduced to the propositional case either by using a standard definition of unification, or by adding a clause

$$demo(KB, P') \leftarrow demo(KB, forall(X, P))$$
$$\wedge \; substitute(X, P, Y, P')$$

as discussed by Kowalski [Kow90]. The predicate $substitute(X, P, Y, P')$ holds when P' results from substituting the *term* Y for the variable X in P.

The definition is non-deterministic because it is not determined, in the first clause of the definition, what axiom in the knowledge base, having conclusion P, might be needed to demonstrate P. The search for the necessary axiom is performed by the non-deterministic "inference engine" which executes the definition rather than by the definition itself.

We can add an extra argument to indicate the resources needed to construct a proof.

$$demo(KB, P, R+1) \leftarrow axiom(KB, P \leftarrow Q) \wedge demo(KB, Q, R)$$
$$demo(KB, P \wedge Q, R+S) \leftarrow demo(KB, P, R) \wedge demo(KB, Q, S)$$
$$demo(KB, true, 0)$$

However, this argument counts only the number of steps in a proof rather than the number of steps in a search for a proof.

To count the number of steps in the search for a proof, it is necessary to represent the search space explicitly. This can be done by means of a deterministic definition of the *demo* predicate, in which alternative branches of the search space are represented by means of disjuncts:

$$demo(KB, InGoals, OutGoals, R+1) \leftarrow InGoals \equiv (G \wedge Rest) \vee AltGoals$$
$$\wedge\ definition(KB, G \leftarrow D)$$
$$\wedge\ demo(KB, (D \wedge Rest) \vee AltGoals, OutGoals, R)$$
$$demo(KB, InGoals, OutGoals, 0) \leftarrow InGoals \equiv true \vee AltGoals$$
$$\wedge\ OutGoals = true$$
$$demo(KB, InGoals, OutGoals, 0) \leftarrow \neg InGoals \equiv true \vee AltGoals$$
$$\wedge\ OutGoals = InGoals$$

Here $demo(KB, InGoals, OutGoals, R)$ expresses that the search space of goals, $InGoals$, can be reduced to the search space of subgoals, $Outgoals$, in R steps. For simplicity the definition does not count the resources needed to select the atomic subgoal G and to find its definition $G \leftarrow D$. As we will see in the example definition of $\equiv$ below, the number of steps involved in executing the definition of $\equiv$ may be non-trivial.

The predicate $definition(KB, G \leftarrow D)$ expresses that $G \leftarrow D$ is the complete definition of G in KB. In the general case D is a disjunction of all the conditions of all the clauses having conclusion G. In addition to having its logical meaning as disjunction, $\vee$ can be interpreted as an infix list constructor, terminated by $false$. Thus every disjunction has a final disjunct $false$. In particular, if G is the conclusion of no clause in KB, then D is just $false$. However, if G is abducible, i.e. can be assumed or "made" true, then G has no definition at all. As we will see later, actions which can be performed by the agent are represented by such abducible goals.

Similarly, $\wedge$ can be interpreted as an infix list constructor, terminated by *true*. Thus every conjunction has a final conjunct *true*. In particular, if G is defined by a conditionless clause, then the condition of that clause is taken to be *true* instead.

The infix predicate, $\equiv$, is any predicate which deterministically expresses the logical equivalence of its two arguments. Procedurally, $\equiv$ can be viewed as selecting both a branch $(G \wedge Rest)$ of the search space and a goal G in the branch. Different deterministic definitions of $\equiv$ give rise to different selection strategies, which in turn give rise to different strategies for searching the search space. For example, the following definition gives rise to Prolog-style depth-first search:

$$((D1 \vee D2) \wedge Rest) \vee AltGoals \equiv D' \vee AltGoals' \leftarrow$$
$$(D1 \wedge Rest) \equiv D'$$
$$\wedge (D2 \wedge Rest) \vee AltGoals \equiv AltGoals'$$
$$(false \wedge Rest) \vee AltGoals \equiv AltGoals$$
$$((C1 \wedge C2) \wedge Rest) \equiv (C1 \wedge Rest') \leftarrow (C2 \wedge Rest) \equiv Rest'$$
$$(true \wedge Rest) \equiv Rest$$

Notice that the first two clauses are like the definition of *append* for lists constructed using $\vee$; whereas the last two clauses are like the definition of *append* for lists constructed using $\wedge$.[1]

The infix predicate $=$ is simple identity, defined by the clause $X = X \leftarrow$.

The *demo* predicate has an argument (the third parameter) which records the state of the search space after the resources allocated to goal reduction have been exhausted. This makes it possible for goals to persist from one cycle to the next, and for execution of the *demo* predicate to resume when additional resources are made available in later cycles.

9.3 Knowledge Assimilation, Integrity Constraints and Goals

The conventional assimilate predicate deals with four cases:
- The input can be demonstrated from the knowledge base, in which case the next state of the knowledge base is identical to the current state.

---

[1] Notice that the example definition of $\equiv$ does not distinguish between abducible and non-abducible goals. Abducible goals should be treated in *demo* as though they were true, without actually being replaced by *true*. Catering for abducibles can be done by modifying the definitions of *demo* and/or of $\equiv$. I leave the details to the reader.

- The input together with one part of the knowledge base can be used to demonstrate the remaining part. In this case the next state of the knowledge base is the input together with the first part of the current state of the knowledge base.
- The input is inconsistent with the knowledge base. Together, the input and the knowledge base need to be revised to restore consistency.
- The input and the knowledge base are logically independent. In this case either the input is added to the knowledge base directly, or, if it is more appropriate, an abductive explanation of the input is added instead.

The formalisation of such knowledge assimilation as a meta-logic program is straightforward, Kowalski [Kow79]. Unfortunately it is quite different from the kind of *assimilate* predicate needed by an active, resource-bounded agent.

An agent needs to process its inputs in real time, both to update its knowledge base and to determine whether any immediate action is required to respond to the input. For example, the representation of the sentence

"If it is raining, carry an umbrella"

should result in the agent attempting to carry an umbrella soon after observing that it is raining.

It is such "online" processing of inputs that is required in the top-level cycle of our agent, rather than the "offline" restructuring of the knowledge base with which the conventional notion of knowledge assimilation is concerned.

In the remainder of this paper, we shall assume that "offline" knowledge assimilation is an activity which takes place either in parallel with the top-level cycle of the agent or at times when the rate of input is very low. We shall focus instead on the resource-bounded assimilation of inputs that needs to take place "online".

The example of carrying an umbrella when it rains is typical of the kind of knowledge an active agent would use to relate conditions it observes to actions it performs. Many other examples readily come to mind:

" In an emergency, press the alarm signal button."

" Give up your seat, if someone else needs it more than you do."

" If it is after 10.00pm and there is no good reason to stay awake, go to sleep."

It is natural to formalise such sentences by means of condition-action production rules. In this paper, we shall explore the possibility that they can be formalised by means of integrity constraints instead. [2]

Integrity constraints in database systems express obligations and prohibitions that all states of a database must satisfy. Such conditions can be expressed by sentences of first-order logic. The obligations and prohibitions associated with such first-order sentences are implicit in the semantics of integrity constraints rather than explicit, as they would be if they were written as sentences of deontic logic, e.g. Jones and Sergot [JS93].

Thus, for example, we might formalise the "obligation" to carry an umbrella when it is raining by the integrity constraint:

$$holds(rain, T) \rightarrow holds(carry(self, umbrella), T)$$

If this were an ordinary sentence in the knowledge base, it would allow the agent to conclude that it is carrying an umbrella whenever it rains, whether it is actually carrying an umbrella or not. As an integrity constraint, however, the sentence imposes an obligation on the ordinary sentences to establish that the conclusion holds, independently of any integrity constraints. This can be done by abducing that some event of putting up an umbrella happens when it first starts raining and by preventing any event of putting down the umbrella while it is still raining.

But it is not enough simply to add to the knowledge base a sentence stating that the agent is performing an action. As we will see in the next section, before the assertion can be added to the knowledge base, the agent needs to output the action to the environment, and the environment needs to confirm that the attempted action has been successful.

Although the distinction between integrity constraints and ordinary sentences is intuitively clear, there have been many different attempts to give the "semantics" of integrity constraints a formal characterisation. For deductive databases, these formalisations include the *theoremhood view*, Lloyd and Topor [LT85], that integrity constraints are theorems that should be logical consequences of the completion of the database, the *consistency view*, Sadri and Kowalski [SK87], that they should be consistent with the completion, and the views that integrity constraints should be understood as *epistemic*, Reiter [Rei90] or *metalevel*, Sadri and Kowalski [SK87] statements about what the database "knows" or can demonstrate.

Despite the differences between these formalisations, the proof procedures that have been developed for verifying integrity constraints generally treat them as goals to be satisfied and are similar in practice. Indeed, it is exactly such a relationship between

---

[2]This interpretation of production rules in terms of integrity constraints resembles the transformation of Rashid [Ras94]. The exact relationship, however, between our interpretation and this transformation needs to be investigated further.

integrity constraints and goals which we will take as the operational semantics of integrity constraints used for "online" assimilation of inputs. We will regard integrity constraints as passive goals that become active when they are "triggered" by appropriate inputs.

For the sake of simplicity, we will assume that integrity constraints are stored in the knowledge base, but are distinguished from ordinary sentences by their syntax. We will assume, in particular, that all such integrity constraints are written in the form:

$$I \to C$$

where I is an atomic formula and C is a "complete" conjunction of all the constraints that the knowledge base should satisfy when I holds. Operationally, if an input matches I (unifies with I) then the integrity constraint is "triggered" (resolved with the input) and the appropriate instance of C (resolvent) is added to the current search space of goals. More formally (and more simply, ignoring unification), we can define such online assimilation of inputs by:

$$assimilate(InKB, InGoals, Input, OutKB, OutGoals, T) \leftarrow$$
$$constraint(KB, Input \to C)$$
$$\land \ OutKB = (Input \land InKB)$$
$$\land \ OutGoals = (C \land InGoals)$$

Here the assimilate predicate expresses the relationship that holds between the states of the knowledge base and of the search space of goals, before and after observing $Input$ at time T. To simplify the definition of *assimilate*, as a matter of convention, a lack of input is recorded instead as an input of *true*.

The predicate *constraint* expresses that C is a conjunction of all the constraints that should hold when $Input$ holds. If there are no such constraints then C is assumed to be *true*. As a result of these conventions

$$constraint(KB, true \to true)$$

holds as a special case.

The simplified representation of integrity constraints assumes that they have been "precompiled" so that inputs trigger integrity constraints directly in one step without any intermediate deductions. This assumption simplifies the definition of *assimilate* (and *cycle*) because it means that it is unnecessary to record forward deductions from the input that have not yet resolved with the integrity constraints. It also facilitates the agent's ability to process inputs in real time. As a result of this simplifying assumption, assimilation of the input consumes only one unit of resource rather than the more general m units we assumed would be necessary in the Introduction.

Notice that we have also assumed that integrity constraints can be written as implications which have only a single atomic condition. This assumption can be relaxed in various ways; for example, by allowing conclusions C which contain negative literals or implications. Thus the constraint
$$A \wedge B \to C$$
could be written as
$$A \to (B \to C)$$
and/or
$$B \to (A \to C)$$
whereas
$$\neg(A \wedge B)$$
could be written as
$$A \to \neg B$$
or
$$A \to (B \to false).$$

Any such relaxation would require appropriate extension of the *demo* predicate. A proof procedure which incorporates many of the features required of such an extension has been developed by Fung [Fun93], following an initial proposal of Kowalski [Kow92]. Further discussion of this matter is beyond the scope of this paper.

9.4 The Cycle Predicate Reconsidered

We can now formulate the recursive clause of the *cycle* predicate more precisely and more formally:

$cycle(KB, Goals, Input.InRest, try(Act, Result).OutRest, T) \leftarrow$
    $assimilate(KB, Goals, Input, KB', Goals', T)$
    $\wedge\ demo(KB', Goals', Goals'', R)$
    $\wedge\ R \leq n$
    $\wedge\ try\text{-}action(KB', Goals'', try(Act, Result), KB'', Goals''', T + R + 2)$
    $\wedge\ cycle(KB'', Goals''', InRest, OutRest, T + R + 3)$

Our intention is that the definition should have both procedural and declarative readings. The procedural reading requires the *cycle* predicate to be executed as a process concurrently with an environment process. The declarative reading specifies what sequences of inputs and outputs constitute acceptable behaviour.

Here the inputs and outputs are represented by an input stream and output stream respectively. The first item, *Input*, in the input stream represents the actual input at time T. The first item, $try(Act, Result)$, in the output stream represents the next output, if any, attempted at time $t + R + 2$. If the action Act is not "*nil*". then it is "tried" at the moment the output item is "consumed" by the environment. At the same time, the environment instantiates the variable *Result* indicating whether the attempted action succeeds or fails.

The time parameter, T, is local to the agent and behaves as an internal clock used to "time stamp" inputs and outputs when they are recorded in the knowledge base. For the sake of simplicity, we have assumed that time is measured in terms of inference steps, and that each inference step takes one unit of time. For simplicity, we have also assumed that observing and assimilating the input together take two time units, whereas trying an (atomic) action takes only one.

The constant n is the amount of resource available for goal reduction, whereas R is the actual resource consumed. Notice that, for the *cycle* predicate to execute efficiently as it is written, the condition $R \leq n$ needs to be executed as a constraint or to be coroutined with the execution of the *demo* condition. In a more elaborate version of the *cycle* predicate, n might be computed by the agent, varying in a manner which is appropriate to the circumstances.

The *try-action* predicate analyses the search space of goals $Goals''$ to determine whether it contains any action which the agent can try to execute. There are three cases:

- There is such an action, and the attempt to execute it succeeds. In this case, the agent commits to the branch of the search space containing the action.

- There is such an action, but the attempt to execute it fails. In this case the branch containing the action is discarded.

- There is no such action, in which case the search space is unchanged.

The *try-action* predicate can be defined more formally as follows:

$$try\text{-}action(KB, Goals, try(Act, Result), KB', Goals', T) \leftarrow$$
$$\quad action\text{-}needed(Goals, Act, Rest, AltGoals, T)$$
$$\quad \wedge\ Result = success$$
$$\quad \wedge\ Goals' = Rest$$
$$\quad \wedge\ KB' = (do(self, Act, T) \wedge KB)$$

$$\textit{try-action}(KB, Goals, try(Act, Result), KB', Goals', T) \leftarrow$$
$$\textit{action-needed}(Goals, Act, Rest, AltGoals, T)$$
$$\wedge\ Result = failure$$
$$\wedge\ Goals' = AltGoals$$
$$\wedge\ KB' = KB$$
$$\textit{try-action}(KB, Goals, try(nil, nil), KB', Goals', T) \leftarrow$$
$$\neg \exists Act, Rest, AltGoals\ \textit{action-needed}(Goals, Act, Rest, AltGoals, T)$$
$$\wedge\ Goals' = Goals$$
$$\wedge\ KB' = KB$$
$$\textit{action-needed}(Goals, Act, Rest, AltGoals, T) \leftarrow$$
$$Goals \equiv (do(self, Act, T) \wedge Rest) \vee AltGoals$$

Here the goal, $do(self, Act, T)$, is abducible in the sense that it has no definition in KB (before it is added to the knowledge base, in the case that the attempt to execute Act succeeds). Alternatively, it may be viewed as having a definition (*true* in case $Result = success$, $false$ in case $Result = failure$) which is held externally in the environment. This second view is similar to that of query-the-user, Sergot [Ser83].

Operationally, the calls to the predicate *action-needed* in the first two clauses select one branch of the search space from among all the branches containing an action which the agent can try to execute at that time. This instantiates the variable Act to a concrete value. The partially instantiated term $try(Act, Result)$ where $Result$ is a variable, is now available as an output on the output stream to the environment.

For the *cycle* predicate to behave effectively, it needs to be executed as a process concurrently with other processes corresponding to the environment and to other agents. The agent and the environment communicate with one another by means of terms in input and output streams. The input stream of the agent is an output stream of the environment; and, conversely, the output stream of the agent is an input stream of the environment. As in concurrent logic programming languages, a process can send a message, such as $try(Act, Result)$, containing a variable, which is instantiated by the process receiving the message. In this case, the environment process instantiates the variable, $Result$, to one of the values *success* or *failure*.

If the result of the attempted action is *success*, then the agent commits to the action and discards the alternative branches of the search space. This commitment is similar to the non-deterministic commitment of concurrent logic programming languages. It is tempting to speculate that the declarative reading of the *try-action* predicate definition, embedded within the definition of the *cycle* predicate, might give a (meta)logical semantics to such non-deterministic commitment.

The *try-action* predicate combines committed choice, when actions succeed, with search, when actions fail. The second clause in the definition deals with the case when actions fail. The branch (being a conjunction equivalent to $false$) containing the selected and failed action is discarded and the agent enters the next cycle with the search space consisting of the remaining alternative branches.

Notice that the definitions of the *demo* and *action-needed* predicates are neutral with respect to the search strategy used to select branches in the search space. Thus backtracking upon failure is only one of the many possible search strategies covered by our definitions.

Another possibility is to employ an evaluation function to evaluate alternative courses of action represented by alternative branches of the search space. The same evaluation function could be used both to direct the search towards the most promising part of the search space in the definition of *demo* and to select the most promising next action in the definition of *try-action*. Such use of an evaluation function would need to be taken into account in determining the total amount of resources consumed by the agent within a given cycle.

As we will see in greater detail in the next section, the branch

$$do(self, Act, T) \land Rest,$$

whose first subgoal is selected for attempted execution at time T, represents a partial plan for accomplishing the agent's goals. The action associated with the first subgoal represents the first step of the plan; $Rest$ represents the remainder of the plan. Depending upon how much resource is available for generating the plan before the first action is needed, $Rest$ will contain more or less detail about the rest of the plan. The less resource, the less detail; and the more the agent behaves reactively. The more resource, the more detail; and the more the agent behaves deliberately.

The advantage of deliberation is that it allows the agent to look ahead, compare alternative partial plans and try the most promising alternative. In many cases it can avoid trying an unproductive action by foreseeing that it would eventually lead to failure. The disadvantage is that in many situations the need to perform an action is so urgent that there simply is no time for such deliberation. Moreover, when the future is unpredictable, planning can be a waste of time.

The value of the parameter n determines the balance between deliberation and reactivity. As we have already remarked, it might be useful for this parameter to be computed and for its value to depend upon the circumstances. But the balance between deliberation and reactivity also depends upon the kind of knowledge that is represented in the knowledge base, as we will see in the following section.

9.5 Knowledge Representation Matters

The feasibility of the agent architecture outlined above depends crucially upon the way in which knowledge is represented in the knowledge base. It must be represented, in particular, in such a way that the backward reasoning performed by the *demo* predicate generates plans in a forward direction, starting with an action that can be performed in the current state. Conventional logic-based representations of actions and their effects behave instead in such a way that backward reasoning corresponds to reasoning backwards in time while forward reasoning corresponds to reasoning forwards in time.

Consider, for example, the goal of going from one location to another. A typical logic-based representation, of the kind normally associated with the situation calculus of McCarthy and Hayes [MH69] or the event calculus of Kowalski and Sergot [KS86], might employ, along with frame axioms or persistence axioms, a clause of the following simplified form:

$$holds(loc(Agent, Y), T+1) \leftarrow holds(loc(Agent, X), T)$$
$$\land\ next\text{-}to(X, Y)$$
$$\land\ holds(clear(Y), T)$$
$$\land\ do(Agent, step(X, Y), T)$$

Using such a sentence in the definition of the *demo* predicate to reduce goals to subgoals would generate plans backwards, starting from the last action to the first. The agent would not be able to execute any action until it generates a complete plan.

What we need instead is a representation such as:

$$go(Agent, X, Z, T) \leftarrow holds(loc(Agent, X), T)$$
$$\land\ next\text{-}to(X, Y)$$
$$\land\ holds(clear(Y), T)$$
$$\land\ T' \leq T + n'$$
$$\land\ do(Agent, step(X, Y), T')$$
$$\land\ go(Agent, Y, Z, T')$$
$$go(Agent, X, X, T) \leftarrow holds(loc(Agent, X), T)$$

which includes both the current state and goal state in the same predicate. Here n' is a parameter (sufficiently larger than $n + 3$) regulating the rate of movement from one location to the next. Using such a representation backwards to reduce goals to subgoals generates plans forwards, starting from the first action in a plan. It can be interrupted any time after the first action has been generated, to try executing that action, even if no part of the rest of the plan has yet been generated.

Notice that even with this representation, however, conventional axioms such as

$$holds(loc(Agent, Y), T2) \leftarrow do(Agent, step(X, Y), T1)$$
$$\wedge\ T1 < T2$$
$$\wedge\ \neg \exists T^*[do(Agent, step(Y, Z), T^*) \wedge T1 < T^* < T2]$$

of the kind used in the event calculus, are still required (for example, to solve the condition $holds(loc(Agent, X), T)$ in the definition of go). The crucial matter is that the goal of the agent's going at time T to a destination Z from its current location X should be represented as

$$go(self, X, Z, T)$$

rather than as

$$holds(loc(self, Z), T).$$

Notice, however, that, although the definition of go might work in theory, it gives rise to a brute-force search, starting from the current location X, that will not work in practice. With this representation, unless the value of n is sufficiently large to allow the generation of a complete plan, the agent will move about at random, totally ignoring the destination Z.

For the agent to behave effectively, the choice of the location Y next to X in the definition of go needs to take the destination Z into account. In traditional AI approaches this is done by using heuristic functions to evaluate alternatives in the search space and to select more promising alternatives in preference to less promising ones. In expert system approaches, on the other hand, such knowledge is more commonly incorporated into the object-level knowledge base itself. In our case this can be done simply by adding an extra condition, $towards(X, Y, Z)$, to the definition of go to restrict the choice of next location Y to one whose distance to the destination is closest among all the locations next to X which are clear at time T. This extra condition can be specified in the form:

$$towards(X, Y, Z) \leftarrow \forall Y'[next\text{-}to(X, Y') \wedge holds(clear(Y'), T) \rightarrow$$
$$dist(Y, Z) \leq dist(Y', Z)]$$

which can be converted into conventional logic programming form in standard ways.

9.6 Logic-based Multi-Agent Systems

A preliminary version of the resource-bounded, logic-based agent architecture described above has been implemented in a multi-agent environment by Davila [DQ94]. Several agents are placed at various initial locations on a rectangular grid and are given the goal of going from their initial locations to different destinations.

Given a grid with no obstacles, except for those created by one agent temporarily blocking another, the implementation confirmed our expectation that planning confers no advantages over purely reactive behaviour. This is because, without an agent having a sophisticated model of the behaviour of other agents, in this environment the obstacles created by agents blocking one another are totally unpredictable.

The implementation was written in a combination of Prolog, used for programming the logic-based cycle of the individual agents, and April, for programming the interactions between the agents and the environment. April, McCabe and Clark [MC94], is a process-oriented symbolic language, which has grown out of the experience of using concurrent logic programming languages such as Parlog. The discussion of the previous section of this paper, showing how inputs and outputs can be implemented using input-output streams containing variables, suggests that an implementation combining Prolog and a concurrent logic programming language might be more appropriate from a logical point of view.

Although planning had no value in our simple multi-agent experiment, we anticipate that it will be important in other applications where predictions can be made reliably. To predict the future, an agent needs to model other agents as well as to communicate with other agents both to avoid conflicts and to achieve common goals.

It was in fact for the purpose of modelling other agents that we earlier proposed the use of meta-logic programming to solve the puzzle of the three wise men, Kowalski and Kim [KK91]. More recently, we have begun to investigate the use of argumentation to resolve conflicts between different agents, Kowalski and Toni [KT94].

9.7 Conclusions and Future Work

The proposal outlined in this paper is a first step towards the development of a resource-bound, logic-based agent architecture. Although it is firmly based on the use of logic both at the object level to represent domain knowledge and at the metalevel to control the observation-thought-action cycle, it makes important concessions to the anti-logic, reactive agent school. In particular, it concedes, that when the future is unpredictable, rational planning is not only a waste of time but interferes with the ability to act effectively and in a timely manner.

None the less, compared with purely reactive architectures, our logic-based agent model can exploit reliable knowledge about the future, to avoid short-term actions that ultimately and predictably fail to achieve long term goals. Moreover, it can be extended to make the future more predictable, both by exploiting meta-logic to reason about other agents and by allowing agents to communicate with each other to negotiate co-ordinated plans of action. Integrating such extensions with the simplified agent model outlined in this paper is an important direction for future research.

There are at least two other important extensions to be considered. One is to investigate how several agents can be combined so that to an external observer they behave as though they were a single agent. The other is to investigate how goal-oriented behaviour can emerge as a property of the behaviour of a single agent or collection of agents.

It seems that much of the work needed for the first of these extensions has already been done in the work on using meta-logic for combining theories Brogi et al. [BMPT94, BT95]. This needs to be developed further to take integrity constraints, goals, subgoals and actions into account.

The second of these extensions seems to be related to the use of integrity constraints to obtain the behaviour of condition-action rules. This, in turn, is related to logic programming representations in which conclusions of implications represent goals. These relationships need to be investigated further to take account of the properties that emerge when agents interact in multi-agent systems.

Acknowledgements

This work was partly supported by Fujitsu Laboratories. I am grateful to Krzysztof Apt, Jacinto Davila, Murray Shanahan and Franco Turini for their helpful comments on an earlier draft of this paper, and to Rodney Brooks and Alan Macworth for alerting me to the importance of the problem investigated in this paper.

References

[BMPT94] A. Brogi, P. Mancarella, D. Pedreschi, and F. Turini. Modular logic programming. *ACM Transactions on Programming Languages and Systems*, 16(4):1361–1398, 1994.

[BT95] A. Brogi and F. Turini. Fully abstract compositional semantics for an algebra of logic programs. 1995. To appear in *Theoretical Computer Science*.

[DQ94] J.A. Davila Quintero. *Knowledge assimilation in multi-agent systems*. MSc. Thesis, Imperial College, London, 1994.

[Fun93] T. H. Fung. *Theorem proving approach with constraint handling and its applications on databases*. MSc. Thesis, Imperial College, London, 1993.

[GN87] M.R. Genesereth and N.J. Nilsson. *Logical foundations of artificial intelligence*. Morgan Kaufmann Publishers Inc., 1987.

[Jia94] Y. Jiang. Ambivalent logic as the semantic basis of metalogic programming: I. In *Proc. Eleventh International Conference on Logic Programming*, pages 387–401, 1994.

[JS93]   A. Jones and M. Sergot. On the characterisation of law and computer systems: the normative systems perspective. In J.-J.Ch. Meyer and R.J. Wieringa, editors, *Deontic logic in computer science: normative system specification*, chapter 12. Wiley, 1993.

[KK91]   R.A. Kowalski and J.S. Kim. A metalogic programming approach to multi-agent knowledge and belief. In Lifschitz V., editor, *Artificial intelligence and mathematical theory of computation*, pages 231–246. Academic Press, 1991.

[Kow79]   R.A. Kowalski. *Logic for problem solving*. Elsevier, New York, 1979.

[Kow90]   R.A. Kowalski. Problems and promises of computational logic. In J.W. Lloyd, editor, *Proc. Symposium on Computational Logic*. Springer Verlag Lecture Notes in Computer Science, 1990.

[Kow92]   R.A. Kowalski. A dual form of logic programming. 1992. Lecture Notes, Workshop in Honour of Jack Minker, University of Maryland.

[KS86]   R.A. Kowalski and M. Sergot. A logic-based calculus of events. *New Generation Computing*, 4:67–95, 1986.

[KT94]   R.A. Kowalski and F. Toni. Argument and reconciliation. In *Proc. Workshop on Legal Reasoning, International Symposium on Fifth Generation Computer Systems*, Tokyo, Japan, 1994.

[LT85]   J.W. Lloyd and R.W. Topor. A basis for deductive database system. *Journal of Logic Programming*, 4:93–109, 1985.

[MC94]   F. McCabe and K.L. Clark. April – agent process interaction language. In *Proc. ECAI94 Workshop on Agent Theories, Architectures, and Languages*, 1994. To be published by Springer Verlag.

[MH69]   J. McCarthy and P.J. Hayes. Some philosophical problems from the standpoint of artificial intelligence. *Machine Intelligence*, 4:463–502, 1969.

[Ras94]   L. Rashid. A semantics for a class of stratified production system programs. *Journal of Logic Programming*, 21(1):31–57, 1994.

[Rei90]   R. Reiter. On asking what a database knows. In J.W. Lloyd, editor, *Computational Logic*, pages 96–113. Springer Verlag, Esprit Basic Research Series, 1990.

[Ser83]   M. Sergot. A query-the-user facility for logic programming. In Degano and Sandwell, editors, *Integrated interactive computer systems*, pages 27–41. North Holland Press, 1983.

[SK87]   F. Sadri and R.A. Kowalski. An application of general purpose theorem-proving to database integrity. In J. Minker, editor, *Foundations of deductive databases and logic Programming*, pages 313–362. Morgan Kaufmann, 1987.

10 Modal and Meta Languages: Consistency and Expressiveness

Luigia Carlucci Aiello, Marta Cialdea, Daniele Nardi, Marco Schaerf

Abstract

In knowledge representation several formalisms for reasoning about knowledge in a multi agent scenario have been proposed. More specifically, we can identify a family of languages based on the use of a modal operator and another one based on the use of first-order logic enriched with meta-level capabilities.

In this paper we consider these two approaches by addressing the issues of consistency that arise from self-referentiality, their expressiveness and the methods for translating classical modal systems into meta-level first-order formalisms.

10.1 Introduction

The need to reason about knowledge and reasoning arises in several applications of Artificial Intelligence (AI), where an explicit description of a reasoning process is used by the system (see for example Aiello et al. [ACN91] and Aiello and Levi [AL84]). In the past, there have been several attempts to build meta-level architectures for programming the control strategies of an inference process, thus providing a declarative specification of domain specific heuristics. More recently, there has been a growing interest in modeling the knowledge of an agent, which is aware of other agents and tries to take advantage of its knowledge about them. Knowledge about other agents' knowledge should thus be considered as any other source of information available to the agent for decision making about the actions to take. In particular, an agent may infer information from other agents' knowledge or ignorance of some facts; an agent may be introspective, namely capable of reasoning about its own knowledge and reasoning.

Two are the main approaches to formalize an agent's knowledge and beliefs; they are here indicated as the modal and the meta approach, respectively.

The modal approach has initially been developed by logicians and philosophers and later on it has attracted the interest of researchers in AI. It aims at formalizing the knowledge of an agent by a logic language augmented with a modal operator interpreted as knowledge or belief. The most common modal systems for knowledge and belief are characterized by possible-worlds semantics, which provides a descriptive view: that is to say, knowledge or belief are regarded as propositions expressing the relationship between the agent and the external world.

In meta-level approaches knowledge about knowledge is represented by admitting sentences to be arguments of other sentences, without abandoning the framework of first-order logic. As a result several formal systems can be obtained: in particular, we distinguish amalgamated and separated languages, depending on whether the meta-expressions are seen as an extension of the basic language, or as a distinct language. The meta-level approach has its roots in the work of logicians, but it has also emerged from the attempt to characterize forms of reasoning expressible in several systems based on logic languages, where one could effectively define provability by designing a meta-level interpreter for the language.

In this paper we review both modal and meta languages to represent knowledge and belief, trying to point out their relationship and their effectiveness as knowledge representation formalisms. We are not specifically concerned here with the difference between knowledge and belief; we are rather interested in how such notions can be realized. Therefore, in the paper the terms knowledge and belief are often used interchangeably.

We focus our presentation on the expressivity of the approaches considered, on the consistency problems arising in the formalization of epistemic notions and, in particular, knowledge and belief, and finally, on the possibility of translating modal languages for knowledge and belief in a meta-level setting. We cannot examine here all aspects of meta and modal systems. One important aspect that we do not address is their implementation.

The paper is organized as follows: In the next section, we briefly recall modal logics for knowledge and belief, while meta-level approaches are discussed in Section 3. Section 4 contains a discussion on the differences and merits of the two approaches. Finally, in Section 5 we draw some conclusions.

10.2 Modal Approaches

In this section we present the best known formalisms, based on modal logics, proposed in the AI literature to deal with knowledge and belief. The section only presents the basic notions of modal logics; a complete introduction to the subject can be found in any textbook on modal logic, such as, for example, Hughes and Cresswell [HC68, HC84].

10.2.1 The possible-worlds semantics

Possible-worlds structures were introduced in 1963 by Kripke [Kri63] as a tool to give a formal semantics to modal logics. The technique was subsequently used by Hintikka [Hin62] in the modal analysis of the concepts of knowledge and belief. The language $\mathcal{L}_k$ of a propositional modal system can be defined as follows:

- If α is a propositional constant then $\alpha \in \mathcal{L}_k$.
- If $\alpha, \beta \in \mathcal{L}_k$ then $\alpha \wedge \beta$, $\alpha \vee \beta$, $\alpha \supset \beta$, $\neg \alpha$ and $\mathbf{K}\alpha \in \mathcal{L}_k$.

Modal atoms i.e. sentences like $\mathbf{K}p$, are interpreted as "the agent knows p". Different properties can be reasonably required for the modal operator $\mathbf{K}$, depending on what has to be represented, thus determining a great number of different modal systems. The most common ones (called *normal* modal systems) can be given a formal semantics by means of structures consisting of sets of classical interpretations, called *possible worlds*. Every modal formula is assigned a truth value in each possible world, in such a way that ordinary propositions (those not containing the modal operator $\mathbf{K}$) are given the classical semantics; i.e. the meaning of the connectives is the same as in classical logic. In order to give a meaning to formulas prefixed by a modal operator, a binary relation on the worlds is introduced, called the *accessibility relation*, and a sentence of the form $\mathbf{K}p$ is considered to be true in a given world if and only if p is true in every world accessible from it, intuitively, p is true in every world the agent can imagine as a possible state of affairs.

Formally, a Kripke structure, or model, is a triple $\mathcal{M} = <W, R, V>$, where W is a set of possible worlds, R is the accessibility relation between possible worlds and V is the valuation function, assigning a truth value to every proposition in every world. Truth in a world w of a model $\mathcal{M}$, denoted by $\mathcal{M}, w \models$, is defined by:

- $\mathcal{M}, w \models p$ iff $V(p, w) = true$, where p is an atomic formula
- $\mathcal{M}, w \models \neg \alpha$ iff $\mathcal{M}, w \not\models \alpha$
- $\mathcal{M}, w \models \alpha \wedge \beta$ iff $\mathcal{M}, w \models \alpha$ and $\mathcal{M}, w \models \beta$
- $\mathcal{M}, w \models \alpha \vee \beta$ iff $\mathcal{M}, w \models \alpha$ or $\mathcal{M}, w \models \beta$
- $\mathcal{M}, w \models \mathbf{K}\alpha$ iff for all w_1 s. t. wRw_1, $\mathcal{M}, w_1 \models \alpha$

A formula α is true in a model $\mathcal{M}$, and we write $\mathcal{M} \models \alpha$ if we have $\mathcal{M}, w \models \alpha$ for all $w \in W$. The class of models of a particular modal system is determined by specifying properties that the accessibility relation R must satisfy. A formula α is valid in a modal logic ($\models \alpha$) iff it is true in every model of the corresponding class.

Modal propositional systems are given a Hilbert style proof theory[1] by adding modal axiom schemata and inference rules to the classical system. Normal systems are characterized by the *necessitation* rule:

$$\frac{\vdash \alpha}{\vdash \mathbf{K}\alpha}$$

---

[1] Most modal logics can be given also natural deduction and sequent proof systems, as well as tableaux and resolution style refutation systems.

that allows one to infer that an agent knows all the valid formulas of the language, and by the following schema:

$$K) \ \mathbf{K}(p \supset q) \supset (\mathbf{K}p \supset \mathbf{K}q)$$

Axiom schema K (in honor of Kripke), called distribution axiom, means that the modal operator of knowledge $\mathbf{K}$ distributes over implication.

Other systems are obtained by addition of axioms further constraining the behavior of the modal operator. The most common normal modal systems are characterized by necessitation, the schema K and a subset of the following axioms:

$$T) \ \mathbf{K}p \supset p$$
$$4) \ \mathbf{K}p \supset \mathbf{KK}p$$
$$5) \ \neg \mathbf{K}p \supset \mathbf{K}\neg \mathbf{K}p$$

The schema T (for truth), called knowledge axiom, means that the agent does not know anything false. Axiom schema 4, called positive introspection, means that if the agent knows something, then he knows that he knows it. Axiom schema 5, called negative introspection, means that when the agent does not know something then he knows that he does not know it.

The semantics of such systems is characterized by properties of the accessibility relation R. Let us consider, for example, the system whose axioms are all of the above, called $S5$. In the class of Kripke structures constituting its semantics R is a reflexive (as established by axiom T), transitive (4) and euclidean (5) relation.[2]

Subsystems of $S5$ are characterized by weaker accessibility relations (see Halpern and Moses [HM92] for a quick overview of this family of modal systems). In particular, if we want to model belief rather than knowledge, we must reject the schema T, since we admit that agents may believe false things.

An important consequence of any modal system for knowledge and belief that can be given a possible world semantics like the one that has been sketched above is logical omniscience. In fact, in the presence of necessitation, the agent necessarily believes every valid formula (if α is valid, then $\mathbf{K}\alpha$ holds). Moreover, if the agent knows p and $p \supset q$, that is, if $\mathbf{K}p$ and $\mathbf{K}(p \supset q)$ hold, then by (K) we deduce that the agent knows q i.e.: $\mathbf{K}q$ holds. This amounts to saying that the agent is able to apply modus ponens and therefore it is logically omniscient because it knows all the logical consequences of its knowledge. Note that while the axiom schemata (T), (4), (5) are consequences of the properties of the accessibility relation R, the axiom schema (K) holds in every Kripke

---

[2]A relation R is euclidean if for every worlds w, w_1, w_2 wRw_1 and wRw_2 imply w_1Rw_2. Notice that a relation that is reflexive, transitive and euclidean is an equivalence relation.

structure and the necessitation rule is correct no matter which the properties of the accessibility relation are (i.e. they are both implicit in the possible worlds semantics).

Till this point we have only considered the case of a single agent. The formal systems can be generalized to deal with multiple agents. In this case we have a number of modal operators, $\mathbf{K}_{a_1}, ..., \mathbf{K}_{a_m}$, one for each agent. The possible-world semantics is then defined by $\mathcal{M} = <W, R_1, ..., R_m, V>$, where $R_1, ..., R_m$ are m accessibility relations between possible-worlds, one for each agent. Even though in most cases the same modal system is associated with all the agents (see for example Halpern and Moses [HM92]), in principle it is possible to model situations where different agents are given accessibility relations with different properties.

10.2.2 Extension to a first-order language

As we have seen, modal propositional logic is a very interesting formalism to represent reasoning about knowledge and belief; however, a propositional language is not powerful enough and it seems necessary to extend the base language to a full first-order one. The semantics for a language admitting only closed first-order formulas in the scope of modal operators is straightforward; conversely, the extension to first-order formulas with free variables that are possibly quantified across the scope of modal operators raises several questions, some of which are briefly addressed in this section.

The most general way to define a first-order modal language is by means of the following formation rules (assuming the usual formation rules for the set of terms):

- $p(t_1, t_2, ..., t_n)$ is an atomic wff iff p is a predicate symbol of arity n and $t_1, t_2, ..., t_n$ are terms.

- if α and β are wffs so are $\alpha \wedge \beta$, $\alpha \vee \beta$, $\alpha \supset \beta$, $\neg \alpha$, $\exists x \alpha$, $\forall x \alpha$ and $\mathbf{K}\alpha$.

The possible world semantics can be extended as follows. Let $\mathcal{M} = (W, R, D, V)$, where W and R are the same as in the propositional case, D is a domain of objects and V is a standard valuation function, assigning a value, in every possible world, to both terms and predicates. The definition of truth is naturally obtained like in classical logic, by means of variable assignments and by addition of the classical clauses for the definition of satisfaction of quantified formulae.

An immediate consequence of such a definition is that the two formulae $\mathbf{K}\exists x p(x)$ and $\exists x \mathbf{K} p(x)$ are given a different meaning. The first formula predicates that in every possible world there is an individual (possibly different in different worlds) which has the property p, while the latter requires the existence of an individual (the same in all possible worlds) satisfying the property p. This corresponds to the difference between saying "The agent knows that there exists a murderer" (if someone was killed) and "There exists someone such that the agent knows he is the murderer", i.e. the agent knows who was the killer.

Kripke in his original work [Kri63] (as well as the sketchy definition above) assumes that the domain of quantification D is the same in every possible world. In the class of Kripke structures that satisfy such property the following formula is valid:

$$\forall x \mathbf{K} p(x) \supset \mathbf{K} \forall x \ p(x)$$

The above formula, called Barcan formula, excludes the case that the agent, although believing $p(x)$ for every object x, does not conclude $\forall x \ p(x)$ because it does not realize that the objects it has considered correspond to their totality. If, on the contrary, this is to be allowed and, consequently, the Barcan formula rejected, the fixed domain assumption is dropped in favor of a set of domains: a domain of individuals D_i is associated with every possible world w_i of a Kripke structure. This allows one to deal with situations where there are some objects to be considered only in some possible worlds.

A logical consequence of any system obtained by extending first-order logic with the axiom schema K and the rule of necessitation is the converse of Barcan formula:

$$\mathbf{K} \forall x \ p(x) \supset \forall x \mathbf{K} p(x)$$

From the semantical point of view, this means that the domains of the possible worlds structure are nested, i.e. if there is a connection between the two worlds w_1 and w_2 ($w_1 R w_2$) then the corresponding domains D_1 and D_2 are in the relation $D_1 \subseteq D_2$. The interpretation of a universally quantified formula in a world w is given by letting the variable range over the domain associated with w. The hypothesis of non decreasing domains is crucial in that it assures that the interpretation of a formula of the form $\forall x \ \mathbf{K} p(x)$ in a world w_i is well defined: if x is assigned an object $d \in D_i$, it can be checked whether $p(x)$ is satisfied by d in any world w_j that is accessible from w, provided that $d \in D_j$.

In order to relax the assumption on domain inclusion, a semantics based on 3-valued logics has been proposed, where all predicates over terms denoting objects not in the domain are assigned the truth value *undefined*: if a formula, containing a term a denoting the object d, is evaluated in a world whose domain does not contain d, then this formula is assigned the truth value *undefined*.

In the previous subsection we have mentioned the problem of logical omniscience raised by the modal treatment of knowledge. Another aspect where the modal treatment apparently conflicts with intuition is in relation with *referential opacity*, i.e. the fact that Leibnitz's law of equality substitution:

$$\forall x \forall y (x = y \supset (\alpha[x] \equiv \alpha[y]))$$

should not hold in the scope of a modal operator. For example, if we know that *Zeus* is a God ($\mathbf{K} God(Zeus)$), and it is the case that *Zeus* and *Jupiter* are two names for the same person ($Zeus = Jupiter$), then it does not necessarily follow that we know that $Jupiter$ is a God (**K**$God(Jupiter)$); in fact it follows only if we also know such equality i.e. **K**($Zeus = Jupiter$). However, if the usual axioms for first-order theories with identity are added to a modal system, referential opacity no longer holds. In order to preserve this property, the law of substitution has to be weakened. The semantical counterpart of this is a modification of the way terms are interpreted. In the above definition of first-order semantics we have considered the case where individual and functional constants have the same denotation in all possible worlds. This feature is called *rigid designation*. However, in order to preserve referential opacity the interpretation of terms must be *non-rigid*: individual and functional constants may denote different objects and functions in different worlds, i.e. terms are interpreted intensionally. Note however that when non-rigid designation is adopted in conjunction with variable domain semantics, the classically valid sentence:

$$\forall x \alpha[x] \supset \alpha[t]$$

does not hold in general, but it can be assumed only when x does not occur in a modal context in α.

10.3 Meta-Level Approaches

The distinction between language and meta-language has been a fundamental issue in the history of mathematical logic. It has been introduced in order to overcome the paradoxes caused by self-reference (the most famous one is probably the impossibility of assigning a truth value to the sentence "this sentence is false"). An informal distinction between language and meta-language has been used by many mathematicians already in the 19th century, and the first discussions about this topic arise with the first formal systems introduced by Frege [Fre85]. His aim was to have a universal language where objects and properties of objects have the same status. Thus predicates of the language were also objects and therefore could be replaced for variables. Russell [Rus05] proved that Frege's system was inconsistent, consequently he proposed a hierarchical system where each level has a different language and formulas at one level can only quantify over objects at the lower level. He also showed that such a system is consistent.

This same distinction has then been deeply analyzed by researchers in Artificial Intelligence who are interested in representing information in a structured framework where it is possible to define different levels of description of the problem, or situation, at hand. Typically, the information needed in knowledge-based systems is of two kinds: object-level or knowledge about the problem and meta-level or knowledge about the object-level (for instance control knowledge on how to process object-level information). The various uses of meta, either as the controller of the object-level or as a declarative specification of properties of the object-level, has been pursued effectively in many AI systems (see for example Aiello and Levi [AL84]). The need to formally represent these two different kinds of knowledge has led to the study of meta-level architectures, where both object-level and meta-level knowledge are explicitly represented.

Meta-level architectures, as well as the formal systems developed in logics, can be classified in two main categories, here referred to as *amalgamated* and *separated*. Roughly speaking, a system is amalgamated when there is no distinction between object and meta language; in particular, this is the case when the formal system is characterized as a single theory. We can consider Frege's system as the first amalgamated system and Russell's system as the first separated one. In separated systems, also called hierarchical, sentences of the meta-level and object-level language are kept separate, so that the object-level theory is an ordinary (first-order) theory, while the meta-theory contains sentences about the object-level theory, using names to refer to symbols and expressions of the object level.

Below we sketch the main issues concerning meta-level architectures. For this purpose we consider a logical theory as composed of two parts: a set of formulas in some language $\mathcal{L}$ and a deductive apparatus. If we call Δ the set of axioms and R the set of the inference rules, then a theory T is defined as:

$$T = \{\phi | \phi \in \mathcal{L} \text{ and } \Delta \vdash_R \phi\}$$

where $\vdash_R$ denotes provability using the rules of inference R.

If one wants to assert properties of the theory T, the language $\mathcal{L}$ must be extended in such a way that facts about T and hence about formulas of $\mathcal{L}$, can be stated. Therefore, we rename $\mathcal{L}$ as $\mathcal{OL}$ (object-level language) and introduce the meta-language $\mathcal{ML}$ (meta-level language), as the language used to state properties about T. Consequently, we use the symbols $\Delta_\mathcal{O}$ and $\Delta_\mathcal{M}$ to distinguish the axioms of the object-level and meta-level, respectively, while $R_\mathcal{O}$ and $R_\mathcal{M}$ denote the inference rules of the two levels, object and meta, respectively. Finally, we denote provability at the object-level as $\vdash_\mathcal{O}$ and provability at the meta-level as $\vdash_\mathcal{M}$.

The distinction between amalgamated and separated systems is originated by a different method of defining the meta-language: in the case of amalgamation $\mathcal{ML}$ is regarded as an extension of $\mathcal{OL}$, while in the case of separation $\mathcal{ML}$ is taken as the language of a theory distinguished from the object-level one. Before discussing the details of these two kinds of systems, we describe the basic features of a meta-level architecture:

- the *naming* relation, providing names at the meta-level for the expressions of the object-level theory;

- the meta-level formalization of object-level properties, such as truth, provability, knowledge, etc.;
- the *linking rules*, that establish the connection between object-level and meta-level in the inference process.

As already noted, the meta-language should be expressive enough to refer to the symbols and sentences occurring in $\mathcal{OL}$, since predicates in $\mathcal{ML}$ may take expressions of the object-level as arguments. This is achieved by establishing a relationship between meta-level symbols and object-level expressions (*naming*), that provides names at the meta-level for object-level expressions. The most common naming relation is called ground representation in Lloyd [Llo88]; it maps constant, variable, function and predicate symbols of the object-level into meta-level constants of different sorts[3]. Object-level terms, atoms and wffs are then represented by meta-level terms using constructors for function application, predicate application, logical connectives, respectively. For example the object-level atom $P(a)$ can be represented as $mkappl1(P^*, a^*)$, where P^* is a meta-level constant of sort predicate symbol representing P and a^* is a meta-level constant of sort constant symbol representing a. The formal properties of the naming relation are studied by van Harmelen [vH92], who notices that the naming relation is normally assumed to be total, functional and injective.

A useful abbreviation that is often used to provide names to object-level expressions is by means of a quoting mechanism. For example '$P(a)$' may be used to denote a meta-level constant symbol naming the object level atom $P(a)$. The situation is slightly more complex if we want to express quantified expressions at the meta-level. To this end $P(a)$ is represented as 'P'('a'), thus making it possible to use meta-level variables inside quoted expressions.

The reason for introducing a meta-level description is to formalize properties of the object-level theory. Therefore, at the meta-level we find the axioms that account for notions such as truth, provability, knowledge and belief. The notion of truth is typically formalized through a one place predicate $TRUE(`\alpha`)$, which states the truth of an object-level formula α. The formalization of truth has been studied mostly by logicians and philosophers.

Provability has a more direct interest for AI applications since it may be used to provide a system with some knowledge about its own reasoning process and is therefore related to the problems of inference control. Provability is usually represented by a meta-level

---

[3]Lloyd [Llo88] points out that the naïve PROLOG meta-interpreter, uses a different representation technique, called non ground, since it represents object-level variables as meta-level variables allowing one to use the basic unification mechanism of PROLOG. Although very useful for some forms of meta-level programming, the non ground representation is not general and it will not be further considered here.

predicate $DEMO(\text{`}T\text{'}, \text{`}\alpha\text{'})$, which denotes the derivability of the object-level formula α, named 'α' at the meta-level, in the object-level theory T represented by 'T'.

The distinguished meta-level predicates, expressing properties of the object-level, are related to the object-level itself by *linking rules*. The most widely used ones are the so-called reflection principles, which take the form of inference rules:

$$\frac{T \vdash_{\mathcal{O}} \alpha}{PR \vdash_{\mathcal{M}} DEMO(\text{`}T\text{'}, \text{`}\alpha\text{'})}$$

$$\frac{PR \vdash_{\mathcal{M}} DEMO(\text{`}T\text{'}, \text{`}\alpha\text{'})}{T \vdash_{\mathcal{O}} \alpha}$$

where PR is a meta-level description of provability at the object-level. The first one, known as Upward Reflection, allows one to assert the provability of $DEMO(\text{`}T\text{'}, \text{`}\alpha\text{'})$ in the meta-theory when α can be proved in the object-theory T. The second reflection principle allows one to assert the provability of α in the object-level theory whenever it is possible to prove $DEMO(\text{`}T\text{'}, \text{`}\alpha\text{'})$ in the meta-theory. Other linking rules can be adopted, depending on the relation between levels that has to be modeled. Giunchiglia and Serafini [GS94] propose different sets of linking rules in a hierarchical multi-level structure and prove their correspondence with some modal systems.

Many attempts to formalize knowledge and belief using meta-level systems have been proposed for AI applications (see, for example, Aiello et al. [ANS91], Brogi and Turini [BT91], Haas [Haa86], Kim and Kowalski [KK91], Konolige [Kon81] and McCarthy [McC79]). In this framework knowledge and belief are usually represented as two place predicates $K(a, \text{`}\alpha\text{'})$ and $B(a, \text{`}\alpha\text{'})$, respectively, where a denotes an agent and 'α' is the proposition known, or believed by the agent. In this case the meta-level theory is used to represent the beliefs about other agents' belief, or the introspective beliefs, i.e. beliefs about belief, whereas the object-level theory represents the agent's own beliefs.

10.3.1 Amalgamated approaches

The idea of the amalgamated approach is to integrate object-language and meta-language in a single framework. More precisely this means that:

$$(\mathcal{OL} \cap \mathcal{ML}) \neq \emptyset$$

As a special case we have:

$$\mathcal{OL} = \mathcal{ML}$$

namely there is no distinction between object and meta-language. Therefore mixed sentences, i.e. those combining object-level and meta-level expressions are allowed. The term amalgamation was introduced by Bowen and Kowalski [BK82] in a logic programming setting. They argue that a necessary condition for amalgamation is that the meta-language is powerful enough to express interesting notions, such as provability at the object-level. Bowen and Kowalski show how Horn Clause languages satisfy this requirement and provide a meta-level description of provability for a Horn Clause language, which can be regarded as a meta-level interpreter for Horn Clauses. Therefore, they propose a system where $\mathcal{OL} = \mathcal{ML}$ and the language is Horn Clauses.

Their system is then completed by a ground naming relation, the meta-level clauses defining a two-place predicate $DEMO$ to represent $\vdash_\mathcal{O}$ and the reflection principles as above. They discuss several examples to show the expressivity of mixed sentences such as:

$$innocent(x) \leftarrow Person(x),$$
$$\neg Demo(facts, `Guilty(x)'),$$
$$Relevant(facts).$$

stating that a person is innocent if he or she cannot be proved guilty by analyzing some relevant facts.

The formalization of epistemic notions in a meta-level amalgamated style has been pursued in the work by Kim and Kowalski [KK91]. In the paper they show how it is possible to capture some of the properties of modal logics via an extended $DEMO$ predicate. As an example, they propose a formalization and solution of the three wisemen puzzle.

McCarthy [McC79], starting from the original aim of Frege of distinguishing between sense and denotation, proposes a formalization based on first-order logic where *concepts* are used to formalize the sense of propositions and *denotations* provide the interpretation of concepts in direct contexts. Concepts and denotations are related by an explicitly defined denotation function and are amalgamated without relying on an explicit object-level meta-level organization. In this framework McCarthy shows how to formalize the truth of propositions, as well as the epistemic notions of knowledge and belief. More recently McCarthy [McC93] further developed the notion of context, pointing out several relations that hold among contexts.

Another amalgamated language has been proposed within the OMEGA system of Attardi and Simi [AS84, SM88]. OMEGA is based on the notion of description, that resembles a frame but is characterized by set-theoretic semantics. The meta-level representation of the object-level is introduced in OMEGA by means of meta-descriptions and is used in the formalization of epistemic notions. The amalgamation is very similar to that of Bowen and Kowalski, but for the use of descriptions. The main distinction between the two systems is in the reflection principles, that in OMEGA take the following form:

$$\frac{T_1 \cup T_2 \vdash \alpha}{T_1 \cup PR \vdash DEMO(`T_2`, `\alpha`)}$$

$$\frac{T_1 \cup PR \vdash DEMO(`T_2`, `\alpha`)}{T_1 \cup T_2 \vdash \alpha}$$

The point is the dependence on a background theory. In the reflection principles of OMEGA the object-level theory is T_2, and the other axioms of the amalgamated language are in T_1. Bowen and Kowalski reflection principles only use T_2 (i.e. the object-level axioms).

This difference is of great importance and in order to explain it we need to recall the notion of conservative system. We say that a system is *conservative* with respect to a set of inference rules, when every theorem provable with the addition of those rules, is also provable without them. In other words, the system is conservative if the reflection principles are derivable in it, otherwise it is non-conservative. In fact, the system of Bowen and Kowalski is conservative, w.r.t. the introduction of the reflection principles, while OMEGA is not because of the different formulation used.

As an example taken from Attardi and Simi [AS91], in OMEGA the following statement is valid:

$$DEMO(`A`, `B`) \supset (A \supset B)$$

and it extends the original theory. Note that this same statement does not hold in Bowen and Kowalski's system.

As we have already seen in the previous subsection, one of the main reasons for the use of meta has been the development of theories of truth. There are, however, strict links between theories of truth and theories of knowledge and belief. Turner [Tur90] has considered these similarities, reviewing the most important approaches to theories of truth and applied these results to define syntactic, meta-level treatment of modalities, including modal logics for knowledge and belief. In his book he presents several interesting systems for truth and modality, pointing out which restrictions are necessary to preserve the consistency of the system.

While Turner's work is concerned with truth and modality in general, Davies [Dav90] extends it to specifically deal with the concepts of knowledge and belief. Davies proposes two meta-logics that are extensions of the modal systems T and $S4$, where the modal operator is replaced by a meta-predicate. In order to preserve consistency the necessitation rule is weakened so that it only applies to tautological formulas of the object language and not of the amalgamated language. This restriction is somehow unnatural, since his system does not validate the formula $Know(`Know(A \vee \neg A)`)$ since, even if $Know(`A \vee \neg A`)$ is a tautology of the amalgamated language, it is not a tautology of the object language.

Using the results obtained by Davies and Turner, Attardi and Simi [AS91] prove that the system proposed by Bowen and Kowalski is consistent, while OMEGA is not. In fact, the more general form of reflection adopted in OMEGA makes $DEMO$ a truth predicate, i. e. $DEMO(\{\}, `B`) \equiv B$.

To overcome this negative result, they propose two different reflection principles which ensure consistency, which are the following ones:

$$\frac{T \vdash_{PC} \alpha}{PR \vdash DEMO(`T`, `\alpha`)}$$

$$\frac{T_1 \cup PR \vdash DEMO(`T_2`, `\alpha`)}{T_1 \cup T_2 \vdash \alpha}$$

These revised reflection principles are more restricted in the possibility of applying the reflection upwards, since it only applies to situations where no underlying theory is present and only the classical rules of inference are used ($\vdash_{PC}$ denotes provability with only classical rules, not the reflection principles). Hence, this system suffers from the problem mentioned for Davies's system with restricted necessitation. Attardi and Simi [AS94] further develop their approach pointing out some difficulties arising in the arbitrary use of reflection principles.

10.3.2 Separated approaches

The idea of separating the meta-level from the object-level is the simpler way to avoid the paradoxes originating from self-reference. This is achieved by defining a meta-level language distinct from the object level language and by defining two distinct theories, the object-theory and the meta-theory. Therefore we have:

$$\mathcal{OL} \cap \mathcal{ML} = \emptyset$$

In this way it is not possible to write sentences mixing object-level and meta-level expressions and the connection between meta-theory and object-theory is provided only by the naming relation and by the reflection principles. Obviously, since the meta-theory is itself a theory we can have a distinct meta-meta-theory and so on. Therefore, we can consider separated systems as multi-language or multi-theory.

The above features can be found in the FOL system by Weyhrauch [Wey80]. FOL is based on the notion of context, which should be regarded as a finite representation of a first-order theory. Each context has its own language, a set of axioms and a partial model of the theory[4]. Meta-theory and object-theory are thus defined as two separate contexts. The naming relation between the meta-context and the object-context is established by specifying the interpretation (in the partial model of the meta-context) of meta-level symbols as object-level structures. In fact, the object-context is viewed as a partial model of the meta-context.

In FOL the reflection principles can be applied in a restricted form that allows one to make a deduction step in the object-context by temporarily switching the attention to the meta-context (reflection up), then doing a proof in the meta-context and finally getting the result in the object-context (reflection down).

The FOL system has been used in the formalization of epistemic reasoning in a scenario with multiple agents in the work by Aiello et al. [ANS88, ANS91]. In this work an agent is represented as a structure of contexts, which makes it possible to formalize reasoning about other agents' knowledge and reasoning. Although in a separated framework it is not possible to represent the axiom schemata of the modal systems for knowledge and belief, the work shows that by a suitable structure of contexts and a clever application of the reflection principles complex reasoning tasks can be accomplished.

Early work by Konolige [Kon81], where he tries to formalize epistemic notions such as knowledge and belief in a meta-level framework, can also be regarded as a separated approach. As for his subsequent work on the sentential model of belief, his aim is to describe the beliefs of an agent through a deductive structure computing the consequences of a basic belief set. The connection between the object and the meta-level is achieved by means of a form of semantic attachment.

The explicit use of meta-level specifications in separated architectures for logic programming has been advocated by several authors (Aiello et al. [ACS86], Costantini and Lanzarone [CL89] and Hill and Lloyd [HL94]); this work is mainly concerned with inference control and the proposed systems are not well suited to the formalization of complex epistemic notions.

The separation of languages has several consequences on the expressivity of a meta-level system. Let us first consider the naming relation. For example $name(P, `P`)$ says

---

[4]A partial model is an ordinary model but partially specified, that is it can be undefined for some expressions. A more detailed description of this issue can be found in Nardi [Nar88] and Talcott and Weyhrauch [WT90]

that 'P' is the meta-level symbol representing the object-level predicate P. In this form the naming relation is not directly expressible in a separated language, because it is indeed a mixed sentence, since P belongs to the object-level language and 'P' belongs to the meta-level language. However, if we build an additional meta-level M', which can refer to the previous meta-level M and to the previous object-level O, we can write in M', $name(P^M, P^O)$, where P^M is the name in M' of the name in M of P and P^O is the name in M' of P. This construction, besides being relatively complex, needs to be repeated if one wants to describe the naming relation for M'.

Similar problems arise when we try to express facts that hold at every nesting depth of a meta construction. This criticism is developed by Perlis [Per85], who considers two examples one involving universal quantification and the other existential quantification. Let us consider the sentence $\forall x(Bel(a,x) \supset Not\text{-}religious\text{-}belief(x))$, expressing that a has no religious beliefs. If we consider a language with separate levels of meta representation we can write the sentence $\forall x_i(Bel_{i+1}(a,x_i) \supset Not\text{-}religious\text{-}belief_{i+1}(x_i))$, which is restricted to the levels i and $i+1$ and is not easily generalizable for all levels. This observation also applies to the axiom schemata of the modal systems of knowledge and belief that cross the levels and thus cannot be directly expressed in a separated framework.

Let us now consider a meta-level separated architecture for representing a situation with two agents a_1 and a_2. The representation of this scenario requires that each agent (a_1 for example) is equipped with a meta-level context M_1 and two object-level contexts T_{11} and T_{12}. In this way it is possible to express the introspective beliefs of a_1, through T_{11} and the beliefs of a_1 about a_2, within T_{12}. For example $\mathbf{B}_{a_1}\mathbf{B}_{a_2}p$ corresponds to having asserted p in T_{12}. The process of reasoning about knowledge is then described by the meta-level axioms and is accomplished by applying the reflection principles. However, such an architecture is limited to express reasoning of one level of nesting and assertions of the kind $\mathbf{B}_{a_1}\mathbf{B}_{a_2}\mathbf{B}_{a_1}p$ are not directly expressible without adding more structure. Moreover, in this way one cannot represent disjunctive or negative information as object-level beliefs (see for example the points raised by Moore [Moo77]). For example the sentence $\mathbf{B}_{a_1}\neg\mathbf{B}_{a_2}p$, cannot be explicitly represented in an object-level theory containing the beliefs of a_1 about a_2.

The main advantage of separated architectures is ease of implementation, although in many cases this is done for the simple case of two levels: object and meta.

10.4 Modal and Meta Languages: A Discussion

The discussion of the relationships between modal and meta languages in the formalization of knowledge and belief addresses three issues: expressivity, consistency and translations.

10.4.1 Expressivity

In this subsection we consider some of the representational issues addressed by the modal formalism: referential opacity, logical omniscience and quantifying-in. In addition, we show other forms of quantification at the meta-level, that make it possible to combine reasoning about knowledge and belief with abduction and other forms of nonmonotonic reasoning.

As already seen, referential opacity is a fundamental property if we want to realistically formalize the notions of knowledge and belief. In fact, given this property, the substitutivity of equals within the scope of modal operators is not allowed and this avoids the unacceptable assumption that agents can always substitute equals in every context. In general, if it is possible to distinguish between object-level and meta-level, referential opacity can be enforced. In fact, the equality of two object-level individuals, is not confused with the equality of their meta-level representations. To this end it is useful to use sorts to distinguish the levels of meta-representation. There are several ways to obtain the same result in amalgamated languages; an interesting proposal has been presented by Arbab [Arb89]. In his work he describes a system where every variable and constant is suitably indexed according to the position where it appears. This more complex naming mechanism allows one to distinguish the same constant when it appears in different contexts, hence allowing the substitution of equals only in equivalent contexts. This solution is actually very radical, since there is almost no relation between the various occurrences of the same term. To relax this assumption, we can define a more sophisticated naming mechanism, where we can still allow substitutions in different, but somehow related contexts, where this notion of similarity can be encoded in the naming function.

The other distinguishing property of classical (possible-world) modal approaches is logical omniscience. Indeed this property is sometimes considered problematic, because it corresponds to the often unrealistic assumption that agents are represented as ideal reasoners, capable to infer all the logical consequences of their basic beliefs. In a meta-level logic it is possible to model agents with limited reasoning capabilities by describing a weak deductive (object-level) system at the meta-level, like in the logic of explicit beliefs of Levesque [Lev84].

Up to this point, we have considered a propositional object language and to move to a first-order language we need to consider the representation of $\forall x \alpha(x)$ and $\exists x \alpha(x)$.

The ground representation requires a meta-level constant symbol, that we call mcx, to represent the object-level variable x, so that $\forall x \alpha(x)$ is represented by the meta-level term $forall(mcx, \alpha'(mcx))$ where $\alpha'(mcx)$ denotes the representation of the formula $\alpha(x)$ where the free occurrences of the symbol x have been represented by the meta-level constant mcx. This representation of object-level variables requires an explicit treatment of quantifiers at the meta-level.

Quantifying inside the scope of a modal operator should be regarded as a form of quantification at the meta-level. In fact, if we consider $\forall x \mathbf{B}\alpha(x)$, the object-level theory is associated with what is believed, that is the formula inside the scope of $\mathbf{B}$. Thus the free occurrences of the variable x are interpreted in terms of a meta-level quantifier and the variable x is defined as a meta-level variable ranging over the individuals of the object-level domain. In this framework it is possible to represent different interpretations of quantifying-in. In particular, following McCarthy [McC79], we can define truth in a possible world, thus admitting an explicit representation of the interpretation of quantifiers within possible worlds.

Finally we sketch some ideas about the use of meta-level logics to combine reasoning about knowledge and belief with other forms of common sense reasoning. In particular, we focus on abduction, since reasoning about other agents' knowledge often has a hypothetical nature, very close to the idea of abductive reasoning. For example, if we want to discover the explanation of a certain information that can be attributed to an agent, we can use abduction. Abductive explanations of a given fact φ are usually characterized as those sentences α's that, in conjunction with background knowledge Θ, logically imply φ and enjoy some additional properties that make them preferred over other explanations. Among such properties, consistency with Θ and minimality in the set $\{\beta : \Theta \cup \{\beta\} \models \varphi\}$ are required. Levesque [Lev89] characterizes abduction in terms of belief, thus making it dependent upon the logical consequence relation modeled by the chosen notion of belief. In other words, abductive reasoning is related to the logic of belief modeling the reasoning capabilities of the agent. If $\mathbf{B}$ is a belief operator and e an "epistemic state", i.e. a set of background beliefs determining which sentences are believed, the notation $e \models \mathbf{B}\alpha$ says that α is believed at the epistemic state e. Abductive explanations for a fact φ in the epistemic state e are chosen among those $\alpha's$ belonging to:

$$\{\alpha : e \models \mathbf{B}(\alpha \supset \varphi) \land \neg \mathbf{B}\neg \alpha\}$$

Good explanations are required to be minimal in the above set, according to a simplicity criterium taking into account the literals occurring in them.

In a meta-level system we can easily define a predicate $Explains(\alpha, \varphi, e)$ which is true iff α explains φ in e according to the above definition. A similar definition cannot be simply embodied within modal systems. A simple example is discussed by Bowen and Kowalski [BK82].

10.4.2 Consistency

Here we briefly address the consistency of the systems that arise in the formalization of epistemic notions. A more extensive discussion of these issues is outside the scope of this paper, we refer the reader to Perlis [Per85, Per88] and Turner [Tur90] for more details.

Amalgamated logics for knowledge and belief must be carefully designed in order to avoid inconsistency. In fact, let T be a first-order theory with naming capabilities and a provability predicate, i. e. a predicate $demo$ satisfying:

$D1$. If $T \vdash \phi$ then $T \vdash demo(`\phi')$;

$D2$. $T \vdash demo(`\phi') \wedge demo(`\phi \supset \psi') \supset demo(`\psi')$;

$D3$. $T \vdash demo(`\phi') \supset demo(`demo(`\psi')')$.

Let us assume, moreover, that T satisfies the following diagonalization property:

Diag. Let $\psi(x)$ have only x free. There is a sentence ϕ such that $T \vdash \phi \equiv \psi(`\phi')$.

Note that this property is satisfied by any sufficiently expressive amalgamated metalanguage. If the schema $demo(`\psi') \supset \psi$ is added for any $\psi$ to the above theory $T$ – i. e. $demo$ represents a sound provability relation – then the resulting theory is inconsistent (Löb's theorem). In particular, $demo$ cannot be a truth predicate satisfying the schema $demo(`\psi') \equiv \psi$.

In a modal approach, instead of the $demo$ predicate a modal operator $\Box$ is used. The modal counterpart of $demo(`\phi')$ is the formula $\Box\phi$, thus the modal version of $D1$, $D2$ and $D3$ are respectively the necessitation rule and the axiom schemata $K$ and 4. In a modal setting the addition of the schema $T$ (modal counterpart of $demo(`\psi') \supset \psi$) does not lead to inconsistency because the modal version of *Diag* does not hold in general, i. e. the following property does not hold for $S4$:

MDiag. Let $\psi(p)$ be a modal formula where the proposition p occurs. There is a sentence ϕ such that $T \vdash \phi \equiv \psi(\phi)$, where $\psi(\phi)$ stands for the formula obtained by substituting ϕ for each occurrence of p.

Modal theories of provability have been studied where *MDiag* holds and, obviously, the modal axiom T does not hold. For a discussion of this subject see Smorynski [Smo85].

It follows from the above considerations that a straightforward translation of the modal axioms into a meta-level amalgamated system produces an inconsistent result, as shown by Montague [Mon63]. On the other hand, the addition of self-referential capabilities to a modal system can make it inconsistent, see Perlis [Per88].

10.4.3 Translation

In this subsection we address the attempts made of directly translating modal theories for knowledge and belief into first-order theories where the modal operator is replaced by a (meta) predicate and the axiom schemata by other schemata. Translations of modal systems into first-order theories can also be based on the possible-world semantics. In this case worlds are explicitly mentioned in the formulas and the accessibility relation is axiomatized as a binary predicate (see Ohlbach [Ohl93] for a recent overview). In the following we consider syntactic translations where the modal operators – which represent meta-level notions – are translated into meta-level predicates. Montague [Mon63] has shown that any theory obtained through this translation is inconsistent if it contains, at least, the system T. Further results along the same line have been obtained by Thomason [Tho80]. For a while these results discouraged any attempt of treating modalities as syntactical objects, but recent work on this topic overcomes the above mentioned problems.

Des Rivieres and Levesque [dRL88] point out that the causes of inconsistency in the theories resulting from the translation of a modal system are in the fact that they are defined on a broader set of formulas. More precisely, the set of formulas to which we can apply the translated axiom schemata, in the first-order case, does not contain only the translation of the formulas allowed in the modal case, but also includes some other formulas. They prove that we can have a consistent first-order theory if the axiom schemata can apply only to formulas which are a translation of modal ones; called *regular* formulas. In fact, in the modal case, the variables that appear free within the scope of a modal operator range over individuals of the interpretation domain, while in the meta-level translation they can range, within the scope of the meta-level predicates representing the modal operators, over the meta-domain, which includes names of first-order expressions. Consequently, the translation of a modal axiom applies to (the name of) any formula of the language, including, for example, $\exists x\ knows(x)$, that is not the translation of any modal formula. Des Rivieres and Levesque's result ensures consistency of the first-order treatment, whenever the meta-variables range is restricted to regular formulae, i.e. formulae that have a modal counterpart.

Starting from a different perspective, the work of Turner [Tur90] also deals with the problem of defining consistent systems for the syntactic treatment of modality. However, he is not interested in a direct translation of the modal systems, but in systems with similar properties that still retain the extra expressive power of the meta-approaches. His systems T' and $S4'$ only restrict the use of the rule of necessitation to formulas satisfying a property of regularity. This restriction is similar to the one imposed by des Rivieres and Levesque, but it does not apply to axiom schemata. Hence, Turner's systems are more expressive than their modal counterpart, the only drawback of this approach is that it is not general. For example, the modal system $S5$ cannot be simulated in this fashion, because it would produce an inconsistent system. Along the same lines, Davies [Dav90] introduces other systems for knowledge and belief, but with a more restricted rule of necessitation.

The results obtained by Turner and Davies are very encouraging for the possibility of syntactically treating modalities, but the generalization of their work to the treatment of any modal system is not obvious, probably more complex restrictions are needed. For example, as far as $S5$ is concerned, we already know that a consistent translation exists if we restrict our language to regular formulas, but probably we can still retain some of the extra expressive power of the meta-language without losing consistency.

Giunchiglia and Serafini [GS94] map the most common (normal) propositional logics onto hierarchical structures of classical systems, where an infinity of systems is needed in order to account for the inherent amalgamated feature of modal languages. Each system in the hierarchy has a propositional language augmented with a unary predicate, that is used as a meta provability/belief predicate for the system below it, and a set of individual constants that are names for formulae. Even if meta-languages can be extensions of the respective object-languages, consistency is ensured by strong restrictions on the language containing neither variables nor quantifiers.

10.5 Conclusions

We have discussed the two main approaches to the formalization of knowledge and belief: modal and meta-level. In the modal approaches the main emphasis is on semantical issues, in fact special interpretations are defined for the modal operators denoting the attitudes of knowledge and belief; the possible-worlds semantics is the most important example of semantical construction. However, such a semantics commits itself to the strong assumption of logical omniscience. A number of proposals have been presented in the literature to overcome this shortcoming, but, although interesting, none of them seems to solve the problem completely.

Meta-level approaches do not attribute any special semantic property to the meta-level predicates denoting either knowledge or belief, but their properties are formalized via axiom schemata and inference rules, such as reflection principles. The main concern of meta-level systems is their expressive power, which can lead to the definition of more general frameworks, than their modal counterparts.

During the last years, some very important results have been obtained in the consistency of the syntactical treatment of modality and the negative results of Montague and

Thomason, that discouraged the use of meta-logics for the treatment of modalities, have been balanced. In fact, we are now in a position where we can build consistent meta-level systems for knowledge and belief and verify their practical usefulness.

There are still several open issues, here we list some of the important ones:

1. The exact boundary between the modal logics that can be treated syntactically and those that cannot has not been clearly defined. Through des Rivieres and Levesque's work we can provide a direct translation of any modal logic, but nothing of the extra-expressive power of the meta-level is preserved. Conversely, the work of Turner and Davies provides a consistent meta-level syntactic treatment only for some of the modal logics (e.g. not for $S5$). We believe that a broader class of modal systems can be dealt with in a meta-level framework, without sacrificing the extra-expressive power.

2. The expressive power of meta-level systems can make it feasible the simultaneous treatment of different forms of common sense reasoning. There are several forms of reasoning which can be naturally formalized in a meta-level system, such as non-monotonic, abductive or inductive reasoning. As already shown by Bowen and Kowalski, abductive reasoning can be easily formalized in a meta-level framework, even if we adopt more general definitions of abduction. Moreover, a meta-level approach can integrate control information which may help improving the solution of the different tasks. The integration of all the above aspects, although it seems possible in principle, needs further investigation.

Acknowledgments

We thank the referees for their comments and suggestions.

References

[ACN91] L. Carlucci Aiello, M. Cialdea, and D. Nardi. Reasoning about student knowledge and reasoning. In *Proceedings of the Twelfth International Joint Conference on Artificial Intelligence (IJCAI-91)*, pages 1087–1094. Morgan Kaufmann, Los Altos, 1991.

[ACS86] L. Carlucci Aiello, C. Cecchi, and D. Sartini. Representation and use of metaknowledge. *Proceedings of IEEE*, 74(10):1304–1321, 1986.

[AL84] L. Carlucci Aiello and G. Levi. The uses of metaknowledge in AI systems. In *Proceedings of the Sixth European Conference on Artificial Intelligence (ECAI-84)*, pages 705–717. North-Holland Publ. Co., Amsterdam, 1984.

[ANS88] L. Carlucci Aiello, D. Nardi, and M. Schaerf. Reasoning about knowledge and ignorance. In *Proceedings of the International Conference on Fifth Generation Computer Systems (FGCS-88)*, pages 618–627. ICOT, 1988.

[ANS91] L. Carlucci Aiello, D. Nardi, and M. Schaerf. Reasoning about knowledge and reasoning in a meta-level architecture. *Applied Intelligence*, 1(1):55–57, 1991.

[Arb89] B. Arbab. How to represent opaque sentences in first order logic. In *Proceedings of the Eleventh International Joint Conference on Artificial Intelligence (IJCAI-89)*, pages 458–462. Morgan Kaufmann, Los Altos, 1989.

[AS84] G. Attardi and M. Simi. Meta-level reasoning across viewpoints. In *Proceedings of the Sixth European Conference on Artificial Intelligence (ECAI-84)*, pages 315–324. North-Holland Publ. Co., Amsterdam, 1984.

[AS91] G. Attardi and M. Simi. Reflections about reflection. In *Proceedings of the Second International Conference on the Principles of Knowledge Representation and Reasoning (KR-91)*, pages 22–31. Morgan Kaufmann, Los Altos, 1991.

[AS94] G. Attardi and M. Simi. Proofs in context. In *Proceedings of the Fourth International Conference on the Principles of Knowledge Representation and Reasoning (KR-94)*, pages 15–26. Morgan Kaufmann, Los Altos, 1994.

[BK82] K. A. Bowen and R. A. Kowalski. Amalgamating language and metalanguage. In S. A. Tarnlund, editor, *Logic Programming*, pages 153–173. Academic Press, New York, 1982.

[BT91] A. Brogi and F. Turini. Metalogic for knowledge representation. In *Proceedings of the Second International Conference on the Principles of Knowledge Representation and Reasoning (KR-91)*, pages 61–69. Morgan Kaufmann, Los Altos, 1991.

[CL89] S. Costantini and G. A. Lanzarone. A metalogic programming language. In *Proceedings of the Sixth International Conference on Logic Programming (ICLP-89)*, pages 218–233. The MIT Press, 1989.

[Dav90] N. Davies. Towards a first order theory of reasoning agents. In *Proceedings of the Ninth European Conference on Artificial Intelligence (ECAI-90)*, pages 195–200. Pitman, 1990.

[dRL88] J. des Rivieres and H. J. Levesque. The consistency of syntactical treatments of knowledge (how to compile quantificational modal logics into classical FOL). *Computational Intelligence*, 4:31–41, 1988.

[Fre85] G. Frege. On sense and meaning. In A. P. Martininch, editor, *The Philosophy of Language*, pages 212–220. Oxford University Press, 1985. Originally published in 1892.

[GS94] F. Giunchiglia and L. Serafini. Multilanguage hierarchical logics, or: how can we do without modal logics. *Artificial Intelligence Journal*, 65(1):29–70, 1994.

[Haa86] A. R. Haas. A syntactic theory of belief and action. *Artificial Intelligence Journal*, 28:245–292, 1986.

[HC68] G. E. Hughes and M. J. Cresswell. *An Introduction to Modal Logic*. Methuen University Press, London, 1968.

[HC84] G. E. Hughes and M. J. Cresswell. *A Companion to Modal Logic*. Methuen University Press, London, 1984.

[Hin62] J. Hintikka. *Knowledge and belief*. Cornell University Press, Ithaca, New York, 1962.

[HL94] P. M. Hill and J. W. Lloyd. *The Gödel Programming Language*. The MIT Press, Reading, Massachusetts, 1994.

[HM92] J. Halpern and Y. Moses. A guide to completeness and complexity for modal logics of knowledge and belief. *Artificial Intelligence Journal*, 54(3):319–379, 1992.

[KK91] J. S. Kim and R. A. Kowalski. A metalogic programming approach to multi-agent knowledge and belief. In *Artificial Intelligence and Mathematical Theory of Computation*, pages 231–246. Academic Press, 1991. Papers in Honor of John McCarthy.

[Kon81] K. Konolige. A first-order formalization of knowledge and action for a multiagent planning system. In *Machine Intelligence*, volume 10, pages 503–508. John Wiley & Sons, 1981.

[Kri63] S. A. Kripke. Semantical considerations on modal logic. *Acta Philosophica Fennica*, 16:83–94, 1963.

[Lev84] H. J. Levesque. A logic of implicit and explicit belief. In *Proceedings of the Fourth National Conference on Artificial Intelligence (AAAI-84)*, pages 198–202. Morgan Kaufmann, Los Altos, 1984.

[Lev89] H. J. Levesque. A knowledge-level account of abduction. In *Proceedings of the Eleventh International Joint Conference on Artificial Intelligence (IJCAI-89)*, pages 1061–1067. Morgan Kaufmann, Los Altos, 1989.

[Llo88] J. W. Lloyd. Directions for meta programming. In *Proceedings of the International Conference on Fifth Generation Computer Systems (FGCS-88)*, pages 609–617. ICOT, 1988.

[McC79] J. McCarthy. First-order theories of individual concepts and propositions. In *Machine Intelligence*, volume 9, pages 120–147. Univ. of Edinburgh Press, 1979.

[McC93] J. McCarthy. Notes on formalizing contexts. In *Proceedings of the Thirteenth International Joint Conference on Artificial Intelligence (IJCAI-93)*, pages 555–560. Morgan Kaufmann, Los Altos, 1993.

[MN88] P. Maes and D. Nardi, editors. *Meta-level Architectures and Reflection*. North-Holland Publ. Co., Amsterdam, 1988.

[Mon63] R. Montague. Syntactical treatment of modality, with corollaries on reflexion principles and finite axiomatizability. *Acta Philosophica Fennica*, 16:153–167, 1963.

[Moo77] R. Moore. Reasoning about knowledge and action. In *Proceedings of the Fifth International Joint Conference on Artificial Intelligence (IJCAI-77)*, pages 223–227, 1977.

[Nar88] D. Nardi. Evaluation and reflection in FOL. pages 195–207. 1988. In [MN88].

[Ohl93] H. J. Ohlbach. Translation methods for non-classical logics: an overview. *Bulletin of the IGPL*, 1(1):69–90, 1993.

[Per85] D. Perlis. Languages with self-references I. *Artificial Intelligence Journal*, 25:301–322, 1985.

[Per88] D. Perlis. Languages with self-references II. *Artificial Intelligence Journal*, 34(2):179–212, 1988.

[Rus05] B. Russell. On denoting. *Mind*, 14:479–493, 1905.

[SM88] M. Simi and E. Motta. OMEGA: an integrated reflective framework. pages 209–226. 1988. In [MN88].

[Smo85] C. Smorynski. *Self-reference and Modal Logic*. Springer-Verlag, 1985.

[Tho80] R. Thomason. A note on syntactical treatments of modality. *Synthese*, 44:391–395, 1980.

[Tur90] R. Turner. *Truth and Modality for Knowledge Representation*. Pitman Pub., London, 1990.

[vH92] F. van Harmelen. Definable naming relations in meta-level systems. In *Proceedings of the International Conference on Meta-Programming (META-92)*, pages 89–104. Springer-Verlag, 1992.

[Wey80] R. W. Weyhrauch. Prolegomena to a theory of mechanized formal reasoning. *Artificial Intelligence Journal*, 13(1):133–170, 1980.

[WT90] R. W. Weyhrauch and C. Talcott. Towards a theory of mechanizable theories: I (FOL contexts — the extensional view). In *Proceedings of the Ninth European Conference on Artificial Intelligence (ECAI-90)*, pages 634–639. Pitman, 1990.

11 Model-based Diagnosis Preferences and Strategies Representation with Logic Meta-Programming

Carlos V. Damásio, Wolfgang Nejdl, Luís M. Pereira, Michael Schroeder

Abstract

Preferences and strategies are fundamental to model-based diagnosis, for specifying preferred and fall-back approaches to the diagnosis task, both to capture general and domain specific criteria, but also to tackle the complexity issue by employing heuristics. A formal framework based on extended logic programming and meta-programs is provided to represent preferences and strategies required by model-based diagnosis. This framework is clearer and more expressive than other approaches that have addressed these problems. We show how the concepts of preferences and strategies are directly programmed and captured by logic meta-programming and meta-reasoning methods, and their implementation techniques.

The paper is intended as proof-of-principle that all concepts needed by a model-based diagnosis system can represented declaratively and captured by a logic meta-program. Specialized more efficient algorithms can be substituted for the simpler proof-of-principle ones we include, and are the subject of ongoing work.

11.1 Introduction

Model-based diagnosis has been an increasingly active area for the last 5 to 10 years, and a lot of advances have been made during this period. Conceived specifically to tackle the disadvantages of heuristic diagnosis methods, such as difficult expandability and maintainability, model-based diagnosis has focused on the declarative representation of system descriptions, and also their use by an independent diagnosis algorithm to analyze discrepancies between observed and predicted behaviour in order to pinpoint suspect components. Research has focused both on efficient diagnosis algorithms and better expressibility of system descriptions, using fault models, physical impossibility axioms, hierarchies and others. A representative set of papers has been published in Hamscher, Console, and de Kleer [HCd92], in Console, and Friedrich [CF94], in the annual workshops on Principles of Diagnosis and AI conferences such as KR, ECAI, AAAI and IJCAI. A schematic overview of the main principle of model-based diagnosis, i.e. the computation of diagnoses based on the analysis of predicted behaviour and actual observations, is shown in Figure 11.1.

Recently, some of our work has focused on the declarative representation of the diagnosis process itself within a model-based diagnosis system, a topic which until now has only been discussed by a few authors (Struss [Str92b, Str92a], Damásio, Nejdl and

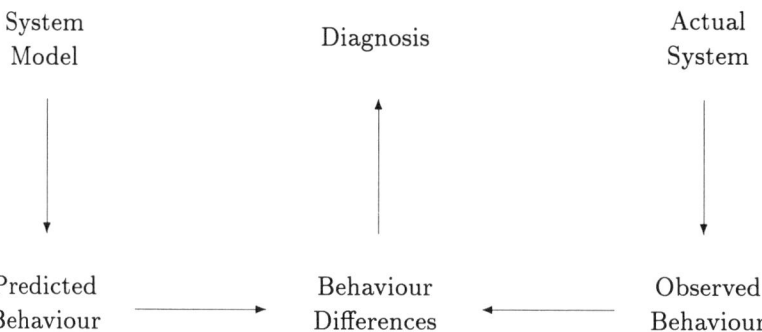

Figure 11.1
Model-based Diagnosis

Pereira [DNP94], and Fröhlich, Nejdl, and Schroeder [FNS94]). A declarative representation of the diagnosis process including the preferences and strategies used during this process is very important, as it gives the user greater power over important aspects of the diagnosis process which have remained implicit in most current diagnosis systems.

In this paper we are able to give a complete description of this diagnosis process by using logic meta-programming concepts and methods. We show how to declaratively represent the diagnosis process itself, extending the model-based approach of declarative modelling to the meta-layer, i.e. the diagnosis process. This allows the user to express properties of the diagnosis process (user preferences and diagnosis strategies) in the same declarative way as the system description, which is an important prerequisite for the diagnosis of large and complicated systems.

In the following we use a 4-layer model to describe the diagnosis process.

| The Program Layer |
| The Revision Layer |
| The Preference Layer |
| The Strategy Layer |

The first layer (the program layer) corresponds to the system description, modeled as a logic program, including any assumptions made (for example about the correctness of components). Predictions are computed in this layer. The second is the revision layer, which reacts to external observations contradicting the predicted values and revises the assumptions accordingly. In the third layer we implement the concept of preferred revisions, specifying which assumptions should be retracted, retained or changed in order to compute consistent revisions. Finally, the specification of diagnosis strategies in the fourth, higher, layer represents the actual decision making process of the diagnostic agent in cases where a set of preferred revisions[1] has been computed, but where further actions, measurements, model refinements etc. are necessary in order to arrive at truly satisfactory diagnoses.

We will define all four layers using meta-programming methods as follows:

1. The program layer. To work on this layer, we define the two-argument `demo` and ¬demo meta-predicates, expressing provability.

2. The revision layer. We define meta-predicates `revisions` and `minimal_revisions`, expressing the concept of revisions by changing assumptions, regarding correctness of components, their fault modes, strategic assumptions, etc.

3. The preference layer. Essentially there is a single meta-predicate designated `preferred_revisions`, which takes preferences on possible revisions of assumptions into account.

4. The strategy layer. Two meta-predicates are provided, namely the `coherent_WH` and `adopted_WH`. In this layer, the strategic decisions during the diagnosis process are represented by a meta-program.

A detailed discussion of the model-based concepts involved follows in a later part of the paper. A short introduction will be given in the next section using a digital circuit example.

Higher layers are defined using lower layers. In the paper we provide declarative and naïve algorithms for each of the layers. The main contribution of this paper is the connection and definition of these layers, as well as their high level meta-programming description. An actual system would redefine these meta-predicates for more specific and efficient implementations. This is ongoing work.

The syntax, semantics and programming style were mainly motivated by the works of Bowen and Kowalski [BK82], Hill and Lloyd [HL89, HL94], and Subrahmanian [Sub89]. Our meta-programs are written in `typewriter` style and object-programs in *math* style.

All the results presented assume the consistency-based approach to diagnosis (e.g. de Kleer and Williams [dKW87, dKW89], Reiter [Rei87], and Struss [SD89]). For the abductive view of Poole [Poo89], Console and Torasso [CT90, CT91], and Friedrich, Gottlob, and Nejdl [FGN90] the basic approach of this paper remains valid but will require some extensions/modifications.

In the remainder we often use variables as shorthand for their ground instances, which are covered by the underlying theory. It is not the purpose of the paper to address nor resolve any issues relative to the problems pertaining to the representation and manipulation of general first-order theories at a meta-level in logic programming. A whole gamut

---

[1] Revisions of assumptions about correctness of components and diagnoses are equivalent.

of research in the last 10 years exists, and is still underway, tackling such issues. These are by no means of easy, nor have generally accepted solutions been proposed. Indeed, other more specialized papers in this book address just such issues. Accordingly, our stance is to keep to the ground case, with the expectancy that others will, in due course, show how, and under what circumstances and restrictions, the generalizing move from the ground to the non-ground case can be made. In any case, we believe we have made our representational conventions clear. Another reason for the restriction concerns the correctness of the procedures used for executing programs and meta-programs: they can only be guaranteed correct for the ground case. Their generalization and implementation for the non-ground case (which involves tabulation and constructive negation) is again ongoing work not reported here.

11.2 Model-based Diagnosis

As mentioned previously, the heart of a model-based diagnosis system consists of a model of the system to be diagnosed. In this section we show how these models can be represented by extended logic programs. A formal definition of the language used is introduced later. Basically we extend logic programs by allowing explicitly negated literals. Negation by default[2] is denoted by *"not"* and explicit negation by *"¬"*. The intended semantics will be evident from the examples, and defined in section 11.3. Additionally, we will informally discuss the role of preferences and strategies during the diagnosis process.

11.2.1 Introduction to preferences and strategies

The main principle of model-based diagnosis is to explicitly describe the system to be diagnosed. In this paper (and in the related paper Fröhlich, Nejdl, and Schroeder [FNS94]) we extend this principle to the diagnosis process itself.

So far, in most current diagnostic systems, diagnosis is considered as a static problem, i.e. the system description and the diagnosis goals are fixed and are not changed during the diagnosis process. As already observed in recent work such as Struss [Str92a], Damásio, Nejdl and Pereira [DNP94], and Böttcher and Dressler [BD93b, BD94], this is a big disadvantage when diagnosing complex systems. All information such as different specifications, modelling assumptions, abstractions, and specializations, are active during the whole diagnosis process, thus limiting the complexity of systems that can be diagnosed as shown in Davis [Dav84], Mozetic [Moz92], Struss [Str92a], and Böttcher and Dressler [BD93b, BD94], or requiring ad hoc methods of controlling the diagnosis process.

---

[2]Negation by default is also known as negation by failure.

Struss [Str92a, Str92b] coined the concept of *diagnosis as a process* to express the idea that the diagnosis process involves diagnostic decisions, which should be modeled as well within a model-based diagnosis system. Subsequent work has focused on formalizing (part of) this notion. Friedrich [Fri93] covers the principle of abstraction, Damásio, Nejdl, and Pereira [DNP94] focus on preference relations for diagnoses. The most general approach up to now has been defined by Böttcher and Dressler [BD93b, BD94], including a set of preferences and strategies, respectively, to guide the diagnosis process. Unfortunately, their description and semantics is explicitly based on iteration, resulting in a procedural semantics. They show how to compute diagnoses but cannot characterize their results in a declarative way. Meta-programming is especially apt to secure diagnosis from this drawback.

Based on the idea of working hypotheses guiding the diagnosis process we develop a syntax and semantics of a strategy language. The semantics of the language provides a declarative characterization of diagnoses as the result of a diagnosis process. The companion paper Fröhlich, Nejdl, and Schroeder [FNS94] focuses on a modal-logic-based semantics of such a strategy language.

In the next section we will use an example diagnosis of a calculator and its underlying digital circuits to describe the concepts informally. We will come back to this example throughout the whole paper, to exemplify the formal concepts introduced.

11.2.2 Description of the diagnosis process of an adder

Description at the abstract level Assume you have a calculator on your desk, and type in the expression "0+0=" just to make sure that your calculator can evaluate the most basic expressions. Imagine your astonishment, when the calculator does not output "0", but instead outputs "256". In the following we will discuss your diagnostic reasoning process for this (admittedly quite unlikely) event to show the diagnostic concepts involved.

Model-based diagnosis uses a component-based model of the device. Our calculator can be described at an abstract level as being composed of a keyboard, a processing unit and a display, as per figure 11.2.

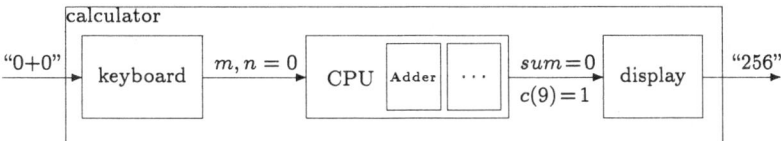

Figure 11.2
Abstract view of a calculator.

Assuming simple high level models of the components of our calculator, we come to the conclusion that "0+0" should evaluated to "0", not to the actual observation "256". The discrepancy between predicted value and observed value makes us realize that some component of the calculator has to be faulty. An immediate repair would be to replace the whole calculator by the new one. Considering the price of calculators nowadays, this might not be a bad decision. However, with a view to enlarging this example, we will not take that decision. Instead, we will assume that a finer grained repair will be cheaper and continue diagnosis on a less abstract level. This would be somewhat problematic if we could not take any additional measurements, thus having no clue about which component should be suspected and further decomposed for analysis. Fortunately (examples in stories and papers are always beneficial), we can reduce the problem to one component if we first measure the connections among the components and thus rule out two of the three components. In our case the measurement of registers m, n and sum shows them all zero, while the overflow bit is set. As the keyboard reacts to the input of "0+0" by setting both registers m and n to 0, we can exonerate it. Similar arguments hold for the display, so we come to the conclusion that the processing unit is faulty and has to be decomposed to pinpoint the fault.

The decomposition of the processing unit leads to an 8-bit-adder. From now on, we will have to rely on more sophisticated diagnostic reasoning, as no measurements are possible inside the 8-bit-adder (a quite realistic assumption, which allows us to discuss our diagnostic abilities in further detail). We will therefore have to use different models and assumptions, plus the already existing measurements of registers m, n and sum and overflow bit $c(9)$ to come to a finer diagnostic conclusion.

Description of the functional view of the adder There are at least two ways to describe an 8-bit-adder: a functional view and a physical view. The former stresses a hierarchical structure in which the 8-bit-adder can be decomposed into eight full adders, each of them again consisting of two and-gates, two xor-gates and an or-gate. The connections of the gates are shown in figure 11.3.

Description of the physical view The physical view considers the actual implementation of the adder, which is realized by a set of micro chips implementing the logical functions "and", "or" and "xor". Each of these chips consists of 16 gates (see figure 11.4).

Within the physical view we do not focus on hierarchies, but on possible failure behavior, including fault statistics. One piece of information could tell us that "xor" chips will be faulted quite often in the failure mode stuck-at-1 (output 1 independently of inputs). We will assume possible behaviour modes of chips to be ok, abnormal and specifically stuck-at-1 and stuck-at-0. Other "unknown" failure modes might be possible.

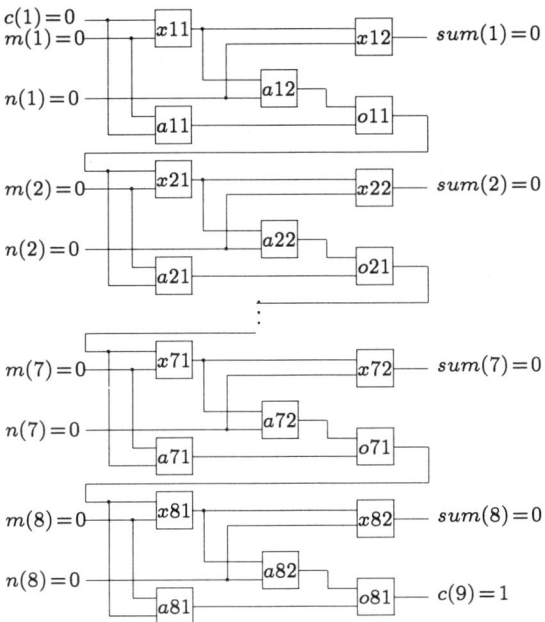

Figure 11.3
8-bit-adder composed of eight full adders

Description of the diagnosis process A possible description of the diagnostic strategies used during the diagnosis process might be as follows:

1. Diagnosis of the calculator will be step by step. Our general diagnosis strategy will be to start on an abstract level and decompose suspected components. We might consider decomposing components which are included in every plausible diagnosis.

2. After decomposing the calculator into different components (keyboard, adder and display), we can use additional measurements to discriminate among different diagnoses. This is possible whenever different diagnoses predict different values for a given measurement.

3. We can illustrate preferences at the level of the 8-bit-adder: First, we will use just the functional model and look for single faults. Only when this fails, we will also accept double faults. If no diagnosis can be found under these criteria, we switch to the physical model, but again look only for single faults. Finally, when even this fails, we consider the possibility that our behavioural model is incomplete and that

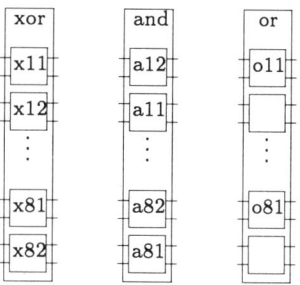

Figure 11.4
Physical view of an 8-bit-adder

an unknown fault has occured.

On the basis of these informal considerations, we recognize that we need to solve the following problems:

1. How to model the different components, such as adder, 8-bit-adder, full adder, and-chip, and-gate, etc. and their composition.

2. How to model the different views such as abstract and detailed descriptions and functional/physical specifications.

3. How to switch between different views and specifications and to express preferences on diagnoses to explore first.

While the second and third questions lead us to the concepts of preferences and strategies, the first one is the most basic problem of how to model the system to be diagnosed. In the next section we will discuss this modelling issue in detail and show how system models can be expressed using recent work on extended logic programming. This will correspond to our first layer, the *program layer*.

11.2.3 Modelling diagnosis problems in extended logic programming

In this section we concentrate on the representation of system models by extended logic programs. Besides basic modelling issues we discuss structure, hierarchies and multiple specifications. Other concepts can be defined analogously. Obviously, how to obtain correct models for specific systems is still a topic to be explored, but at least our representation gives some hints about which constructs to use for various meta-concepts such as hierarchies and multiple specifications, which are essential for large models.

The basic system model consists of a description of the system components and their connections. Components are characterized by their behaviour, which is described by a relation.

Definition 1 Component's Behaviour
To specify the behaviour of a component in a specific mode, let $dom(B)$ be the domain of behavioural modes, $dom(In_k)$ be the domain of input variable In_k and $dom(Out_l)$ be the domain of output variable Out_l. A *component's behaviour* is a relation

$$r_{beh} \subseteq dom(B) \times dom(In_1) \times \ldots \times dom(In_n) \times dom(Out_1) \times \ldots \times dom(Out_m)$$

with no two different values for any output, given a behavioural mode and the input. This basically realizes a functional description from inputs to outputs.

Example 2 Behaviour of an and-gate and of an adder
Assume an and-gate with $dom(B) = \{ok, ab, s_0, s_1, unk\}$, where s_0 and s_1 stand for the failure modes stuck-at-0 and stuck-at-1. Mode unk represents unknown behaviour, and explains any relation between input and output.

The gate has two binary input variables and one binary output variable. The behaviour of the and-gate is expressed by the relation

$$and_{beh}(ok, 0, \_, 0). \quad and_{beh}(ab, 0, \_, 1). \quad and_{beh}(s_0, \_, \_, 0).$$
$$and_{beh}(ok, \_, 0, 0). \quad and_{beh}(ab, \_, 0, 1). \quad and_{beh}(s_1, \_, \_, 1).$$
$$and_{beh}(ok, 1, 1, 1). \quad and_{beh}(ab, 1, 1, 0).$$

The behaviour for the unknown mode in not specified as we do not know how values are related. In general, the behaviour does not have to be given explicitly as above, but can also be coded implicitly like for the adder as follows:

$$adder_{beh}(ok, X, Y, Z) \leftarrow Z = X + Y.$$
$$adder_{beh}(ab, X, Y, Z) \leftarrow Z \neq X + Y.$$

The output values of a component depend on the relation describing the component's behaviour, its current mode and the input values. Two other important aspects are:

- To what problem specification does the component belong?
- To what part of the overall system structure does the component belong?

Usually we do not want to solve the diagnosis problem using all specifications simultaneously, but want to activate and deactivate the specifications depending on some conditions (as discussed in our example of the 8-bit-adder, where we switched from functional to physical specification). Because specifications will be expressed as rules of an extended logic program, we can activate and deactivated these rules by means of specification assumptions. This is implemented in the following way: All rules belonging to a problem specification contain a literal $spec(name)$ in their body, where $name$ indicates to what specification the rule belongs. Simply by assuming $spec(name)$ we can activate the rules describing the specification *name*. The assumption of specifications can be defeated non-monotonically.

Within such specifications or views, devices are usually structured in hierarchies. Again, we need information about which part of the hierarchy is active at the moment. The granularity of components within a problem specification is adjusted by means of the literal $refined(Comp)$. As the diagnosis develops, the granularity of the system becomes finer and the corresponding component rules become activated.

Definition 3 Computing output values of a component

In order to compute the output values of a component the system description contains rule instances of the schema of the form:

$$value(out(Comp, k), Out_k) \leftarrow$$
$$\quad spec(i), refined(Super\_comp), not\ refined(Comp),$$
$$\quad mode(Comp, Mode),$$
$$\quad value(in(Comp, 1), In_1), \ldots, value(in(Comp, n), In_n),$$
$$\quad comp_{beh}(Mode, In_1, \ldots, In_n, Out_1, \ldots, Out_k, \ldots, Out_m).$$

where $value(in(Comp, j), In_j)$ codes that In_j is the value of input j to component $Comp$; and similarly for $value(out(Comp, k), Out_k)$.

Example 4 continued

Let $and(I, 1)$ be the first of the two Ith and-gates of a full adder $fad(I)$. A rule to compute the output of gate $and(I, 1)$ has the form:

$$value(out(and(I, 1), 1), Out_1) \leftarrow$$
$$\quad spec(functional), refined(fad(I)), not\ refined(and(I, 1)),$$
$$\quad mode(and(I, 1), Mode),$$
$$\quad value(in(and(I, 1), 1), In_1), value(in(and(I, 1), 2), In_2),$$
$$\quad and_{beh}(Mode, In_1, In_2, Out_1).$$

In a similar way, we can compute the output of the adder. Since the adder has no super component, the first $refined$ literal is left out.

$$value(out(adder, 1), Out_1) \leftarrow$$
$$\quad spec(functional), not\ refined(adder),$$
$$\quad mode(adder, Mode),$$
$$\quad value(in(adder, 1), In_1), value(in(adder, 2), In_2),$$
$$\quad adder_{beh}(Mode, In_1, In_2, Out_1).$$

The rules above show how to compute an output value when a component's inputs and a relation describing the component's behaviour are given. There are two more cases. A value may be directly observed, or propagated along a causal connection.

Definition 5 Computing observed and propagated values
We assume that the causal connections of a system are coded by the relation $conn(N, M)$, where N and M are nodes. A node value results from an observation or is propagated from another connected node:

$$value(N, V) \leftarrow obs(N, V) \qquad value(N, V) \leftarrow conn(N, M), value(M, V)$$
$$\neg value(N, V) \leftarrow \neg obs(N, V) \qquad \neg value(N, V) \leftarrow conn(M, N), \neg value(M, V)$$

Note that $\neg obs(N, V)$ means V was not observed at N, not that there was no observation. The symbol $\neg$ refers to the explicit negation of logic programs.

The problem of diagnosis is to find an assignment of behavioural modes to the components that explains the observed behaviour. This assignment does not consider all combinations of modes. Normally we prefer the *okay* mode and assume components to be *okay* by default. Only if this leads to inconsistencies do we assume one or more of the components' fault modes. When doing this we also have to be aware that any specified fault modes might not be complete, i.e. that we may have forgotten to specify all possible fault modes. The unknown mode may be used as a catch-all mode, when everything else has failed.

Definition 6
Let c be a component with possible behaviours $dom(B) = \{ok, fm_1, \ldots fm_n\}$, where fm_i is a fault mode. By default we assume that the component is *okay*:

$$mode(c, ok) \leftarrow not\ mode(c, fm_1), \ldots, not\ mode(c, fm_n).$$

Example 7 continued
For the and-gate $and(I, 1)$ the *okay* mode is assumed by default via the rule:

$$mode(and(I, 1), ok) \leftarrow\ not\ mode(and(I, 1), ab), not\ mode(and(I, 1), s_0),$$
$$not\ mode(and(I, 1), s_1), not\ mode(and(I, 1), unk).$$

In the remainder, we will refer to the above rules concerning the and-gate as SD_{and}.

11.3 The Program Layer: Extended Logic Programs

11.3.1 Well-founded semantics with explicit negation

In the previous section we used extended logic programs (*ELPs*) to represent our system models in the program layer, appealing to the intuitive understanding of the reader.

The evolution of logic programming semantics has included the introduction of an explicit form of negation, beside the older implicit (or default) negation typical of logic programming, cf. Gelfond and Lifschitz [GL90a], Pereira and Alferes [PA92], and Wagner [Wag91]. The richer language has been shown adequate for a spate of knowledge representation and reasoning forms by Pereira, Aparício, and Alferes [PAA93], Alferes, Damásio, and Pereira [ADP94a, PDA93b], and Baral and Gelfond [BG94]. In this section we briefly review the formal definition of extended logic programs under well-founded semantics comprising explicit negation.

Our choice of Well-founded Semantics with Explicit Negation as the base semantics is justified by its structural properties. In particular, these are paramount for top-down query evaluation. Because of these properties, which other semantics do not fully enjoy, *WFSX* is a natural candidate for being the semantics of choice for *ELPs*.

WFSX exhibits the structural properties of well-foundedness, cumulativity, rationality, relevance, and partial evaluation. By well-foundedness we mean that it can be characterized (without recourse to three-valued logic) by the least fixpoint of two iterative operators (cf. Alferes, Damásio, and Pereira [ADP94a, ADP94b]). The other properties were introduced by Dix in [Dix91] and [Dix92]. By cumulativity, we refer to the efficiency related ability of using lemmas, i.e. the addition of lemmas does not change the semantics of a program. By rationality, we refer to the ability to add the negation of a non-provable conclusion without changing the semantics; this is important for efficient default reasoning. By relevance, we mean that the top-down evaluation of a literals truth-value is made possible because it requires only the program's call-graph below it. By partial evaluation we mean that the semantics of a partially evaluated program is equivalent to the original program semantics.[3] Additionally, it is amenable to both top-down and bottom-up procedures, and its complexity for finite DATALOG programs is polynomial. These and other properties of *WFSX* are detailed and proven in Alferes [Alf93].

An extended program is a (possibly infinite) set of ground rules of the form

$$L_0 \leftarrow L_1, \ldots, L_m, not\ L_{m+1}, \ldots, not\ L_n\ (0 \leq m \leq n)$$

where each L_i is an objective literal ($0 \leq i \leq n$). An objective literal is either an atom A or its explicit negation $\neg A$. In the sequel, we also use $\neg$ to denote complementary literals wrt the explicit negation, so that $\neg\neg A = A$. The set of all objective literals of a program P is called the extended Herbrand base of P and denoted by $\mathcal{H}(P)$. To syntactic objects of the form *not L* we call default literals. Literals are either objective or default literals.

WFSX follows from Well-founded semantics of van Gelder, Ross, and Schlipf [vGRS91] for normal programs plus the coherence requirement relating the two forms of negation:

---

[3]Stable model based approaches, such as answer-sets, enjoy neither cumulativity, nor rationality, nor relevance.

"For any objective literal L, if $\neg L$ is entailed by the semantics then not L is also entailed".

This requirement states that whenever some literal is explicitly false then it must per force be assumed false by default.

We will use some Gödel-language constructs and meta-programming facilities, proposed by Hill and Lloyd [HL94], to simplify the writing of meta-programs. They will be introduced and shortly explained as they are needed by our meta-programs.

11.3.2 Query language

Having defined our basic language, to be used both as object and meta-language, we proceed to define the object formula language used for the goals to be given to programs.

Definition 8 Object Formula

Let P be an extended logic program and $\mathcal{L}_P$ its associated language.

1. If L is a literal of $\mathcal{L}_P$ then $\lceil L \rceil$ is an object formula;
2. If $\lceil O \rceil$ and $\lceil O' \rceil$ are object formulas then $\lceil O \wedge O' \rceil$ and $\lceil O \vee O' \rceil$ are object formulas;
3. Nothing else is an object formula.

The Language $\mathcal{L}_{Obj}$ is the set of all object formulas.

We use the notation $\lceil P \rceil$ for representing an object extended logic program P in the meta-level. We assume only that this representation is a set of rules in some format. To access, in the meta-level, the representation of a rule $H \leftarrow B$ of object program P we will make use of the ternary meta-predicate rule($\lceil P \rceil, \lceil H \rceil, \lceil B \rceil$). Additional meta-predicates to handle the representation of goals and atoms will also be employed. Other more particular details are left open for the implementation.

Definition 9 Query Language

Let P be an extended logic program and $\mathcal{L}_{Obj}$ its associated object formulas language. The query language is formed by the predicates demo($\lceil P \rceil, \lceil O \rceil$) and $\neg$demo($\lceil P \rceil, \lceil O \rceil$), where $\lceil O \rceil$ is an object formula and $\lceil P \rceil$ is the representation of P. The intended interpretation of the demo predicate is given by:

- demo($\lceil P \rceil, \lceil O \rceil$) is true iff $P \models_{WFSX} O$;
- $\neg$demo($\lceil P \rceil, \lceil O \rceil$) is true iff $P \not\models_{WFSX} O$;

Note that when $P \not\models_{WFSX} O$ then O may be either false or undefined.

The query language provides the means to access the object programs within an extended logic meta-program. After establishing this relation we proceed to study properties of the demo predicate. Because our extended logic programs do not allow disjunctions in the heads the following simple properties hold:

Proposition 10
Let P be an extended logic program. Then,

$$\begin{aligned}
\text{demo}(\ulcorner P \urcorner, \ulcorner O \wedge O' \urcorner) &\equiv \text{demo}(\ulcorner P \urcorner, \ulcorner O \urcorner) \wedge \text{demo}(\ulcorner P \urcorner, \ulcorner O' \urcorner) \\
\text{demo}(\ulcorner P \urcorner, \ulcorner O \vee O' \urcorner) &\equiv \text{demo}(\ulcorner P \urcorner, \ulcorner O \urcorner) \vee \text{demo}(\ulcorner P \urcorner, \ulcorner O' \urcorner) \\
\neg\text{demo}(\ulcorner P \urcorner, \ulcorner O \wedge O' \urcorner) &\equiv \neg\text{demo}(\ulcorner P \urcorner, \ulcorner O \urcorner) \vee \neg\text{demo}(\ulcorner P \urcorner, \ulcorner O' \urcorner) \\
\neg\text{demo}(\ulcorner P \urcorner, \ulcorner O \vee O' \urcorner) &\equiv \neg\text{demo}(\ulcorner P \urcorner, \ulcorner O \urcorner) \wedge \neg\text{demo}(\ulcorner P \urcorner, \ulcorner O' \urcorner)
\end{aligned}$$

The demo predicate definition is based on the results presented by Alferes, Damásio, and Pereira [ADP94b, ADP94c], where a top-down derivation procedure (SLX) for $WFSX$ was first presented. The novelty of the algorithms for ground programs in this paper is the use of a syntax independent representation, similar to the work of Hill and Lloyd [HL89, HL94], for handling object level theories and goals in well-founded semantics with explicit negation. The meta-interpreter we produce implements SLX, and so is defined and correct only for ground programs.

11.3.3 *WFSX* meta-interpreter

The definition of SLX is directly inspired by the semantic tree characterization of Alferes, Damásio, and Pereira [ADP94b], and relies on SLDNF-like definitions of derivation, refutation, and failure.

It is known that the main issues in the definition of top-down procedures for WFS are infinite positive recursion, and infinite recursion through negation by default, as discussed by Bol and Degerstedt [BD93a], Chan and Warren [CW92, CW93], Pereira, Aparício, and Alferes [PAA91], and Ross [Ros92]. The former gives rise to the truth value false, so that, for some L involved in the recursion, there must exist a refutation for *not* L, and no refutation for L; the latter, to the truth value "undefined" so that both L and *not* L must have no refutation.

For recursion through negation by default we want both L and *not* L to have no refutations, which seems to violate the negation as failure rule. Indeed, in SLDNF that rule states that if there is no refutation for L then *not* L should succeed, and if L has a refutation then *not* L should fail. This is so because SLDNF relies on a two-valued semantics, and so failure of verity means falsity and vice-versa. In WFS, being a three valued semantics, the same cannot apply. In fact, for WFS a failure of L simply means that L is not true, i.e. it can be either false or undefined.

Rather than considering an extra status for literals in derivations as Ross [Ros92] does, we distinguish instead two kinds of derivations: SLX-T-derivations that prove verity relative to the well-founded model (WFM), and SLX-TU-derivations (where TU stands for "true or undefined") that prove non-falsity relative to the WFM. Now, for any L, the verity of *not L* succeeds iff there is no SLX-TU-refutation for L (i.e. L is false), and the non-falsity of *not L* succeeds iff there is no SLX-T-refutation for L (i.e. L is not true).

The main difference in the generalization to extended programs resides in the treatment of negation by default. In order to fulfill the coherence requirement an additional way must be allowed to refute a literal *not L*. In fact *not L* is true if $\neg L$ is also true. This is achieved by imposing, in SLX-TU-derivations, the verification of a precondition for the application of resolution on objective literals. Namely: "In a SLX-TU-derivation resolution can only be applied to a selected objective literal L *if there is no SLX-T-refutation for* $\neg L$".

To guarantee termination by eliminating both positive cyclic recursion and negative cyclic recursion (i.e. through negation by default), we need to access two types of ancestors: *local ancestors* are assigned to all literals appearing in derivations, and are used for detecting positive cyclic recursion; *global ancestors* are assigned to whole derivations, and are used to detect negative cyclic recursion.

Accordingly, to implement the meta-interpreter, the first argument of demo/2 is a theory, i.e. the meta-level representation of an extended logic program, and the second argument the goal to be proved. The actual definition of demo/2 is made in terms of a five argument `demo( M, Th, Goal, LAnc, GAnc)`. Mode argument M states which is the current context, "true" or "true or undefined" abbreviated to t and tu. Arguments Th and `Goal` are the representations of the theory and of the goal object formula. The last two arguments are, respectively, the set of local ancestors and the set of global ancestors which are used to guarantee termination for finite propositional programs. For details and a formal description the reader is referred to the articles by Alferes, Damásio, and Pereira [ADP94b, ADP94c].

demo(Th, Goal) ←
    demo(t,Th,Goal,{},{}).

demo(t, _, Goal, _, _) ← true(Goal).
demo(tu, _, Goal, _, _) ← true(Goal).
demo(M, Th, Goal, LAnc, GAnc) ← and(G1, G2, Goal),
    demo(M, Th, G1, LAnc, GAnc),
    demo(M, Th, G2, LAnc, GAnc).
demo(M, Th, Goal, LAnc, GAnc) ← or(G1, _, Goal),
    demo(M, Th, G1, LAnc, GAnc).
demo(M, Th, Goal, LAnc, GAnc) ← or(_, G2, Goal),
    demo(M, Th, G2, LAnc, GAnc).

demo(t, Th, Goal, LAnc, GAnc) ← dneg(G1,Goal),
    ¬demo(tu, Th, G1, {}, LAnc ∪ GAnc).

demo(tu, Th, Goal, LAnc, GAnc) ← dneg(G1, Goal),
    ¬demo(t, Th, G1, {}, LAnc ∪ GAnc).

demo(t, Th, Goal, LAnc, GAnc) ← atom(Goal),
    not (Goal in LAnc ∪ GAnc),
    rule(Th,Goal,Body),
    demo(t, Th, Body, {Goal} ∪ LAnc, GAnc).

demo(tu, Th, Goal, LAnc, GAnc) ← atom(Goal),
    not (Goal in LAnc),
    xneg(Goal,G1),
    ¬demo(t, Th, G1, {}, LAnc ∪ GAnc),
    rule(Th, Goal, Body),
    demo(tu, Th, Body, {Goal} ∪ LAnc, GAnc).

The ¬demo predicate may be given two definitions. The first is gotten by applying CWA to the above rules. We obtain a sound and complete proof-procedure for $WFSX$ just in case the goal is syntactically correct, i.e. if the first argument of demo/2 is a theory and the second one an object formula, as per definition 8:

¬demo(M, Th, Goal, LAnc, GAnc) ←
    not demo(M, Th, Goal, LAnc, GAnc).

Alternatively, an explicit definition of the ¬demo predicate can be given, obtainable from the one above. But, as the method of Pereira, Aparício, and Alferes [PAA91], it suffers from several deficiencies, mainly due to the potential exponential rewriting of predicate definitions. For a predicate with n rules, each with m literals, the procedural execution generates m^n successors in the search space. Therefore, the use of the above definition is preferable to the explicit one, if syntactic correctness is guaranteed from the start (this can be easily checked before calling the demo predicate).

In the rest of the paper we only use explicit negation at the meta-level in meta-predicate ¬demo. Other uses of this form of negation at the meta-level can be envisaged like, for instance, representing taxonomies of meta-concepts allowing default rules with exceptions and exceptions to exceptions. For details see Pereira, Aparício, and Alferes [PAA93].

The above algorithm is just a "vanilla" meta-interpreter for $WFSX$. Of course, it is assumed that the following predicates are correctly defined, with the obvious interpretations: **and/3**, **or/3**, **rule/3**, **dneg/2**, **xneg/2**, **atom/1**, **false/1**, and **true/1**.

For instance, the three argument **and** predicate constructs the representation of object formula $[G1 \wedge G2]$ in its last argument, from the representation of object formulas

⌈G1⌉ and ⌈G2⌉ in the first and second arguments. The language used was inspired by the *Gödel* logic programming language of Hill and Lloyd [HL94]. For ease of presentation we use Gödel extensional and intensional finite set terms. In the above programs we make use only of the extensional sets Gödel's bracket notation, the union (∪) and set membership (*in*) operations. With the appropriate modifications and additions, all programs presented in this paper can be run either in *Prolog* or *Gödel*.

11.4 Revisions and Preferences

We now turn to the next two layers, the *revision* and the *preference* ones. The revision layer is responsible for revising a logic program (respectively its assumptions) if inconsistencies arise; the preference layer is used to express preferences on how to choose among possible revisions.

11.4.1 Integrity constraints

The problem of diagnosis can be reduced to revision of an extended logic program as shown by Pereira, Damásio, and Alferes [PDA93b, PDA93a], and Damásio, Nejdl, and Pereira [DNP94]. The modeling concepts discussed in the previous sections allow us to code information about the system to be diagnosed in the language of such programs. We are also able to express the assumption that all components are working fine by default. What is still missing is how to express that these assumptions may lead to contradiction and how the program is to be revised in order to remove contradiction, so that specific revisions will then be equivalent to the diagnoses of the system.

Example 11 continued
Consider the system description SD_{and} of section 11.2.3. The and-gate is part of a full adder $fad(8)$, i.e. the eighth full adder. Together with the observations (according to the coding convention of def. 1)

$$\{obs(in(and(8,1),1),1), obs(in(and(8,1),2),1), obs(out(and(8,1),1),0)\}$$

and the assumption $H = \{spec(functional), refined(fad(8))\}$, the well-founded model of *WFSX* entails both $value(out(and(8,1),1),0)$, which is derived from the observed output, and $value(out(and(8,1),1),1)$, which is computed from the observed inputs.

We need to express that it is contradictory to have two different values at any one node. We capture this with an integrity constraint.

Definition 12 Integrity constraints
An integrity constraint of an extended logic program P has the following form, where

each L_i is a literal belonging to the language of P:

$$L_1 \vee L_2 \vee \ldots \vee L_n \Leftarrow L_{n+1} \wedge L_{n+2} \wedge \ldots \wedge L_{n+m} \qquad (n+m \geq 0)$$

To an empty consequent (head) we associate the symbol **f** and to an empty antecedent (body) the symbol **t**. An integrity theory is a set of integrity constraints.

Notice that the literals appearing in the constraints can be objective or default ones.

Example 13 continued

Now we can express that only one value can be assigned to a node with the integrity constraint

$$V_1 = V_2 \Leftarrow value(N, V_1), value(N, V_2).$$

Furthermore, we can express that a component is in only one mode at a time with

$$M_1 = M_2 \Leftarrow mode(C, M_1), mode(C, M_2).$$

The use of non-ground integrity constraints is for ease of presentation. Theoretically speaking, we always consider the ground instances of the constraints as the underlying intended constraint theory. The equality theory used is CET of Clark [Cla78], syntactical equality in the Herbrand universe. When the two terms in the equality relation are both ground then we can easily substitute this relation by either **t** or **f**.

We can extend the object language of definition 8 with a new kind of object formulas corresponding to the integrity constraints. This is accomplished by adding the following item to the definition:

- If $\lceil O \rceil$ and $\lceil O' \rceil$ are object formulas then $\lceil O \Leftarrow O' \rceil$ is an object formula;

This extension is syntactically more general than definition 12 because the head and body of an integrity constraint is an arbitrary object formula. However, Alferes, Damásio and Pereira have shown in [ADP94a] that the restricted form of def. 12 is powerful enough to equivalently define any kind of constraint imposed on a three-valued model of an extended logic program.

Because our language does not allow disjunctive heads in object level program rules, we adhere to the theorem view of constraint satisfaction defined in the following manner:

Definition 14 Constraint violation and satisfaction

Given a set of literals $\mathcal{I}$[4], a ground integrity constraint

$$L_1 \vee L_2 \vee \ldots \vee L_n \Leftarrow L_{n+1} \wedge L_{n+2} \wedge \ldots \wedge L_{n+m} \qquad (n+m \geq 0)$$

is violated by $\mathcal{I}$ iff every literal $L_{n+1}, \ldots, L_{n+m} \in \mathcal{I}$ and none of the literals $L_1, \ldots, L_n \in \mathcal{I}$. Otherwise, the constraint is satisfied by $\mathcal{I}$.

---

[4]Notice that the notion of constraint satisfaction refers to an arbitrary set of literals $\mathcal{I}$, whether contradictory or not.

Example 15 continued
Consider the ground integrity constraint

$$\Leftarrow value(out(and(8,1),1),0), value(out(and(8,1),1),1).$$

and $\mathcal{I} = \{value(out(and(8,1),1),0), value(out(and(8,1),1),1)\}$. Then the constraint is not satisfied in $\mathcal{I}$.

In terms of our demo meta-predicate, constraint satisfaction is expressed and verified by:

$$\text{demo}(\lceil P \rceil, \lceil L_1 \vee \ldots \vee L_n \Leftarrow L_{n+1} \wedge \ldots \wedge L_{n+m} \rceil)$$
$$\equiv$$
$$\text{demo}(\lceil P \rceil, \lceil L_1 \rceil) \vee \ldots \vee \text{demo}(\lceil P \rceil, \lceil L_n \rceil)$$
$$\vee$$
$$\neg\text{demo}(\lceil P \rceil, \lceil L_{n+1} \rceil) \vee \ldots \vee \neg\text{demo}(\lceil P \rceil, \lceil L_{n+m} \rceil)$$

We can "meta–program" the notion of satisfaction and violation of integrity constraints:

demo(T, IC) ←
    is_implied_by(_, Body, IC),
    ¬demo(T, Body).
demo(T, IC) ←
    is_implied_by(Head, _, IC),
    demo(T, Head).

¬demo(T, IC) ←
    is_implied_by(Head, Body, IC),
    demo(T, Body),
    ¬demo(T, Head).

Now the definition of a contradictory program follows:

Definition 16 Contradictory program wrt integrity constraints
Let P be an extended logic program and IC a set of integrity constraints. Program P is contradictory wrt IC iff there is an integrity constraint belonging to IC which is violated in the well-founded model of P.

Contradiction with respect to explicit negation can be rendered by introducing integrity constraints $\mathbf{f} \Leftarrow L, \neg L$, for each objective literal L. Of course, if the Herbrand base is infinite we'll have an infinite number of integrity rules to satisfy. Infinite ground programs might be reduce to (or generated by) a finite non-ground one. In such cases, it suffices to extend our algorithms to the non-ground case, whenever these programs are "well behaved" (i.e. having a finite number of finite derivations). This is not a restriction in practice in model-based diagnosis because such systems have a limited finite number of components, observation points, etc... So, by properly assigning sorts or types to our predicates symbols we will be able to deal with tthe aforesaid contradiction ICs.

11.4.2 The revision layer: revision and diagnosis

We are now in a position to discuss the second of our four layers, the *revision layer*, which is concerned with revisions of the assumptions in force, so as to avoid contradiction. We consider changing some of the assumed beliefs (i.e. logic programming closed-world assumptions) to avoid the contradiction. It is natural in diagnosis to revise the basic beliefs instead of derived ones since the former correspond to deeper causes. Thus we need to be able to specify which literals are revisable. These intuitive notions are captured by the concept of open literals:

Definition 17 Open literals

The open literals O_P of a program P are a subset of the set of literals having no rules for them in P.

Example 18 continued

For SD_{and} a subset of open literals is given by

$$\{ mode(and(\_,1), ok), \quad mode(and(\_,1), ab), \quad mode(and(\_,1), s_0),$$
$$mode(and(\_,1), s_1), \quad mode(and(\_,1), unk) \}$$

To declaratively define how the program should be revised in order to avoid contradiction, we consider the result of revising a program in all possible ways. This is accomplished by adding to the original program combinations of open literals (such that their truth value is changed from false to true[5]) and then retaining only the non-contradictory programs obtained in this manner.

Definition 19 Revisions and diagnosis

Let P be a logic program, IC a set of integrity constraints and O_P the set of open literals of P. A revision or diagnosis of P wrt O_P and IC is a subset $R \subseteq O_P$ such that $P \cup R$ is non-contradictory wrt IC.

Example 20 continued

The and-gate is assumed to be ok by default. Since the mode-literals are open ones, we

---

[5] Alferes, Damásio, and Pereira [ADP94a] describes a more general method, by going back on the values of open literals, by changing them from false to either undefined or true. In this paper, for simplicity, we adopt the two–valued revision approach. For a mixed two- and three-valued approach see Pereira, Alferes, and Aparício [PAA94], and the above cited paper.

can switch their truth value from false to true, i.e. we can assume the and-gate to be abnormal, stuck at 0 or 1, or in an unknown fault mode. It turns out that $SD_{and} \cup OBS \cup H$ is not contradictory wrt $IC = \{\Leftarrow value(out(and(I,1),1),0), value(out(and(I,1),1),1)\}$ when H is either $mode(and(I,1), ab)$, or $mode(and(I,1), s_0)$ or $mode(and(I,1), unk)$.

By using a simple generate and test mechanism we can find the revisions of a program:

revision(T, ICs, R) ←
    $\forall_I$(I in ICs → demo(T ∪ R, I)).

revisions(T, ICs, Op, Rs) ←
    Rs = { R : subset(R, Op), revision(T, ICs, R) }.

We make use of the *Gödel* intensional set notation in the *revisions* predicate. A declarative semantics to this construction is given in Naish [BK86], and Hill and Lloyd [HL94] and corresponds to the *Prolog* `findall` meta-predicate. If we perform the Lloyd–Topor [Llo87] transformation to eliminate the universal quantifier on the first rule, we get the equivalent program:

revision(T, ICs, R) ←
    not violates_one_ic(T, ICs, R).

violates_one_ic(T, ICs, R) ←
    I in ICs,
    ¬demo(T ∪ R, I).

revisions(T, ICs, Op, Rs) ←
    Rs = { R : subset(R, Op), revision(T, ICs, R) }.

Obviously this is not an efficient algorithm to compute the revisions of a logic program. Model-generation techniques are necessary to compute revisions of programs. It is possible to design an algorithm that computes a set of formulas on open literals whose models are the revisions. The algorithm is based on the meta-interpreter presented in section 11.3.3. The revisions can be found by applying a tableaux method to the formulas obtained. In this paper we only provide proof-of-principle that all the concepts needed by a model-based diagnosis system can readily be expressed and captured by a meta-program. Specialized efficient algorithms have to be substituted for the above definitions in a real world diagnostic system. Similar comments apply to other algorithms in the rest of the paper.

11.4.3 Minimal revisions

Using the analogue of Reiter's [Rei87] minimal diagnoses, we focus on minimal revisions wrt set inclusion:

Definition 21 Minimal revisions
Let P be a logic program, IC a set of integrity constraints, and O_P the set of open literals of P. A revision R of P wrt O_P and IC is minimal iff there is no other revision $R' \subseteq R$.

The corresponding meta-program is:

minimal_revision(T, ICs, R) ←
   revision(T, ICs, R),
   $\forall_{R'}$ (subset(R', R), revision(T, ICs, R') → R = R').

minimal_revisions(T, ICs, Op, Rs) ←
   Rs = { R : subset(R, Op), minimal_revision(T, ICs, R) }.

Example 22
Consider the full adder shown in figure 11.5. The system description SD_{fad} contains similar rules to those mentioned in SD_{and} for the components $and(I,1)$, $and(I,2)$, $xor(I,1)$, $xor(I,2)$ and $or(I,1)$, where I denotes the order of the full adder. Additionally we have the facts representing the connections among nodes, for instance

$$conn(m(I), in(xor(I,1), 2)) \text{ and } conn(out(and(I,2)), 1), in(or(I,1), 1))$$

Together with the observations shown in figure 11.5 and the assumption that the full adder model $fad(8)$ is *refined* we obtain the minimal revisions

$\{mode(and(8,1), ab)\}$, $\{mode(and(8,2), ab)\}$, $\{mode(or(8,1), ab)\}$,
$\{mode(and(8,1), s_1)\}$, $\{mode(and(8,2), s_1)\}$, $\{mode(or(8,1), s_1)\}$,
$\{mode(and(8,1), unk)\}$, $\{mode(and(8,2), unk)\}$, $\{mode(or(8,1), unk)\}$

with respect to the open literals $\{mode(\_,\_)\}$.

Though we do not consider all possible revisions, but only minimal ones, the number of solutions is still too large. Minimality by set inclusion is too coarse and we need to refine it. In general we want to express that a revision with a given property is preferred to another one. For example, we might prefer certain problem specifications or revisions with minimal cardinality. We will therefore introduce a preference relation over revisions, which captures properties of desired revisions. A preferred revision is to be set inclusion

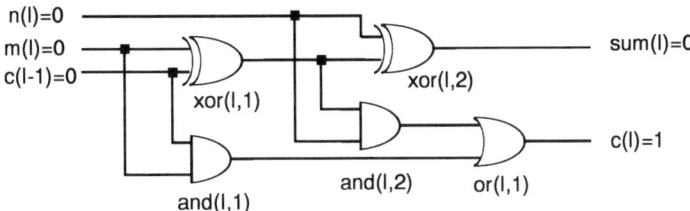

Figure 11.5
A full adder $fad(I)$.

minimal as well as minimal with reference to the preference relation. The definition of preferred revision constitute the core of our third layer, the preference layer.

It is interesting to note that in the database community the concept of intentional updates, for instance see Bry [Bry90], and Guessoum and Lloyd [GL90b], deals with a problem similar to model-based diagnosis. These database approaches are in general abductive based, so they can produce all explanations for new facts/observations/updates as well as making the program consistent (as in consistency-based diagnosis and the current paper). To realize an intensional update, they insert/remove basic or derived facts. In terms of model-based diagnosis, the formalization of Bry [Bry90] for example takes as possible abducibles all base facts and computes all set-minimal explanations consisting of these abducibles.

In model-based diagnosis the use of an explanation based approach is preferred if behaviour of faulty components can be described, while consistency based diagnosis can deal with models consisting of behaviour descriptions of the correct behaviour of the device. If all behavioural descriptions (correct and faulty) are completely given, the results of the consistency-based approach and the explanation based approach become the same (see also Console and Torasso [CT91], and Konolige [Kon92]).

The concepts developed in the remainder of this paper (definitions of preferred revisions and formalization of strategic decisions during the revision process) are independent of our specific definition of minimal revision given in this section. Thus, they can be adopted to various other revision concepts, such as abductive diagnosis or abductive formalizations of intensional updates. The principles and concepts behind intensional updates and model-based diagnosis being similar, it would be an interesting research topic to investigate the use of our concepts in the database area and vice-versa, for example building on work such as by Bry [Bry90], and Guessoum and Lloyd [GL90b],

11.4.4 The preference layer: specification of preferences

Motivation The first formal characterization of consistency based diagnosis by Reiter [Rei87] introduced the concept of set-theoretically minimal diagnoses to describe which consistent revisions (diagnoses) should be preferred to others. Subsequent systems have usually employed a preference relation based on probabilities. These approaches are all special cases of a more general one, which ascribes to each diagnosis a set of properties, and specifies how diagnoses with certain properties are preferred to diagnoses with certain other properties.

For example, a diagnosis containing a single fault might be considered better than one with more than one fault. Diagnoses explaining observations without assuming an unknown fault mode might be preferable to ones using such an unknown fault mode. A component might have different fault modes, leading to preferences based on whether certain fault modes are present or not, or are known to be more frequent than others. We will define a generalized preference concept based on properties of single diagnosis.

The minimal revisions are computed with respect to a set of open literals and a set of integrity constraints. We can use both these sets to influence the minimal revisions.

Example 23 continued

In the example of the full adder from example 22 we obtain nine minimal revisions concerning three different gates. We may want to refine the notion of minimality by set inclusion by stating that a revision based on faulty or-gates is preferred to the other ones. We can express this in the following way:

A minimal revision using open literals $\{mode(or(\_,\_),\_)\}$ is preferable to one using other open literals $\{mode(\_,\_)\}$.

Example 24 continued

Single faults are expressed by the following integrity constraint:

$$C_1 = C_2 \Leftarrow single\_faults, not\ mode(C_1, ok), not\ mode(C_2, ok).$$

We can enforce single faults by using the integrity constraint:

$$single\_faults \Leftarrow$$

where $single\_faults$ is an open literal.

The preference graph Sets of open literals and the integrity constraints are ordered in a directed preference graph of nodes containing them. This induces a partial order on all possible minimal revisions.

Definition 25 Preference Graph
A finite, directed, and acyclic graph $G = (V, E)$, where V is a set of nodes and E a set of edges, is called a preference graph of an underlying program P with open literals O_P, iff each node $v \in V$ has associated to it

- a unique label;
- a set of node integrity constraints $I(v)$;
- a set of node open literals $O_P(v) \subseteq O_P$.

Example 26
Consider a system description of an 8-bit-adder. In a functional specification the adder is composed of 8 full adders, each of which consists of two and-gates, two xor-gates and one or-gate. In this functional view hierarchies are stressed, while different fault modes are ignored. In order to detect single faults in this view, a node v of the preference graph has the form:

$$O_P(v) = \{ \quad mode(\_, ab), spec(\_), single\_faults \quad \}$$
$$I(v) = \{ \quad spec(functional) \Leftarrow,$$
$$single\_faults \Leftarrow,$$
$$C_1 = C_2 \Leftarrow single\_faults, not\ mode(C_1, ok), not\ mode(C_2, ok) \quad \}$$

Similar to the integrity constraints above, we are also able to check for double faults or to enforce the physical specification. In contrast to the functional view, the physical specification stresses a large variety of fault modes and neglects hierarchical structure. Analogously to node v we define nodes w and x. Node w captures double faults in the functional specification and x single faults in the physical specification.

The preference graph $G = (\{v, w, x, y\}, \{(v, w), (v, x), (w, y), (x, y)\})$ shown in figure 11.6, prefers revisions with single faults using the functional view over double faults using the functional view, and over single faults using the physical view. These in turn are preferred over arbitrary revisions.

Intuitively, a preference graph expresses the covering relation of a partial order:

Definition 27 Induced partial order
Let $G = (V, E)$ be a preference graph and $v, w \in V$; then w is reachable from v, iff there is a path
$$\{(v_0, v_1), (v_1, v_2), \ldots, (v_{k-1}, v_k)\}, \quad (v_i, v_{i+1}) \in E,$$
such that $v_0 = v$, $v_k = w$.

We say $v <_p w$ iff w is reachable from v whereas v is not reachable from w.

$$
\begin{aligned}
O_P(v) &= \{\ mode(\_,ab), spec(\_), single\_faults\ \} \\
I(v) &= \{\ spec(functional) \Leftarrow, single\_faults \Leftarrow, C_1 = C_2 \Leftarrow \\
&\quad single\_faults, not\ mode(C_1,ok), not\ mode(C_2,ok)\ \}
\end{aligned}
$$

$$
\begin{aligned}
O_P(w) &= \{\ mode(\_,ab), spec(\_), double\_faults\ \} \\
I(w) &= \{\ spec(functional) \Leftarrow, double\_faults \Leftarrow, \\
&\quad C_1 = C_2 \vee C_1 = C_3 \vee C_2 = C_3 \Leftarrow double\_faults, \\
&\quad not\ mode(C_1,ok), not\ mode(C_2,ok), not\ mode(C_3,ok)\ \}
\end{aligned}
$$

$$
\begin{aligned}
O_P(x) &= \{\ mode(\_,\_), spec(\_), single\_faults\ \} \\
I(x) &= \{\ spec(physical) \Leftarrow, single\_faults \Leftarrow, C_1 = C_2 \Leftarrow \\
&\quad single\_faults, not\ mode(C_1,ok), not\ mode(C_2,ok)\ \}
\end{aligned}
$$

$$
\begin{aligned}
O_P(y) &= \{\ mode(\_,\_), spec(\_)\ \} \\
I(y) &= \{\ \}
\end{aligned}
$$

Figure 11.6
A preference graph.

Preferred revisions The extended definition of a preferred revision demands not only minimality by set inclusion, but minimality with respect to the partial ordering induced by the preference graph too.

Definition 28 Preferred Revision
Let P and $G = (V, E)$ denote an extended logic program and its associated preference graph.

Let $v \in V$ and let R be a minimal revision of P wrt $O_P(v)$ and $I(v)$. If there is no revision we say node v is invalid. Otherwise R is a minimal revision of node v.

A *preferred revision* of P with respect to G is a revision of a $<_p$-minimal node of the set of valid nodes of G, according to the induced partial order.

Example 29
A minimal revision wrt $O_P(v)$ and $I(v)$ is

$$r_v = \{spec(functional), single\_faults, mode(or(8,1), ab)\}$$

Thus node v is valid. As v is $<_p$-minimal, r_v is a preferred revision.

A meta-program to compute preferred revisions is the following one:

node_revisions(P, G, V, Rs) ←
    node_in_graph(G, V),
    node_constraints(G, V, Iv),
    node_open_literals(G, V, Ov),

minimal_revisions(P, Iv, Ov, Rs).

invalid_node(P, G, V) ←
    node_revisions(P, G, V, {}).

preferred_revision(P, G, V, R) ←
    node_in_graph(G, V),
    $\forall_U$(reachable(G, U, V) → invalid_node(P, G, U)),
    node_revisions(P, G, V, R), R ≠ {}.

reachable(G, U, V) ←
    edge_in_graph(G, U, V).
reachable(G, U, W) ←
    edge_in_graph(G, U, V),
    reachable(G, V, W).

The intended semantics of the graph manipulating operations (node_in_graph/2 and edge_in_graph/3) is evident. The set of preferred revisions obtained using a new predicate preferred_revisions(P,G,Rs) can be computed using the intensional sets operation on predicate preferred_revision above described.

11.4.5 Expressing commonly used preferences

Below we have added some useful integrity constraints to specify properties for various commonly used preferences in diagnosis problems. The corresponding property literals (such as $spec(i)$, max_n_faults, $same\_typed\_faults$, $only\_mode$, $faults\_complete(C)$) can then be used in the preference graph.

- **Choice of specification.**
  There are different specifications for the problem denoted by $spec(\_)$. We need a general integrity constraint to express that only one specification is active at a time

$$I = J \Leftarrow spec(I), spec(J).$$

and to activate a particular one

$$spec(i) \Leftarrow .$$

If we try to find diagnoses with specification I and want to consider specification J only, when reasoning with specification I does not lead to a diagnosis, we have to specify $spec(I)$ as more preferable than $spec(J)$ in the preference graph. This is a useful strategy in order to avoid reasoning with overly complex specifications, if simple ones are already sufficient to find suitable diagnoses.

- **Number of faults.**
  A diagnosis must not contain more than n faults. The head of the integrity constraint imposes that two of the $n+1$ variables are equal.

$$\bigvee_{i,j=1,\, i\neq j}^{n+1} C_i = C_j \Leftarrow max\_n\_faults, not\, mode(C_1, ok), \ldots, not\, mode(C_{n+1}, ok).$$

Again, in the preference graph we specify that literal max_i_faults is preferable to max_i+1_faults. This computes higher cardinality diagnoses only, if lower ones are not available.

- **Preference of homogeneous diagnoses.**
  If more than one component is faulty the diagnoses containing only diagnoses of the same type are preferred to mixed ones.

$$T_1 = T_2 \Leftarrow same\_typed\_faults,\quad not\, mode(C_1, ok), not\, mode(C_2, ok),\\ type(C_1, T_1), type(C_2, T_2).$$

where $type/2$ is a predicate denoting a component's type.

- **Preference of fault modes.**
  For a given component c with possible fault modes fm_1 to fm_n the fault mode fm_i may be the most plausible one. By assuming $only\_mode(c, fm_i)$ modes other than fm_i are not considered.

$$M_2 = ok \vee M_1 = M_2 \Leftarrow only\_mode(C, M_1), mode(C, M_2).$$

- **Faults complete.**
  If the specified fault modes of a component are assumed to be complete then the unknown fault mode unk is forbidden. This is expressed by the integrity constraint

$$\Leftarrow faults\_complete(C), mode(C, unk).$$

Using property literals and property conditions in our revision process, we have to model property literals as *open literals*: In order to use the properties listed above the literals

$$O_{prop} = \{\ mode(\_,\_), spec(\_), max\_n\_faults, same\_typed\_faults,\\ only\_mode(\_,\_), faults\_complete\ \}$$

has to be the set of open literals.

If we want to model *complex properties*, we can employ logical combinations of the simpler ones.

11.5 The Strategy Layer: Strategies

Finally, we specify our fourth layer, the *strategy layer*. Strategic diagnostic decisions are taken whenever we have found a set of consistent revisions (diagnoses), and have to contemplate whether or not we wish to continue the diagnostic process, and how. Therefore we need to consider conditions depending on sets of diagnoses, whereas preferences depend only on properties of single diagnoses.

The main contribution of this chapter is the definition of a syntax and a semantics for the strategy language. Based on it we then define the diagnostic process as the process to find a satisfactory set of working hypotheses obeying some given strategy formula and the corresponding diagnoses. But first, let us provide additional motivation for the strategy layer.

11.5.1 Motivation

We have already discussed the need for strategic decisions in our initial example. Using a hierarchical description of the system, diagnoses are initially found on a very abstract level. Since these diagnoses (or revisions) make the program consistent again, work at the preference level stops after having found a set of preferred diagnoses according to the preference graph. In the initial example for instance, we want to continue refining suspected components. One way of defining suspect components might be those which occur in every diagnosis found.

Example 30 Structural refinement (Böttcher and Dressler [BD93b, BD94])
If an abstract system component c occurs in all diagnoses we then prefer to activate a more detailed model for c.

This "structural refinement" strategy needs to access all current diagnoses before choosing to activate more detailed models. Another strategy, called "behavioural refinement", involves a condition on two diagnoses:

Example 31 Behavioural refinement (Böttcher and Dressler [BD93b, BD94])
If there are two diagnoses assigning distinct fault modes to a component then, in order to discriminate between the two, a more detailed model for the faults is activated. This is behavioural refinement.

A third strategy is called "measurements". The condition for the strategy "measurements" involves the models resulting from a combination of system description, observations, and a diagnosis. "Measurements" checks which observations are predicted by a certain diagnosis. Given a set of diagnoses, we check whether at least two diagnoses predict incompatible values at some observation point. By performing this observation we can discriminate between the two diagnoses, and drop (at least) one of them.

Example 32 Measurements
Given a diagnosis we can make predictions using the system description and observations. Having obtained at least two diagnoses their predictions may contradict each other. In this case we want to make additional observations in order to distinguish which of these diagnoses remain possible.

This can be extended to model in a declarative way a complete measurement strategy based on entropy, the one used in most diagnosis systems (e.g. de Kleer and Williams [dKW87]). Entropy-based measurement selection generalizes our example to more than two diagnoses and to achieve maximum discrimination from each measurement. Entropy measurement can do a look-ahead, given the possible outcomes of measurements. Specifying such a strategy explicitly makes the term "model-based system" even more appropriate, by effectively including the various diagnosis and measurement strategies declaratively. Other measurement strategies are defined by McIlraith [McI94].

After this informal discussion of diagnostic strategies, let us define the formalism and language needed to express such strategies.

11.5.2 Working hypotheses

A basic concept for defining strategies is that of "working hypotheses" as defined by Struss [Str92b, Str92b], and Böttcher and Dressler [BD93b, BD94]. For a strategy such as "structural refinement" the working hypotheses express the granularity of a component's description. If a component is abnormal in all reported diagnoses it should be refined. This is achieved by adding a suitable working hypothesis to the system description.

Definition 33 Working Hypotheses
The set of working hypotheses WH of a program P is a subset of the open literals of P.

Example 34
In definition 3 we defined rules to compute a component's output values. Because the component is part of the hierarchical structure, all rules must contain $refined(\_)$-literals in their body, so as to assign a granularity level to every component. Literals $refined(\_)$ are open ones,
$$WH = \{refined(\_)\}$$
is a set of working hypotheses for structural refinement.

Now we have to define how a set of working hypotheses is associated with the current set of diagnoses:

Definition 35 Preferred Revisions under Working Hypotheses
Let WH be a set of working hypotheses for program P and a preference graph G, with the restriction that for each node v in G sets $O(v)$ and WH are disjoint. The preferred revisions under working hypotheses $H \subseteq WH$ of P, denoted by $\mathcal{R}_{\langle P,G,WH \rangle}(H)$, are the preferred revisions of $P \cup H$ wrt G.

If the context is clear, we refer to $\mathcal{R}_{\langle P,G,WH \rangle}(H)$ by $\mathcal{R}(H)$.

Example 36 continued
Recall again our calculator of figure 11.2, displaying the result "256" for the input "0+0". The calculator consists of keyboard, CPU and display. The CPU contains an 8-bit-adder composed of 8 full adders. A set of working hypotheses $H \subseteq WH$ reflecting part of the hierarchical structure of the calculator is

$$H = \{refined(calculator), refined(cpu), refined(adder), refined(fad(8))\}$$

Additionally, consider the preference graph G of figure 11.6, to compute the single (preferred) faults of the functional specification, where we obtain

$$\mathcal{R}_{\langle SD_{calculator} \cup OBS, G, WH \rangle}(H) =$$

$$\{\{mode(or(8,1), ab)\}, \{mode(and(8,1), ab)\}, \{mode(and(8,2), ab)\}\}$$

Working hypotheses are used to control the diagnostic process, i.e. to define when we become satisfied with a set of diagnoses or whether and which further actions or assumptions have to be adopted to obtain a more satisfying set. In the next section we define a language to control the diagnosis process.

11.5.3 Syntax of the strategy language

In order to express strategies, in particular for model-based diagnosis, a language is necessary with enough expressive power to capture statements of the form:

"Some properties hold in all/some models determined by revisions".

It is important to realize that such a statement cannot be expressed in the object level language. The revisions have to be at hand to evaluate resulting models, and to check if some properties hold in all/some of them. The object level contains a system description, observations, and a set of working hypotheses, and the meta-level comprises the strategy formulas. The two levels form a cycle (see figure 11.7), where the object level passes revisions, whose calculation is based on the working hypotheses, to the meta-level, while

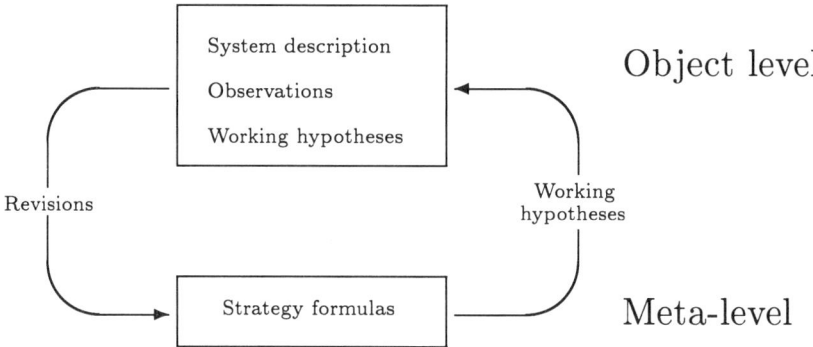

Figure 11.7
Circle of object- and meta-level reasoning.

the meta-level uses the revisions to evaluate the strategy formulas and to pass back new working hypotheses to the object level.

A strategy formula relates conditions that hold in the last cycle step of the iteration, expressed in its premises, to conditions that are made to hold in the next cycle step, expressed in their conclusion.

Thus, the strategy formulas have to be powerful enough to express proving a goal given a program at the object level, and to reason about all programs obtained from a set of revisions.

Consider again structural refinement (c.f. example 30). In logical terms, for a refinable component c, a theory T and a set of integrity constraints ICs, what is intended is the satisfiability of the following formula:

$$\forall_R \, [\texttt{minimal\_revision}(T, ICs, R) \to \texttt{demo}(T \cup R, \lceil ab(c) \wedge refinable(c) \rceil)] \to$$
$$\forall_{R'} \, [\texttt{minimal\_revision}(T, ICs, R') \to \texttt{demo}(T \cup R', \lceil refined(c) \rceil)]$$

So, if component c is abnormal in all current minimal diagnoses of theory T wrt to integrity constraints ICs, then it should be refined in all revisions. Of course, instead of only considering minimal revisions, preferred revisions might be used.

In order to simplify the writing of strategies we omit the program and the revisions from formulas, and obtain a modal-like syntax for representing strategy formulas. We also extend the language with an existential quantifier ranging over revisions and propositional connectives:

Definition 37 Strategy formulas

1. The constants **f** and **t** are strategy formulas;
2. If $F \in \mathcal{L}_{Obj}$ then $demo(F)$ is a strategy formula;
3. If S and S' are strategy formulas, then their boolean combinations are strategy formulas.
4. If S is a strategy formula, then $\forall_R S$ and $\exists_R S$ are strategy formulas;
5. Nothing else is a strategy formula.

The Language $\mathcal{L}_{Strat}$ is the set of all strategy formulas.

The definition above deals only with ground literals of the object program. In the remainder we sometimes use an extension of the strategy language allowing quantification of the form $\forall$ and $\exists$ of variables ranging over finite domain types of the object program. These quantifiers are just used as shorthand for the non-ground representations of formulas. They should not be confused with the quantifiers over revisions, denoted by $\forall_R$ and $\exists_R$. These quantifiers over revisions, by the way, are similar to the usual modal operators $\Box$ and $\Diamond$ (see Chellas [Che80]), specifying, what is true in all worlds (diagnoses) or in at least one world (diagnosis), and are required inasmuch we want to evaluate conditions over sets of diagnosis (comparable to modal logic $S5$).

Accordingly, we express the structural refinement strategy as follows:

$$S_{sr}(c) = \forall_R \, demo(not \, mode(c, ok) \land refinable(c)) \rightarrow \forall_R \, demo(refined(c))$$

Example 38 Behavioural Refinement

Behavioural refinement is a little more complex than structural refinement, and is coded as a conjunction of the following formulas, obtained for each distinct pair of fault modes[6] fm_1 and fm_2 of component c:

$$\exists_R (demo(faultmode(c, fm_1))) \land \exists_R (demo(faultmode(c, fm_2))) \rightarrow$$
$$\forall_R \, demo(refined\_fm(c))$$

A more concise representation exists if the first order extension of the strategy language is used. The propositional strategy constraints used above are obtained from their first-order versions by first substituting the variables by their ground instances and then dropping the quantifiers.

$$\forall C \colon \forall_R \, demo(not \, mode(C, ok) \land refinable(C)) \rightarrow \forall_R \, demo(refined(C))$$

and

$$\forall C \colon \forall F_1 \colon \forall F_2 \colon \quad \exists_R (demo(faultmode(C, F_1))) \land$$
$$\exists_R (demo(faultmode(C, F_2))) \land F_1 \neq F_2)$$
$$\rightarrow \forall_R \, demo(refined\_fm(C))$$

---

[6] In order to distinguish between the *okay* mode and other fault modes, the system description contains the rule $faultmode(C, M) \leftarrow mode(C, M), M \neq ok$.

11.5.4 Semantics of the strategy language

The semantics of strategy formulas is defined as a function assigning truth values to the well-formed formulas of language $\mathcal{L}_{Strat}$. The meaning of such formulas is defined with respect to a set of preferred revisions $\mathcal{R}$ under some working hypotheses. We start by explicitly characterizing the truth conditions of the consequence operator

$$\mathcal{R}, \rho \models S$$

with the intended meaning "S is true at ρ from $\mathcal{R}$", where S is a strategy formula and ρ a revision from $\mathcal{R}$.

The program generating the revisions is parameterized by the set of working hypotheses:

Definition 39 Semantics of strategy formulas
Let $H \subseteq WH \subseteq O_P$ be a set of working hypotheses for program P with open literals O_P and preference graph G. Let $\mathcal{R}$ be a set of preferred revisions under working hypotheses H:

(1) $\mathcal{R}, \rho \models \mathbf{t}$
(2) not $\mathcal{R}, \rho \models \mathbf{f}$
(3) $\mathcal{R}, \rho \models demo(F)$ iff $P \cup H \cup \rho \models_{WFSX} F$
(4) $\mathcal{R}, \rho \models \neg S$ iff not $\mathcal{R}, \rho \models S$
(5) $\mathcal{R}, \rho \models S \wedge S'$ iff $\mathcal{R}, \rho \models S$ and $\mathcal{R}, \rho \models S'$
(6) $\mathcal{R}, \rho \models S \vee S'$ iff $\mathcal{R}, \rho \models S$ or $\mathcal{R}, \rho \models S'$
(7) $\mathcal{R}, \rho \models \forall_R S$ iff $\mathcal{R}, \gamma \models S$ for every γ in $\mathcal{R}$
(8) $\mathcal{R}, \rho \models \exists_R S$ iff $\mathcal{R}, \gamma \models S$ for some γ in $\mathcal{R}$

The truth-value of formulas $S \rightarrow S'$, $S \leftarrow S'$, and $S \leftrightarrow S'$ are defined as usual.

Additionally, all formulas have to be in the scope of a quantifier over revisions. This guarantees that the current revision ρ of a formula is not needed. We can then write $\mathcal{R} \models S$ instead of $\mathcal{R}, \rho \models S$.

Notice that nested $\forall_R F$ and $\exists_R F$ quantifiers can always be eliminated due to the properties of the semantics of strategy formulas. A flat strategy formula is obtained, i.e. without embedded ocurrences of $\forall_R F$ and $\exists_R F$.[7]

Validating strategy constraints is very similar to checking integrity constraints. To define the predicate demo/3 we need the theory T and the strategy constraint to be checked plus a set of revisions Rs. The revisions are needed to define the validity of the quantifiers over revisions.

---

[7]The process is similar to eliminating nested occurences of $\Box$ and $\Diamond$ in modal logic $S5$

demo(T, Rs, SF) ←
    is_implied_by(_, Body, SF),
    ¬demo(T, Rs, Body).
demo(T, Rs, SF) ←
    is_implied_by(Head, _, SF),
    demo(T, Rs, Head).
demo(T, Rs, Conj) ←
    and(C1, C2, Conj),
    demo(T, Rs, C1),
    demo(T, Rs, C2).
demo(T, Rs, Disj) ←
    or(D1, _, Disj),
    demo(T, Rs, D1).
demo(T, Rs, Disj) ←
    or(_, D2, Disj),
    demo(T, Rs, D2).
demo(T, Rs, Nec) ←
    forall_rev(SF, Nec),
    $\forall_R$(R in Rs → demo(T ∪ R, SF)).
demo(T, Rs, Ex) ←
    exists_rev(SF, Ex),
    R in Rs, demo(T ∪ R, SF)).

The predicate ¬demo/3 is defined analogously. Notice that we have hidden the current set of working hypotheses within theory T. We also haven't described how set Rs was obtained. This is to allow for generality and modularity of our meta-programs.

11.5.5 Selecting sets of working hypotheses

Using strategy formulas we are able to formulate a wide range of strategies useful in the diagnosis process. Strategy formulas allow us to express whether a set of diagnoses satisfies certain conditions. The aim of the diagnosis process is to identify a set of diagnoses satisfying all desired conditions. Satisfaction of formulas can only be controlled by the set of working hypotheses. Therefore, we need to find a set H of working hypotheses satisfying all strategy formulas, i.e. meeting two requirements:

R1: All necessary working hypotheses are adopted.

R2: No superfluous working hypothesis is adopted.

If a set H of working hypotheses satisfies a strategy formula S, requirement R1 is met. It is trickier to meet requirement R2. Notice the similarities among the rationale behind conditions $R1$ and $R2$ and the definitions of revision and minimal revision. Clearly, the difference is that satisfying strategy formulas is done on a higher (meta) level. The concepts and ideas are the same. So we define:

Definition 40 Coherent set of working hypotheses
A set of working hypotheses $H \subseteq WH$ of a logic program P with associated preference graph G is called a *coherent set of working hypotheses* with respect to a set of strategy constraints SCs iff

$$\mathcal{R}_{\langle P, G, WH \rangle}(H) \models S$$

for every $S \in SCs$.

The definition above complies with requirement $R1$. To guarantee $R2$ we define:

Definition 41 Adopted set of working hypotheses
A coherent set of working hypotheses $H \subseteq WH$ of a logic program P with associated preference graph G is adopted iff it is inclusion minimal among all coherent sets of working hypotheses.

Example 42
Consider the calculator given in figure 11.2 and the strategy of structural refinement as defined in example 38. We will discuss the strategy to refine the calculator's structure when necessary:

$$S_{sr} = S_{sr}(calculator) \wedge S_{sr}(keyboard) \wedge \ldots \wedge S_{sr}(fad(8))$$

These strategy formulas formalize the intuitions given in section 11.2. At the end, the set

$$H = \{refined(calculator), refined(cpu), refined(adder), refined(fad(8))\}$$

will be adopted, because first the calculator should be refined, afterwards the cpu and the adder, and finally the 8-bit full adder.

Additionally, we could think about preferred adopted sets of working hypotheses and introduce a working hypotheses preference graph resembling the notion of preference graph. Afterwards, we could move to an higher level considering conditions on sets of working hypotheses. We could also associate a different preference graph to each set of working hypotheses, to gain more generality, etc.

The meta-program to compute a coherent set of working hypotheses is simply:

coherent_WH(T, G, SFs, H) ←
    not violates_one_sf(T ∪ H, G, SFs).

violates_one_sf(T, G, SFs) ←
    S in SFs,
    preferred_revisions(T, G, Rs),
    ¬demo(T, Rs, S).

To determine an adopted set, a meta-predicate adopted_WH/5 along the lines of meta-predicate minimal_revisions/4 should be defined. We leave that to the reader and go on with an overview of strategies.

11.5.6 Overview of strategies

The following is a set of useful diagnosis strategies, expressed in the language defined above. The strategies structural refinement and behavioural refinement are directly adapted from Böttcher and Dressler [BD93b, BD94]. The strategies physical negation and measurements are also mentioned in the works by Böttcher and Dressler but interpreted differently. Our system directly allows the declaration of these strategies in the form below, while they have to be specified procedurally in the system of Böttcher and Dressler [BD94].

- **Structural Refinement**

    If abstract component c is contained in all diagnoses of the system we structurely decompose c.

    $$\forall_R \, demo(not \; mode(c, ok) \land refinable(c)) \rightarrow \forall_R \, demo(refined(c))$$

- **Behavioural Refinement**

    If different consistent models of the system predict different behavioural modes for a component c, then we refine the behavioural model of c in order to discriminate among them.

    $$\exists_R \, (demo(faultmode(c, fm_1))) \land \exists_R \, (demo(faultmode(c, fm_2))) \rightarrow \\ \forall_R \, demo(refined\_fm(c))$$

    In the system description, the more complicated rules for fault models of component c will include $refine\_fm(c)$ in their premise.

- **Measurements**

    If different consistent models of the system predict different values v_1 and v_2 for some component c we can discriminate between these values by making a measurement:

    $$\exists_R \, demo(value(n, v_1)) \wedge \exists_R \, demo(value(n, v_2)) \rightarrow \forall_R \, demo(measure(n))$$

    The adopted set of working hypotheses indicates measurement points which discriminate diagnoses.

- **Introducing physical negation**

    If under some diagnosis component c is assigned a known behavioural mode fm, then we avoid assigning the unknown mode to c.

    $$\exists_R \, demo(faultmode(c, fm)) \rightarrow \forall_R \, demo(fm\_complete(c))$$

    If $fm\_complete$ is active for a component c we can ensure that c is assigned a known fault mode by adding the following rule to the system description:

    $$\neg mode(c, unk) \leftarrow fm\_complete(c)$$

11.6 Conclusion

Declarative specification of the system description has been a major advantage of model-based diagnosis compared to earlier heuristic diagnosis systems. We have extended this principle to the diagnosis process itself, and have shown how to specify important aspects of the diagnostic decision making process in a declarative way. This includes preferences used during the computation of diagnoses, and diagnosis strategies used by the agent to refine and expand diagnoses. A suitable formalization has been developed in this paper supported by extended logic programs and meta-programming.

We have been able to show how the concepts of preferences and strategies can directly be programmed and captured by meta-programming, meta-reasoning methods and implementation techniques using logic programming. While the current paper is mostly intended as proof-of-principle to show that all concepts needed by a model-based diagnosis system can be readily expressed by a meta-program, we think that more specialized and efficient algorithms can be substituted into the same framework without undue conceptual problems. This will yield a very expressive and efficient logic programming environment suitable for diagnostic reasoning, based on models both of the system to be diagnosed and the diagnostic process itself.

Acknowledgments

We thank the referee for his comments and criticisms, which helped to substantially improve the paper. We thank Peter Fröhlich for many fruitful discussions and comments on this paper. We also thank JNICT - Portugal and INIDA-DAAD for their support.

References

[ADP94a]   J. J. Alferes, C. V. Damásio, and L. M. Pereira. A logic programming system for non-monotonic reasoning. Technical report, CRIA, UNINOVA, March 1994. Submitted to the Special Issue of the Journal of Automated Reasoning.

[ADP94b]   J. J. Alferes, C. V. Damásio, and L. M. Pereira. Top–down query evaluation for well-founded semantics with explicit negation. In A. Cohn, editor, *Proc. of the European Conference on Artificial Intelligence'94*, pages 140–144. John Wiley & Sons, August 1994.

[ADP94c]   J. J. Alferes, C. V. Damásio, and L. M. Pereira. A top-down derivation procedure for programs with explicit negation. In M. Bruynooghe, editor, *Proc. of the International Logic Programming Symposium'94*, pages 424–438. MIT Press, November 1994.

[Alf93]   J. J. Alferes. *Semantics of Logic Programs with Explicit Negation*. PhD thesis, Universidade Nova de Lisboa, 1993.

[BD93a]   R. Bol and L. Degerstedt. Tabulated resolution for well founded semantics. In *ILPS'93*. MIT Press, 1993.

[BD93b]   Claudia Böttcher and Oskar Dressler. Diagnosis process dynamics: Holding the diagnostic trackhound in leash. In Ruzena Bajcsy, editor, *Proceedings of the 13th international joint conference on artificial intelligence*, volume 2, pages 1460–1471. Morgan Kaufmann Publishers, Inc., 1993.

[BD94]   Claudia Böttcher and Oskar Dressler. A framework for controlling model-based diagnosis systems with multiple actions. *Annals of Mathematics and Artificial Intelligence, special Issue on Model-based Diagnosis*, 11(1–4), 1994.

[BG94]   C. Baral and M. Gelfond. Logic programming and knowledge representation. *Journal of Logic Programming - Special 10-year issue*, 1994.

[BK82]   K. A. Bowen and R. A. Kowalski. Amalgamating language and metalanguage in logic programming. In K. L. Clark and S.-A. Tarnlund, editors, *Logic Programming*, pages 153–172. Academic Press, 1982.

[BK86]   K. A. Bowen and R. A. Kowalski. Negation and quantifiers in NU-Prolog. In E. Shapiro, editor, *Proc. ICLP'86*, pages 624–634. Lecture Notes in Computer Science 225, Springer-Verlag, 1986.

[Bry90]   Francois Bry. Intensional updates: Abduction via deduction. In *Proceedings of the 7th Intl. Conf. on Logic Programming*, pages 561–575, Jerusalem, June 1990. MIT Press.

[Che80]   Brian F. Chellas. *Modal Logic – An Introduction*. Cambridge University Press, 1980.

[Cla78]   K. Clark. Negation as failure. In H. Gallaire and J. Minker, editors, *Logic and Data Bases*, pages 293–322. Plenum Press, 1978.

[CT90]   Luca Console and Pietro Torasso. Integrating models of correct behavior into abductive diagnosis. In *Proceedings of the European Conference on Artificial Intelligence (ECAI)*, pages 160–166, Stockholm, August 1990. Pitman Publishing.

[CT91]   Luca Console and Pietro Torasso. A spectrum of logical definitions of model-based diagnosis. *Computational Intelligence*, 7(3):133–141, 1991.

[CF94]   Luca Console and Gerhard Friedrich (eds). *Model-based Diagnosis. Annals of Mathematics and Artificial Intelligence*, 11, 1994.

[CW92]   W. Chen and D. S. Warren. A goal–oriented approach to computing well–founded semantics. In *IJCSLP'92*. MIT Press, 1992.

[CW93]   W. Chen and D. S. Warren. Query evaluation under the well founded semantics. In *PODS'93*, 1993.

[Dav84]   Randall Davis. Diagnostic Reasoning Based on Structure and Behavior. *Artificial Intelligence*, 24, 1984.

[Dix91]   J. Dix. Classifying semantics of logic programs. In Nerode et al., editor, *1st workshop on Logic Programming and Non-Monotonic Reasoning*. MIT Press, 1991.

[Dix92]   J. Dix. A framework for representing and characterizing semantics of logic programs. In *3rd KR*. Morgan Kaufmann, 1992.

[dKW87]   Johan de Kleer and Brian C. Williams. Diagnosing multiple faults. *Artificial Intelligence*, 32:97–130, 1987.

[dKW89]   Johan de Kleer and Brian C. Williams. Diagnosis with behavioral modes. In *Proceedings of the International Joint Conference on Artificial Intelligence (IJCAI)*, pages 1324–1330, Detroit, August 1989. Morgan Kaufmann Publishers, Inc.

[DNP94]   Carlos Viegas Damásio, Wolfgang Nejdl, and Luís Moniz Pereira. REVISE: An extended logic programming system for revising knowledge bases. In J. Doyle, E. Sandewall, and P. Torasso, editors, *Knowledge Representation and Reasoning'94*, pages 607–618, Bonn, Germany, May 1994. Morgan Kaufmann.

[FGN90]   Gerhard Friedrich, Georg Gottlob, and Wolfgang Nejdl. Hypothesis classification, abductive diagnosis and therapy. In *First International Workshop on Principles of Diagnosis*, Stanford, July 1990. Also appeared in Proceedings of the International Workshop on Expert Systems in Engineering, Lecture Notes in Artificial Intelligence, Vol. 462, Vienna, September 1990, Springer-Verlag.

[FNS94]   Peter Fröhlich, Wolfgang Nejdl, and Michael Schroeder. A formal semantics for preferences and strategies in model-based diagnosis. In *Working papers of the 5th International Workshop on Principles of Diagnosis (DX-94)*, pages 106–113, New Paltz, NY, October 1994.

[Fri93]   Gerhard Friedrich. Theory diagnoses: a concise characterization of faulty systems. In Ruzena Bajcsy, editor, *13th International Joint Conference on Artificial Intelligence*, volume 2, pages 1466–1471, San Mateo, CA, USA, 1993. Morgan Kaufmann Publishers, Inc.

[GL90a]   M. Gelfond and V. Lifschitz. Logic programs with classical negation. In *ICLP'90*. MIT Press, 1990.

[GL90b]   A. Guessoum and J. W. Lloyd. Updating knowledge bases. *New Generation Computing*, 8(1):71–89, 1990.

[HCd92]   Walter Hamscher, Luca Console, and Johan de Kleer. *Readings in Model-based Diagnosis*. Morgan Kaufmann, 1992.

[HL89]   P.M. Hill and J.W. Lloyd. Analysis of meta-programs. In H.D. Abramson and M. H. Rogers, editors, *Meta-Programming in Logic Programming*, pages 23–52. MIT Press, 1989.

[HL94]   P.M. Hill and J.W. Lloyd. *The Gödel Programming Language*. MIT Press, 1994.

[Kon92]   Kurt Konolige. Abduction versus closure in causal theories. *Artificial Intelligence*, 53:255–272, 1992.

[Llo87]   J. Lloyd. *Foundations of Logic Programming*. Springer–Verlag, second edition, 1987.

[McI94]   Sheila McIlraith. Generating Tests using Abduction. In J. Doyle, E. Sandewall, and P. Torasso, editors, *Knowledge Representation and Reasoning*, pages 449–460, Bonn, Germany, May 1994. Morgan Kaufmann.

[Moz92]   Igor Mozetic. Hierarchical model-based diagnosis. In *[HCd92]*, 1992.

[PA92]   L. M. Pereira and J. J. Alferes. Well founded semantics for logic programs with explicit negation. In *ECAI'92*. Morgan Kaufmann, 1992.

[PAA91]   L. M. Pereira, J. N. Aparício, and J. J. Alferes. A derivation procedure for extended stable models. In *IJCAI'91*. Morgan Kaufmann, 1991.

[PAA93]   L. M. Pereira, J. N. Aparício, and J. J. Alferes. Non–monotonic reasoning with logic programming. *Journal of Logic Programming. Special issue on Nonmonotonic reasoning*, 17(2, 3 & 4), 1993.

[PAA94]   L. M. Pereira, J. J. Alferes, and J. N. Aparício. Contradiction removal semantics with explicit negation. In M. Masuch and L. Pólos, editors, *Knowledge Representation and Reasoning Under Uncertainty*, volume 808 of *LNAI*, pages 91–106. Springer-Verlag, 1994.

[PDA93a] L. M. Pereira, C. Damásio, and J. J. Alferes. Diagnosis and debugging as contradiction removal in logic programs. In L. Damas and M. Filgueiras, editors, *6th Portuguese AI Conf.* Springer–Verlag, 1993.

[PDA93b] L. M. Pereira, C. V. Damásio, and J. J. Alferes. Diagnosis and debugging as contradiction removal. In L. M. Pereira and A. Nerode, editors, *2nd Int. Workshop on Logic Programming and Non-monotonic Reasoning*, pages 334–348, Lisboa, Portugal, June 1993. MIT Press.

[Poo89]   David Poole. Normality and faults in logic-based diagnosis. In *Proceedings of the International Joint Conference on Artificial Intelligence (IJCAI)*, pages 1304–1310, Detroit, August 1989. Morgan Kaufmann Publishers, Inc.

[Rei87]   Raymond Reiter. A theory of diagnosis from first principles. *Artificial Intelligence*, 32(1):57–96, 1987.

[Ros92]   K. A. Ross. A procedural semantics for well-founded negation in logic programs. *Journal of LP*, 13, 1992.

[SD89]   Peter Struss and Oskar Dressler. Physical negation — Integrating fault models into the general diagnostic engine. In *Proceedings of the International Joint Conference on Artificial Intelligence (IJCAI)*, pages 1318–1323, Detroit, August 1989. Morgan Kaufmann Publishers, Inc.

[Str92a] Peter Struss. Diagnosis as a process. In *[HCd92]*, 1992. First appeared in Working Notes of the first International Workshop on Model-based Diagnosis, Paris, 1989.

[Str92b] Peter Struss. What's in SD? towards a theory of modeling in diagnosis. In *[HCd92]*, 1992.

[Sub89]   V. S. Subrahmanian. Foundations of Metalogic Programming. In H.D. Abramson and M. H. Rogers, editors, *Meta-Programming in Logic Programming*, pages 1–14. MIT Press, 1989.

[vGRS91] A. van Gelder, K. A. Ross, and J. S. Schlipf. The well-founded semantics for general logic programs. *Journal of the ACM*, 38(3):620–650, 1991.

[Wag91]   G. Wagner. Logic programming with strong negation and inexact predicates. *J. of Logic and Computation*, 1(6), 1991.

12 The Generalized ChronoBase Temporal Data Model

Suryanarayana M. Sripada and Petra Möller

Abstract

Temporal databases have wide ranging applications from traditional areas such as banking and financial applications to advanced areas such as process and transportation industries. The ChronoBase temporal data model is designed to aid the development of advanced temporal database applications from domains such as IVHS (Intelligent Vehicle Highway Systems), process industry and planning, as well as the more traditional domains. The ChronoBase model introduces new notions such as temporal associations and temporal characteristics which make it much more expressive than other models proposed in the literature. In this paper, we present the ChronoBase temporal data model. The foundations of the ChronoBase model lie in the use of meta-logic programming for temporal knowledge representation.

12.1 Introduction

Meta-Logic programming provides a very powerful framework for knowledge representation and reasoning (Aiello, Nardi, Schaerf [ANS88], Brogi and Turini [BT91], Kowalski and Kim [KK92]). Meta-logic has also been used for advanced temporal knowledge representation and reasoning (Sripada [Sri93a]). In this paper, we describe how meta-logic programming has been applied in the development of an advanced temporal data model, ChronoBase.

Temporal databases have wide ranging applications from traditional areas such as banking to advanced areas such as process and transportation industries (Sripada, Rosser, Bedford, and Kowalski [SRBK94]). As a consequence, research in the area of temporal databases has grown steadily during the last decade. Several interesting temporal data models, temporal query languages, and low-level storage and access mechanisms for temporal data have been proposed in the literature (Soo [Soo91], Kline [Kli93]). While the proposals demonstrate several important features, most of the proposals cover only a small spectrum of the application domains. Motivated by the need for a temporal data model that allows the development of advanced temporal database applications, the ChronoBase temporal data model has been designed. The targeted applications for ChronoBase include domains such as IVHS (Intelligent Vehicle Highway Systems), process industries and planning in addition to the more traditional domains such as administrative databases and banking. The expressiveness of the ChronoBase model is far greater than any other model proposed in the literature so far.

The foundations of the ChronoBase model are grounded in the use of meta-logic programming for temporal knowledge representation and reasoning. Central to the ChronoBase model is the notion of *temporal characteristics* which are defined using metalogic. Temporal characteristics provide a rich temporal knowledge representation formalism for representing a wide range of temporal phenomena such as processes, events and temporal propositions found in real life situations. Associated with temporal characteristics are a set of temporal relational algebra operators that enable inferences to be drawn automatically from the temporal data represented in the database. The aim of this paper is to present the Generalized ChronoBase temporal data model. The ChronoBase temporal database system implemented at ECRC is an instance of the generalized model.

In the following section, we motivate the need for the salient features of the ChronoBase model. The distinguishing features of the ChronoBase model are the notions of *temporal characteristics* and *temporal associations*. We motivate these notions through some examples.

12.2 Motivation

The *first* example illustrates the need for the notion of *temporal characteristics*. This need arises from the fact that different propositions have different temporal properties. Consider the following relation **R1**. The first tuple in R1 denotes that "John worked in the Math department from time[1] 10 to 15".

```
Employee | Department | Time Period
-----------------------------------
  John      Math         [10,15]
  John      Math         [16,20]
```

From the above relation, it is logical to infer the following information

```
  John      Math         [10,20]        I1
  John      Math         [12,17]        I2
  John      Math         [11,14]        I3
```

Now consider the following relation **R2**. The first tuple in R2 denotes that "Bus1 travelled an odd number of miles during the period [10,15]".

```
Vehicle | No.of miles | Time Period
-----------------------------------
  Bus1      odd          [10,15]
  Bus1      odd          [16,20]
```

---

[1] In this example, we assume discrete time and denote closed intervals with $[t_1, t_2]$.

Bus2	even	[10,15]	
Bus2	even	[16,20]	

From the relation R2, the following may be inferred:

Bus2	even	[10,20]	I4

However, the following tuples *should not* be inferred:

Bus2	even	[12,17]	I5
Bus2	even	[11,14]	I6

Notice that while inferences I1 and I4 are similar, and are both allowed, I5 and I6 are not allowed whereas I2 and I3 are. For example, I6 is not allowed since one cannot say whether Bus2 travelled an even number of miles during [11,14] given that Bus2 travelled an even number of miles during [10,15]. Thus the propositions "employment in a department" and "travel an even number of miles" have different temporal properties.

Also notice that while I4 and I5 are similar, I4 is allowed whereas I5 is not. Thus, propositions from a single relation may also have different temporal properties.

The case of Bus1 is even stranger. *None* of the following tuples concerning Bus1 may be inferred:

Bus1	odd	[10,20]	I7
Bus1	odd	[12,17]	I8
Bus1	odd	[11,14]	I9

How would the database know what inferences to make and what not? All the existing temporal data models proposed in the literature *actually make* the *prohibited* inferences concerning Bus1 and Bus2, since there is no way they can distinguish between the different properties of the facts under consideration.

We propose a new notion called *temporal characteristics* by which such distinction can be recorded and acted upon by the DBMS. A temporal characteristic (TC) captures the temporal properties of a fact. We propose that an additional attribute be used to capture the temporal properties of the fact under consideration. For example, the information of R2 may be represented as shown below in relation **R3**:

Vehicle	No. of miles	Time Period	Temporal Characteristics
Bus1	odd	[10,15]	tc1
Bus1	odd	[16,20]	tc1
Bus2	even	[10,15]	tc2
Bus2	even	[16,20]	tc2

The DBMS has to have knowledge concerning the operations that can be performed for each kind of temporal characteristic. For example,

- two tuples of type tc1 with "meeting" time periods *cannot* be merged
- two tuples of type tc2 with "meeting" time periods *can be* merged

and so on. More of this is given in the following sections.

The *second* example motivates the need for the notion of *temporal associations*. It shows that the notion of temporal association is required due to the different kinds of temporal information which may be known about identical propositions. Consider the following relation **R4**. The first tuple denotes that Sue lived in Munich from time 20 to 30. The second tuple denotes that Sue lived in Munich *sometime* during the interval [50,80].

```
Person | City    | Time Period
-------------------------------
Sue      Munich    [20,30]       'during the interval'
Sue      Munich    [50,80]       'sometime during the interval'
```

As a consequence, the following information may be inferred from the first fact:

```
Sue      Munich    [22,28]       'during the interval'
```

However, the following inference *cannot* be made from the second tuple:

```
Sue      Munich    [52,78]       'sometime during the interval'
```

Notice that there is no way of distinguishing between the two tuples of R4 and the inferences that are allowed, without explicitly recording the information 'during the interval' and 'sometime during the interval'.

In fact, the difference between the behaviour of the two tuples arises from the fact that the relationship between the *actual time interval* for which the proposition holds and the corresponding *reference time interval* is different in each case. The first tuple states that "Sue lived in Munich during an interval i1 which is equal to the reference interval [20,30]". The second tuple states that "Sue lived in Munich during an interval i2 which is a subinterval of [50,80]". In the ChronoBase model, temporal associations are used to record such relationships.

A *temporal association* (TA) represents the association between the actual time for which a proposition holds and a reference time element. In general, a temporal association is a constraint on a validity time element.

The above information may be recorded using temporal associations as follows (see relation **R5**):

```
Person | City    | Time Period | Temporal Association
-----------------------------------------------------
Sue      Munich    i1            i1 equals [20,30]
Sue      Munich    i2            i2 subinterval [50,80]
```

The DBMS should ofcourse know that the temporal association "equals" allows inferences over subintervals whereas "subinterval" does not. Therefore, the inference "Sue lived in Munich during [22,28]" is allowed whereas "Sue lived in Munich sometime during [52,78]" is not. On the other hand, "subinterval" allows inferences over super-intervals whereas "equals" does not. Therefore, the inference "Sue lived in Munich sometime during [45,85]" is allowed whereas "Sue lived in Munich during [15,35]" is not allowed[2].

Both of the above situations may occur simultaneously. That is, propositions with different temporal characteristics and different kinds of temporal associations for each proposition may appear in a single relation. The following example illustrates this phenomenon (relation **R6**):

```
Vehicle | No. of miles | Time Period
-----------------------------------------------------------------
Bus2      even           [25,40]      'sometime during the interval'
Bus2      even           [50,70]      'during the interval'
Bus1      odd            [30,60]      'during the interval'
```

Naturally, identifying allowed inferences from R6 is more involved. In addition, temporal associations and temporal characteristics should be taken into account while joining two or more relations. Therefore an appropriate Temporal Relational Algebra (TRA) and a Temporal Query Language are also required.

The above examples illustrate that there is a need for a temporal data model that can capture the temporal properties of different kinds of propositions as well as the different kinds of relationships between the actual time for which a proposition holds and a reference time element. As far as we are aware, the ChronoBase model is the only one that enables the representation of the above kinds of temporal information through the notion of *temporal characteristics* and *temporal associations* and provides a set of TRA operators that can make appropriate inferences automatically. The description of the ChronoBase model is the main focus of this work.

The rest of this paper is organized as follows: Various kinds of Temporal Characteristics are introduced and defined in section 12.3. The definitions, expressed in meta-logic, serve to define the sound and complete inferences that may be drawn from a set of temporal facts. Temporal Associations are defined in section 12.4. The Generalized ChronoBase Temporal Data Model is described in section 12.5. Some of the constructs of the associated Temporal Query Language SQL^C are presented in section 12.6. The ChronoBase advanced temporal data model is outlined in section 12.7. Related work is presented in section 12.8 followed by conclusions in section 12.9.

---
[2]Notice, however, that the inference "Sue lived in Munich sometime during [15,35]" is allowed.

12.3 Temporal Characteristics

The ChronoBase data model aims at the representation of a great variety of temporal phenomena. The definition of complex temporal phenomena requires a powerful knowledge representation and reasoning tool. We have employed meta-logic programming as a tool for this purpose. The nature of temporal phenomena are captured through meta-predicates, which are in turn defined in logic as a meta-language. The temporal phenomena then simply become metalevel facts. A mapping is required that allows these metalevel facts to be represented as tuples in a temporal relational database. In the ChronoBase model, the meta-predicates are mapped into an attribute called *temporal characteristics*.

A tuple in the ChronoBase model denotes a metalevel fact. A ChronoBase tuple consists of a temporal component and a non-temporal component. A temporal characteristic is part of the temporal component of a tuple. It corresponds to the meta-predicate of the metalevel fact. The non-temporal component of a tuple corresponds to the object level term in a metalevel fact. The object level term denotes an object level fact, which we sometimes simply refer to as a *fact* or a *proposition* or a *temporal proposition*.

A temporal characteristic (TC) captures the temporal properties of a proposition. Every TC denotes a unique set of temporal properties. Different TCs imply different sets of properties. The set of temporal properties of a proposition are unique. Therefore, in a ChronoBase tuple, the TC attribute is functionally dependent upon the proposition (i.e. the non-temporal component).

The (set of) temporal properties denoted by a TC are of two kinds. They are called unary and binary temporal properties respectively. A unary property specifies how the truth of a proposition over a time element (point or interval) implies the truth of the proposition over another time element. A binary property specifies how the truth of a proposition over two different time elements implies the truth of the proposition over a third time element. It goes without saying that some of the temporal properties in the set of properties denoted by a TC are unary temporal properties, while the others are binary temporal properties.

The most commonly found TC in the literature is $Holds^t$ (although the notion of a TC is not explicit, the temporal properties denoted by the $Holds^t$ TC are assumed by default). The $Holds^t$ TC states that a proposition holds over a time interval iff it holds at each time point in that interval. This definition of $Holds^t$ implies that, if the proposition is true over an interval, then it is also true at every time point in that interval. The definition also implies that, if the proposition is true over an interval, then it is also true over all sub-intervals of the given interval. These properties of $Holds^t$ are called the unary properties of the $Holds^t$ TC. On the other hand, the definition also implies that, if a proposition is true over two overlapping time intervals, then the proposition is also true over the union of the two time intervals. This is an example of a binary property implied by the TC $Holds^t$.

The proposition "Sue lives in Munich" has the $Holds^t$ TC. The TC that is associated to a proposition as well as the unary and binary properties of that TC are to be known to the temporal database system in order to make automatic inferences and to answer queries appropriately.

An infinite number of TCs are definable in theory. However, in practice, only a finite number of TCs have been discovered. Moreover, there are only a finite set of operations that can be efficiently carried out over time intervals, in order to draw further inferences (using the unary and binary properties). Therefore, a finite set of useful TCs have been identified and pre-defined in the ChronoBase model.

The meta-logical definitions (axioms) corresponding to various TCs serve as specifications for the unary and binary temporal properties of the TCs. They are useful for defining appropriate temporal relational algebra operators for set-oriented inferences in a temporal database. In the rest of this section, we show how the TCs and their properties are defined using logic as a meta-language.

12.3.1 Unary and binary temporal properties of a proposition

We shall adopt the following convention: predicates, function symbols and constants start in the upper case. Variables start in the lower case. Object level facts, i.e. propositions, are named by suitably constructed metalevel terms using a ground representation. Let the meta-predicate $HT^i(r, tc, i)$[3] denote that proposition r is true over interval i with the temporal characteristic tc. Similarly, the meta-predicate $HT^t(r, tc, t)$ denotes that proposition r is true at time point t and has the temporal characteristics tc. We are now set to define the unary and binary temporal properties that may be denoted by a TC.

Unary properties:

o *Interval downward inheritance:*

$$HT^i(r, tc, i) \Rightarrow (\forall i' \; i' \subset i \Rightarrow HT^i(r, tc, i')) \tag{1}$$

o *Point downward inheritance:*

$$HT^i(r, tc, i) \Rightarrow (\forall t \; t \in i \Rightarrow HT^t(r, tc, t)) \tag{2}$$

o *Interval upward inheritance:*

$$(\forall i' \; i' \subseteq i \Rightarrow HT^i(r, tc, i')) \Rightarrow HT^i(r, tc, i) \tag{3}$$

---

[3] HT stands for $HoldsTrue$.

○ *Point upward inheritance:*

$$(\forall t\ t \in i \Rightarrow HT^t(r, tc, t)) \Rightarrow HT^i(r, tc, i) \quad (4)$$

○ *Superinterval inheritance:*

$$HT^i(r, tc, i) \Rightarrow (\forall i'\ i \subset i' \Rightarrow HT^i(r, tc, i')) \quad (5)$$

○ *Subset preclusion:*

$$HT^i(r, tc, i) \Rightarrow (\forall i'\ i' \subset i \wedge i' \neq i \Rightarrow \neg\ HT^i(r, tc, i')) \quad (6)$$

○ *Overlap preclusion:*

$$HT^i(r, tc, i) \Rightarrow (\forall i'\ i'\ overlaps\ i \Rightarrow \neg\ HT^i(r, tc, i')) \quad (7)$$

Binary properties:

○ *concatenation:*

$$i\ meets\ i' \wedge\ HT^i(r, tc, i) \wedge\ HT^i(r, tc, i') \Rightarrow HT^i(r, tc, i \cup i') \quad (8)$$

○ *overlap-union:*

$$i\ overlaps\ i' \wedge\ HT^i(r, tc, i) \wedge\ HT^i(r, tc, i') \Rightarrow HT^i(r, tc, i \cup i') \quad (9)$$

○ *overlap-intersection:*

$$i\ overlaps\ i' \wedge\ HT^i(r, tc, i) \wedge\ HT^i(r, tc, i') \Rightarrow HT^i(r, tc, i \cap i') \quad (10)$$

○ *overlap-difference:*

$$i\ overlaps\ i' \wedge\ HT^i(r, tc, i) \wedge\ HT^i(r, tc, i') \Rightarrow HT^i(r, tc, i \setminus i') \quad (11)$$

○ *starts-difference:*

$$i\ starts\ i' \wedge\ HT^i(r, tc, i) \wedge\ HT^i(r, tc, i') \Rightarrow HT^i(r, tc, i' \setminus i) \quad (12)$$

○ *finishes-difference:*

$$i\ finishes\ i' \wedge\ HT^i(r, tc, i) \wedge\ HT^i(r, tc, i') \Rightarrow HT^i(r, tc, i' \setminus i) \quad (13)$$

12.3.2 Predefined temporal characteristics in ChronoBase

The logical definition of the pre-defined TCs in ChronoBase are given in this subsection.

1. **The Holds_over TC**

    The fact "Peter ran an odd number of miles" is characterized by the $Holds\_over$ TC. The $Holds\_over$ TC states that a proposition r is true over an interval i. No further inferences are possible from this TC.

    $$HT^i(r, Holds\_over, i)$$

2. **The Holds^t TC**

    The fact "John worked in Math" is characterized by the $Holds^t$ TC:

    $$HT^i(r, Holds^t, i) \Leftrightarrow \forall t\ t \in i \Rightarrow HT^t(r, Holds^t, t)$$

    From definition (14) the following properties can be derived (Sripada [Sri91]):
    - *point downward inheritance*
    - *point upward inheritance*
    - *interval downward inheritance*
    - *interval upward inheritance*
    - *concatenation*
    - *overlap − union*
    - *overlap − intersection*
    - *overlap − difference*
    - *starts − difference*
    - *finishes − difference*

3. **The Holds^i TC**

    The fact "Car A travelled at a constant speed" is characterized by the $Holds^i$ TC:

    $$HT^i(r, Holds^i, i) \Leftrightarrow (\forall i'\ i' \subset i \Rightarrow HT^i(r, Holds^i, i')) \wedge$$
    $$(i\ overlaps\ i' \wedge HT^i(r, Holds^i, i) \wedge HT^i(r, Holds^i, i') \Rightarrow$$
    $$HT^i(r, Holds^i, i \cup i'))$$

    The $Holds^i$ TC has the following properties:
    - *interval downward inheritance*
    - *interval upward inheritance*
    - *overlap − union*

- *overlap − intersection*
- *overlap − difference*
- *starts − difference*
- *finishes − difference*

4. **The Holds_event TC**

   The fact "Flight LH758 flew from London to Munich" is characterized by the *Holds_event* TC:

   $$\forall i' \; (i' \subset i \;\lor\; i' \text{ overlaps } i) \Rightarrow \neg HT^i(r, Holds\_event, i')$$

   The *Holds_event* TC has the properties:
   - *subset preclusion*
   - *overlap preclusion*

5. **The Holds_exact TC**

   The fact "Exactly six minutes passed" is characterized by the *Holds_exact* TC:

   $$HT^i(r, Holds\_exact, i) \Rightarrow (\forall i' \; i' \subset i \Rightarrow \neg HT^i(r, Holds\_exact, i'))$$

   A property of the *Holds_exact* TC is:
   - *subset preclusion*

6. **The Holds_up TC**

   The fact "Car B travelled more than 2 miles" is characterized by the *Holds_up* TC:

   $$HT^i(r, Holds\_up, i) \Leftrightarrow \forall i' \; i \subseteq i' \Rightarrow HT^i(r, Holds\_up, i')$$

   Properties of the *Holds_up* TC are:
   - *interval upward inheritance*
   - *super interval inheritance*
   - *concatenation*
   - *overlap − union*

7. **The Holds_down TC**

   The fact "Car C travelled less than 2 miles" is characterized by the *Holds_down* TC:

   $$HT^i(r, Holds\_down, i) \Leftrightarrow \forall i' \; i' \subset i \Rightarrow HT^i(r, Holds\_down, i')$$

   Properties of the *Holds_down* TC are:
   - *interval downward inheritance*

- *overlap − intersection*
- *overlap − difference*
- *starts − difference*
- *finishes − difference*

8. **The Holds_process TC**

    "Processes" are characterized by the *Holds_process* TC with the properties:
    - *concatenation*
    - *overlap − union*

9. **The Holds_even TC**

    The fact "Bus 1 travelled an even number of miles" is characterized by the TC *Holds_even*, with the following properties:
    - *concatenation*
    - *starts − difference*
    - *finishes − difference*

10. **The Holds_diff TC**

    The fact "Car D passed an even number of junctions at a constant speed" is characterized by the *Holds_diff* TC with the properties:
    - *starts − difference*
    - *finishes − difference*

The notion of temporal characteristics also allows one to represent and manipulate a wide variety of temporal phenomena proposed in the AI literature. For example, the TCs *Holds_event* and *Holds_exact* are used to characterize events (Allen [All83], McDermott [McD82]). Allen's events occur over an interval but they do not occur over any subinterval. The non-concatenable propositions of Shoham [Sho87] (eg. travel an odd number of miles) are simply characterized by *Holds_over*[4]. Shoham's concatenable propositions (eg. travel an even number of miles) can be characterized by the *Holds_even* TC.

12.4 Temporal Associations

Temporal associations are relationships between different time elements. A *time element* is either a *time point* or a *time interval*. A time element may be *absolute* (eg. time point 5, time interval [10,20] etc.) or *symbolic* (eg. time point c_{t1}, time interval c_{i1} etc.). Temporal associations may be viewed as constraints between time elements.

---

[4] *Holds_over* corresponds to the temporal association tc1 used in relation R3.

For example, we may know that "John was a lecturer of Physics during a subinterval of [25,50]" but we don't know exactly when. This fact may be represented in the ChronoBase model by giving a symbolic name to the time period when John was a lecturer of Physics. This symbolic time element, say c_{i5}, is associated to the absolute time element [25, 50] through the constraint "c_{i5} subinterval [25, 50]". Notice that, in this framework, constraints may also be expressed between time elements corresponding to different tuples. For example, if "Peter was a professor of Chemistry after John's period of employment as a lecturer of Physics", then Peter's period of employment may be denoted by another symbolic time element c_{i6}. The relationship between c_{i5} and c_{i6} is then represented as the constraint "c_{i6} after c_{i5}".

Single time elements may be denoted by symbolic constants as described above. A set of time elements may be denoted by a symbolic variable. For instance, the temporal association "[50, 60] subinterval $c_{i2} \wedge c_{i2}$ subinterval [30, 80]" denotes a single time interval which is a subinterval of [30, 80] and a superinterval of [50, 60]. The temporal association "[50, 60] subinterval $v_{i2} \wedge v_{i2}$ subinterval [30, 80]" denotes the set of all time intervals which are subintervals of [30, 80] and superintervals of [50, 60]. Thus symbolic constants denote existentially quantified variables (skolem constants or null values) and symbolic variables denote universal quantification. The representation of infinite periodic data is another instance where symbolic variables are useful.

We use $c_{i1}, c_{i2}, \ldots$ to name symbolic constants that denote time intervals and $c_{t1}, c_{t2}, \ldots$ to name symbolic constants that denote time points. Similarly, $v_{i1}, v_{i2}, \ldots$ name symbolic variables that denote sets of time intervals and $v_{t1}, v_{t2}, \ldots$ name symbolic variables that denote sets of time points.

No restrictions are placed on the notion of constraints in the Generalized ChronoBase model. These include all forms of point-point, point-interval and interval-interval relationships as found in, for instance, Allen [All83], Vilian and Kautz [VK86], Kabanza, Stevenne and Wolper [KSW90], Baudinet, Niezette and Wolper [BNW91], Meiri [Mei91], and Koubarakis [Kou93]. Set membership (of finite or infinite domains) as well as equations and inequations involving arithmetic operators are also included. Instances of the generalized model are obtained by restricting the kinds of constraints allowed in a specific implementation of the ChronoBase model.

Definition 1 An *atomic constraint* is any valid relationship between two time elements which does not involve a conjunction or a disjunction. The negation of an atomic constraint is also an atomic constraint.

Definition 2 An atomic constraint between two time elements is a *temporal association*. A conjunction or disjunction of temporal associations is a temporal association.

12.5 The Generalized ChronoBase Temporal Data Model

Having introduced the main features of the ChronoBase model, we are now in a position to define the Generalized ChronoBase model.

In the ChronoBase model, time domains can be either discrete ($DISC$) or continuous ($CONT$). Time is assumed to be linear. A time point is an abstract entity whose position on the time line is unique. Its representation in a database is an approximation subject to the precision of the system. Time elements can be either time points or time intervals. Time intervals denote the set of all time points bounded by an ordered pair of time points $[t_i, t_j]$ along the time line, such that $t_i < t_j$. The ChronoBase model allows the representation of both time points and time intervals since both are required in the ontology for modelling continuous change (Galton [Gal90]). However, a specific application may choose to include only time points or only time intervals in its ontology.

The domains of absolute time elements $\mathcal{D}_{AVT}$ and $\mathcal{D}_{ATT}$ are defined over the time lines for valid time and transaction time respectively.

We define a set of symbolic time elements $\mathcal{D}_{ST}$ and a set of Temporal Characteristics $\mathcal{D}_{TC}$. The domain of Temporal Associations $\mathcal{D}_{TA}$ is the set of sentences in the constraint language $\mathcal{L}$ defined over $\mathcal{D}_{AVT}$ and $\mathcal{D}_{ST}$ and the set of allowed constraint relationships $\mathcal{C}$ together with the logical connectives $\wedge$, $\vee$ and $\neg$. The domain of valid timestamps, $\mathcal{D}_{VT}$ is defined as a subset of $\mathcal{D}_{TC} \times \mathcal{D}_{ST} \times \mathcal{D}_{TA}$.

Transaction time domain has much simpler semantics compared to that of valid time. The temporal associations in the transaction time domain are always "equals" and the temporal characteristics are always $Holds^t$. Therefore, the domain of transaction timestamps, $\mathcal{D}_{TT}$ is defined simply as $\mathcal{D}_{ATT}$.

We also define a set of non-temporal attributes $\{A_1, ..., A_n\}$ and a set of attribute domains $\{D_1, ..., D_n\}$.

A non-temporal relation is called a snapshot relation. This corresponds to the standard relational model. Temporal relations are classified into valid time, transaction time and bitemporal relations according to the presence of valid time or transaction time or both (Ahn and Snodgrass [AS93], Snodgrass et al [S+94a]).

A *snapshot* relation in the ChronoBase model is a subset of the cartesian product of $D_1 \times ... \times D_n$.

A *transaction time* relation is a subset of the cartesian product of $D_1 \times ... \times D_n \times \mathcal{D}_{TT}$.

A *valid time* relation is a subset of the cartesian product of $D_1 \times ... \times D_n \times \mathcal{D}_{VT}$.

A *bitemporal* relation is a subset of the cartesian product of $D_1 \times ... \times D_n \times \mathcal{D}_{VT} \times \mathcal{D}_{TT}$.

EmpDept

sgt	name	dept	validity		
			TC	Ident	TA
1	John	Math	$Holds^t$	c_{i1}	c_{i1} before c_{i2}
2	Tom	Math	$Holds^t$	c_{i2}	c_{i2} subinterval [50, 100]
3	Tim	Physics	$Holds^t$	c_{i3}	duration(c_{i3}) equals duration(c_{i4})
4	Peter	Physics	$Holds^t$	c_{i4}	c_{i4} equals [20, 30]
5	Sue	Physics	$Holds^t$	c_{i5}	duration(c_{i5}) less duration(c_{i2})
6	Charles	Physics	$Holds^t$	c_{i6}	start(c_{i6}) $\in$ [1,5] $\wedge$ end(c_{i6}) $\in$ [20,25]

Figure 12.1
Relative and imprecise information in the ChronoBase model

12.5.1 Valid time relations in ChronoBase

The Generalized ChronoBase model allows one to represent relative and imprecise temporal data in addition to absolute temporal information. For example, the following information is represented as shown in Figure 12.1 (Function *duration* gives the duration of an interval, function *start* gives the start point, and function *end*, the end point).

1. John worked in the Math department before Tom worked in the Math department

2. Tom worked sometime during [50, 100]

3. Tim worked in Physics as long as Peter worked in Physics

4. Sue worked in Physics for a shorter time period than Tom

5. Charles started to work in Physics between 1 and 5 and stopped between 20 and 25

Figure 12.1 shows the valid time interval relation *EmpDept* represented in the ChronoBase conceptual model, with explicit and implicit attributes separated by a double line. The schema specification for a temporal relation includes the list of attributes together with their data types. Furthermore, the relation type ($VALID$, $TRANSACTION$, or $BITEMPORAL$), the kind of time element ($TIMEINTERVAL$ or $TIMEPOINT$), and the time structure ($DISC$ or $CONT$), are also part of the schema specification.

The schema specification for relation *EmpDept* is

    CREATE TABLE EmpDept(name CHAR, dept CHAR)
    AS VALID DISC TIMEINTERVAL.

In the ChronoBase model, temporal information is associated with each tuple. The temporal information associated to a tuple is viewed as a set of implicit attributes. This is equivalent to tuple timestamping. In the presence of surrogates, this model has the

EmpDept

sgt	name	dept	validity			transaction time
			TC	Ident	TA	
1	John	Math	$Holds^t$	c_{i1}	c_{i1} equals [20,50]	[20,24]
1	John	Math	$Holds^t$	c_{i2}	c_{i2} equals [20,30]	[25,now]
2	Tom	Math	$Holds^t$	c_{i3}	c_{i3} subinterval [20,30]	[20,now]
3	Tim	Physics	$Holds^t$	c_{i4}	c_{i4} subinterval [20,30]	[25,now]

MgrDept

sgt	mgr	dept	validity			transaction time
			TC	Ident	TA	
4	Peter	Math	$Holds^t$	c_{i1}	c_{i1} equals [10,25]	[15,now]
5	Anne	Math	$Holds^t$	c_{i2}	c_{i2} subinterval [26,50]	[20,now]
6	Jerry	Physics	$Holds^t$	c_{i3}	c_{i3} equals [15,40]	[15,now]

Figure 12.2
Bitemporal interval relations *EmpDept* and *MgrDept* in ChronoBase

same expressive power as the grouped model of Clifford, Croker and Tuzhilin [CCT93]. In the ChronoBase model, a *surrogate (sgt)* uniquely identifies groups of tuples which correspond to a single object, even when all the attributes of an object vary in time.

The ChronoBase Temporal relational algebra operators take into account temporal characteristics as well as temporal associations to draw inferences from the data represented in the database.

12.5.2 Bitemporal relations in ChronoBase

Figure 12.2 shows the *EmpDept* and *MgrDept* bitemporal relations. They use two pairs of implicit attributes to represent the information related to valid and transaction time respectively.

Relation EmpDept records the department to which an employee belongs. For example, the first tuple states that one believed from time 20 to 24 that John worked in the Math department from 20 to 50. The third tuple states that one believed from time 20 onwards that Tom worked in Math sometime during [20,30].

Relation MgrDept records the managers of the different departments. The first tuple states that one believed from time 15 onwards that Peter managed the Math department from 10 to 25. The second tuple states that one believed from time 20 onwards that Anne managed the Math department sometime during [26,50].

Interesting questions concerning transaction time on the bitemporal relation *EmpDept* are:

- What employment data was recorded late?
- When did the database believe that John worked during [30,40]?

Notice that the first query requires to find those cases in which some time elapsed between the date an employee was employed (valid time) and the date the data was actually stored in the database (transaction time). In other words, ChronoBase allows the combination of valid and transaction time in a single expression. The second query shows how, in contrast to other proposals, the ChronoBase user can query the states of the database.

12.5.3 Temporal interpolation in ChronoBase

Data which involves interpolation can also be represented in ChronoBase. The interpolation functions to be applied may be different for each attribute and for each tuple in a relation. In the schema specification for interpolated relations, the keyword $Ifun$ is added for those attributes for which interpolation functions are to be applied. If the interpolation function to be used is the same for all the tuples, the interpolation function itself may be specified at the schema level. If interpolation functions vary for different tuples, the keyword $Ifun$ is used in the specification, and the actual interpolation function is specified at tuple insertion time.

Although interpolation functions may be arbitrarily complex, the following predefined interpolation functions are supported ($\Re$ denotes the set of real numbers):

- stepwise constant interpolation: $y = c$
- linear interpolation: $y = a * t + b$ with $a, b \in \Re$
- quadratic interpolation: $y = a * t^2 + b * t + c$ with $a, b, c \in \Re$
- exponential interpolation: $y = a * e^{b*t}$ with $a, b \in \Re$

The keywords *stepwise*, *linear*, *quad* or *exp* are used to specify the above mentioned interpolation functions. To compute the value for a time point t, ChronoBase makes use of the last recorded time point before t, the first recorded time point after t, and their respective attribute values. For example, the predefined function to compute the value for time point t using linear interpolation is $f_t = f_{t1} + ((f_{t2} - f_{t1}) * (t - t_1))/(t_2 - t_1)$. Here t_1 refers to the last recorded time point before t, t_2 to the first recorded time point after t; f_{t1} and f_{t2} are the values of the attribute at time t_1 and t_2 respectively.

For transaction time, the interpolation is assumed to be stepwise constant. This reflects the fact that a tuple persists in the database until its (logical) deletion.

Consider a relation *Census* which stores the population of a town. As a census is performed only at certain times, this information is stored in time point relations. An

Census

sgt	city	population		validity
		value	Ifun	
1	Munich	75000	exp	1955
1	Munich	95000	exp	1960
1	Munich	120000	exp	1965
2	London	55000	exp	1950
2	London	90000	exp	1960

Figure 12.3
Interpolation relations *Census* in ChronoBase

exponential interpolation function is used to compute the population at time points which are not explicitly recorded. Figure 12.3 illustrates such a relation. The schema specification for relation *Census* is

  CREATE TABLE Census(city CHAR, population INT Ifun)
  AS VALID DISC TIMEPOINT.

The first tuple of relation *Census* denotes that in 1955 there were 75000 inhabitants in Munich. To compute the number of inhabitants in Munich in 1958, the exponential interpolation function is used.

In the case of the *Census* relation the interpolation function is the same for the whole relation. Therefore the interpolation function can be specified at the schema level by substituting the keyword *Ifun* with the interpolation function to be used, eg. as

  CREATE TABLE Census(city CHAR, population INT exp)
  AS VALID DISC TIMEPOINT.

The specification of the interpolation functions at schema level is useful for optimization purposes. In such a case, the interpolation functions are not recorded at the tuple level.

Interpolation functions may vary for different tuples. Consider a relation *TrafficData* as illustrated in Figure 12.4. The schema specification for relation *TrafficData* is

  CREATE TABLE
  TrafficData(car CHAR, junction CHAR, miles_travelled INT Ifun)
  AS VALID DISC TIMEPOINT.

Relation *TrafficData* records the time point when a car reaches a junction, as well as the miles travelled up to this junction. For time points which are not explicitly recorded, an interpolation function is used to compute the miles travelled. For example, the first tuple of relation *TrafficData* denotes that a VW-Golf reached the junction B4-B7 at 8:30 and had travelled 35 miles. Note that the interpolation function for the BMW525 is

TrafficData

sgt	car	junction	miles_travelled		validity
			value	Ifun	
1	VW-Golf	B4-B7	35	quad	8:30
1	VW-Golf	B7-B8	65	quad	8:40
2	BMW525	B5-B7	50	linear	9:15
2	BMW525	B7-B8	95	linear	9:20

Figure 12.4
Interpolation relation *TrafficData* in ChronoBase

different from the interpolation function for the VW-Golf.

When querying the temporal database, the user is not aware of the existence of interpolation functions. ChronoBase uses them automatically whenever required. For a query like "How many miles were covered by the BMW525 at 9.20?" no interpolation is involved, as the data is explicitly available. On the other hand, for a query like "How many miles were covered by the BMW525 at 9.18?" the linear interpolation function is used to compute the answer.

12.5.4 Instances of the generalized ChronoBase model

Several simplifications of the Generalized ChronoBase model are conceivable. Simplifications result in instances that are computationally more tractable. There are two main categories of simplifications. One class of simplifications result from restricting the kind of temporal associations that are allowed in an instance. The second class of simplifications arise from restricting the kind of time elements that are allowed in an instance. Since both simplifications are orthogonal, combinations of the above mentioned simplifications are possible.

For example, a simplification of the generalized model would be to restrict the time element identifiers *Ident* to symbolic time interval constants, and the TAs to the atomic constraints *equals* and *subinterval*. This results in an instance with polynomial query complexity.

The kind of temporal associations that are allowed in a system may be identified by specifying which of the following conditions hold. Each such condition has an effect on the computational complexity of the query evaluation process:

- **C1** Symbolic constants are allowed as time elements
- **C2** Symbolic variables are allowed as time elements
- **C3** Constraints $<, >$ and $=$ are allowed between time point elements

- **C4** Atomic temporal constraints "equals" and "subinterval" are allowed between time interval elements
- **C5** Interval temporal constraints that are expressible in a time point based algebra are allowed between time interval elements
- **C6** Allen's interval constraints are allowed between time interval elements
- **C7** Conjunctions of atomic constraints are allowed as temporal associations
- **C8** Disjunctions of constraints are allowed as temporal associations
- **C7** Periodic temporal data is represented through constraints on symbolic time elements in temporal associations

For example, allowing condition **C6**, that is, allowing the complete set of Allens interval relationships in the TAs renders query evaluation intractable. However, allowing a subset of Allens relations expressible within a point based algebra (condition **C5**) results in polynomial time complexity for query evaluation (Vilian and Kautz [VK86]). Thus, a number of instances of the Generalized ChronoBase model can be defined, trading expressiveness for computational efficiency. In general, representing constraints in databases increases the expressiveness of a model, but it also warrants new query evaluation mechanisms (Kannelakis et al [KKR90]).

12.6 The Temporal Query Language SQL^C

In this section we outline some of the constructs of the Temporal Query Language SQL^C which has been developed for the ChronoBase system. The full language definition is given in Sripada and Möller [SM93]. SQL^C is a superset of the SQL92 standard (Melton and Simon [MS93], Cannan and Otten [CO93]). Therefore, any valid SQL92 statement is also a valid SQL^C statement. The SQL92 language has been adapted and extended for the temporal domain. Some features of SQL^C are:

- keywords to query and obtain snapshot, valid time, transaction time, and bitemporal relations.
- functions and comparison operators to explicitly specify temporal conditions in the WHERE clause. This includes a symmetric treatment of valid and transaction time.
- valid and transaction *time window* and *time slice* operators to restrict the view of the world to specified time intervals and time points.
- temporal interpolation. SQL^C does not require a special syntax for temporal relations with interpolation functions.

emp	mgr	validity		
		TC	Ident	TA
John	Peter	$Holds^t$	c_{i1}	c_{i1} equals $[20, 25]$
Tim	Jerry	$Holds^t$	c_{i2}	c_{i2} subinterval $[20, 30]$

Figure 12.5
Relation *EmpMgr* for a valid time query

- temporal aggregation and temporal ordering.

In the following, we illustrate some features of SQL^C using example queries on the relations EmpDept and MgrDept (Figure 12.2).

SQL^C*Query* **1** - *Snapshot Query* - *List the managers of all the employees.*

SELECT EmpDept.Name, MgrDept.Mgr
FROM EmpDept, MgrDept
WHERE EmpDept.Dept = MgrDept.Dept

In ChronoBase, any standard SQL92 statement is evaluated by default on a snapshot of the current database state. A snapshot relation is returned which incorporates neither valid nor transaction time.

SQL^C*Query* **2** - *Valid Time Query* - *Who was managed by whom and when?*

SELECT HISTORICAL EmpDept.Name, MgrDept.Mgr
FROM EmpDept, MgrDept
WHERE EmpDept.Dept = MgrDept.Dept

Valid time information is retrieved by adding the keyword *HISTORICAL* to a standard SQL92 query, as illustrated in the SQL^C Query 2. The answer relation will be a valid time relation, as shown in Figure 12.5. In order to use *SELECT HISTORICAL* the relations in the FROM clause of the SELECT statement have to be of type valid time or bitemporal. In the case of bitemporal relations, the latest valid time state is used for the evaluation.

SQL^C*Query* **3** - *Bitemporal Query* - *Who was managed by whom and when, and when was each piece of information believed?*

SELECT BITEMPORAL EmpDept.Name, MgrDept.Mgr
FROM EmpDept, MgrDept
WHERE EmpDept.Dept = MgrDept.Dept

emp	mgr	validity			transaction
		TC	Ident	TA	time
John	Peter	$Holds^t$	c_{i1}	c_{i1} equals [20, 25]	[20, 24]
John	Peter	$Holds^t$	c_{i2}	c_{i2} equals [20, 25]	[25, now]
Tim	Jerry	$Holds^t$	c_{i3}	c_{i3} subinterval [20, 30]	[25, now]

Figure 12.6
Relation *EmpMgr* for a Bitemporal Query

By adding the keyword *BITEMPORAL*, a bitemporal answer relation may be retrieved, as illustrated in Figure 12.6. In this case, it is required that the relations in the FROM clause are of type bitemporal.

SQL^C allows the specification of various temporal conditions explicitly. For this purpose, SQL^C supports constructs to retrieve valid and transaction times associated to a tuple. The function *validity* returns the valid time of the selected tuple in a valid time or bitemporal relation. The function *belief* returns the transaction time of the selected tuple in a transaction time or bitemporal relation. The interval comparison operators *BEFORE, EQUALS, MEETS, OVERLAPS, DURING, STARTS*, and *FINISHES* are supported. The function *duration* returns the duration of a time interval. In order to extract the start and end points of time intervals, functions *start* and *end* are provided.

SQL^C*Query* 4 *What employment data was recorded late in the Math department?*
```
SELECT BITEMPORAL EmpDept.Name
FROM EmpDept
WHERE start(belief(EmpDept.Dept = 'Math')) \=
      start(validity(EmpDept.Dept = 'Math'))
```

Furthermore, SQL^C supports valid and transaction *time window* operators to restrict a selection to a specific time interval. For example, the valid time window operator "*TIMEWINDOW VALIDITY i*" restricts the selection to only those tuples which are valid for at least some time during the specified time interval i.

Similarly, valid and transaction *time slice* operators are offered to restrict a selection to a specific time point.

In order to conveniently express queries like "Count the number of employees per month", SQL^C supports primitives for temporal aggregation and temporal ordering. These include, for example, grouping on valid or transaction time. Primitives are provided to order a group of tuples on valid or transaction time (in ascending or descending order). Purely temporal aggregate functions such as *first* and *latest* access the first or latest tuple within such an ordering.

12.7 Advanced Temporal Data Model

The advanced temporal data model of ChronoBase allows the use of deduction rules (Sripada [Sri92]) to derive new information from explicitly stored information. The deduction rules themselves are stored in the database, and can vary in time (Sripada [Sri93a]).

For example, the employee-manager relationship can be retrieved using the temporal query language SQL^C:

SELECT HISTORICAL EmpDept.Name, MgrDept.Mgr
FROM EmpDept, MgrDept
WHERE EmpDept.Dept = MgrDept.Dept

An alternative is to express the employee-manager relationship in the advanced model as the deduction rule

$EmpMgr(E, M) \leftarrow EmpDept(E, D)$ AND $MgrDept(M, D)$.

Note that non-recursive deduction rules can always be expressed using SQL^C. Using deductive rules facilitates the specification of complex temporal phenomena such as finite state automata. In particular, the modelling of processes, actions, events, causes and effects is facilitated in the advanced data model. Efficient, set-oriented temporal query processing can be supported in the advanced model by using the temporal relational algebra operators of the ChronoBase model.

12.8 Related Work

Several data models have been proposed previously for temporal databases (Ben-Zvi [BZ82], Snodgrass [Sno84], Clifford and Tansel [CT85], Segev and Shoshani [SS87], Gadia [Gad88], Navathe and Ahmed [NA89], Sarda [Sar90], Jensen and Snodgrass [JS93], Lorentzos [Lor93], Jensen, Soo, and Snodgrass [JSS93], Snodgrass et al [S+94b]). All these models represent only the $holds^t$ temporal characteristics, although this assumption is never explicitly stated. The ChronoBase model, on the other hand, allows the representation of many other kinds of temporal characteristics that, to our knowledge, are not representable in any other model.

Some models such as Gadia, Nair, and Poon [GNP92], Maiocchi, Pernici, and Barbic [MPB92], Dyreson and Snodgrass [DS93] deal also with imprecise temporal data. In ChronoBase, imprecise temporal data is represented through temporal associations.

Temporal interpolation functions have been considered by Segev and Shoshani [SS87]. They use time-sequences in which different attributes can have different interpolation functions. The ChronoBase model also supports this feature. Further, the ChronoBase model supports continuous as well as discrete time structures.

Periodic temporal data has been considered by Segev and Shoshani [SS87], Chomicki and Imieliński [CI88], Kabanza, Stevenne, and Wolper [KSW90], Baudinet, Niezette, and Wolper [BNW91], Koubarakis [Kou93]. In the ChronoBase model, there are two ways to represent such data; one way is to represent them through temporal associations in the generalized model, the other way is to represent them through deduction rules in the advanced temporal data model.

Chaudhuri ([Cha88]) was the first to introduce temporal associations into the relational model in the form of Allen's temporal relationships. Temporal tables (Koubarakis [Kou93]) allow the representation of absolute, relative, imprecise and periodic temporal data through the use of constraints. As far as we are aware, these are the only models which allow several kinds of temporal associations in a single relation. The ChronoBase model shares this common feature with Chaudhuri [Cha88] and Koubarakis [Kou93]: each tuple in a relation can have a different temporal association. However, the expressive power of the ChronoBase model is greater because of the notion of temporal characteristics.

In addition to ChronoBase, several other models support both valid time and transaction time (Snodgrass [Sno84], Gadia [Gad88], Navathe and Ahmed [NA89], Sarda [Sar90], Jensen, Soo, and Snodgrass [JSS93], Snodgrass et al [S$^+$94b]). However, these models only allow the representation of the $holds^t$ temporal characteristics. The BCDM model (Jensen, Soo, and Snodgrass [JSS93], also in Jensen and Snodgrass [JS93]) aims at providing a semantics for temporal information in a representation independent manner. The BCDM model also deals only with data representable by the $holds^t$ TC.

Most of the proposals in the literature (Navathe and Ahmed [NA89], Sarda [Sar90]) require some kind of normalization of the temporal relations (eg. maximal intervals) for the temporal relational algebra operators to work correctly. The ChronoBase model does not require such normalisations. We believe that the requirement for normalisation arises out of the fact that the semantics of the $holds^t$ temporal characteristics is not explicitly captured in the model.

The set of temporal relational algebra operators of the ChronoBase model are a superset (and a generalisation) of those found in the literature. This is because TRA operators in the ChronoBase system have to take into account the different temporal associations and temporal characteristics of each tuple, and the operations that can be performed on them in order to draw further inferences.

Several temporal extensions to the SQL query language have been proposed in the literature. Most proposals deal with only one notion of time (valid or transaction time) (Hsu, Jensen, and Snodgrass [HJS93]). Other proposals such as Snodgrass [Sno84], Sarda [Sar90], Ahn [Ahn93], Gadia and Bhargava [GB93], Snodgrass et al [S+94b] include valid as well as transaction time. For example, TQuel (Snodgrass [Sno84]), an extension of the Quel language, allows a restricted form of transaction time querying through the *AS OF* construct which identifies a specific state of the database against which a valid time query is evaluated. SQL^C provides a simple and straightforward way to query and obtain snapshot, valid time, transaction time and bitemporal relations, and to refer to valid and transaction times of a tuple. Constructs for temporal ordering, temporal aggregation, and time window present in SQL^C have been proposed earlier by Navathe and Ahmed [NA89], Sarda [Sar90], Snodgrass, Gomez, and McKenzie [SGM93].

An approach for temporal knowledge representation and reasoning in AI using meta-predicates has been proposed in Sripada [Sri93b]. Meta-programming has also been used in Sripada [Sri92] to formalize the semantics of bitemporal deductive databases where deduction rules can change in time. It was shown in Sripada [Sri88] that valid and transaction times of facts and rules can be derived logically from the Event Calculus (Kowalski and Sergot [KS86]), a theory of time in Horn Clause logic extended with negation-as-failure, given a description of the events that occurred in the real-world and the transactions on the database that recorded the information.

12.9 Conclusions

We have described the ChronoBase temporal data model which is based on meta-logic programming. The ChronoBase model employs the notions of temporal association and temporal characteristics as the realisations of meta-predicates in a database. They serve as a means for representing a wide variety of temporal phenomena in a temporal database.

The temporal characteristics are recorded as an additional attribute in the ChronoBase model together with a schema of the properties of the temporal characteristics which are then used by the temporal relational algebra operators. We believe that in order to represent the rich variety of temporal phenomena such as those considered in this paper, the notion of temporal characteristics is indispensable. We have also briefly described a temporal query language for the ChronoBase model. The ChronoBase model can be extended to deal with probabilistic temporal data, multiple calendars, branching time etc. These issues have not been of main concern in the design of the ChronoBase model.

Historically, the ChronoBase effort started before the design of the TSQL2 proposal (Snodgrass et al. [S+94b]) and progressed in parallel. Many features of the ChronoBase model, such as the notion of Temporal Characteristics and Temporal Associations, as well as several constructs of the SQL^C query language may be used to extend the TSQL2 model and query language thereby making it more expressive.

The generalized ChronoBase model, which combines the power of temporal associations (constraints) with that of temporal characteristics (meta-reasoning), is suitable for representing spatial as well as temporal data. It is worth noting that the analogues of temporal characteristics also arise in the context of spatial reasoning. We are, however, not aware of any spatial data model which can represent the spatial analogues of temporal characteristics. The generalised ChronoBase model could be a good candidate for handling spatio-temporal data.

The ChronoBase data model is geared towards the development of advanced temporal database applications. The ChronoBase model enables the representation of various phenomena proposed in the DB and AI literature such as events, processes and temporal propositions and to draw inferences from such information automatically. Work is in progress at ECRC on implementing several instances of the generalized ChronoBase model in the context of advanced applications from the domains of Transportation and Simulation & Training.

Acknowledgements

The authors are grateful to Wolfgang Nejdl (RWTH, Aachen) and some anonymous referees for many valuable suggestions and comments on earlier drafts of this paper.

References

[Ahn93]   I. Ahn. SQL+T: a Temporal Query Language. In *Proc. of the Int. Workshop on an Infrastructure for Temporal Databases*, Arlington, Texas, June 1993.

[All83]   J. Allen. Maintaining Knowledge about Temporal Intervals. *Communications of the ACM*, 26(11):832–843, 1983.

[ANS88]   L.C. Aiello, D. Nardi, and M. Schaerf. Reasoning about knowledge and ignorance. In *Proceedings FGCS 1988*, 1988.

[AS93]   I. Ahn and R.T. Snodgrass. A Taxonomy of Time in Databases. In *Proc. of ACM SIGMOD International Conference on Management of Data*, New York, 1993.

[BNW91]   M. Baudinet, M. Niezette, and P. Wolper. On the Representation of Infinite Temporal Data and Queries (Extended Abstract). In *Proceedings of the ACM Symposium on Principles of Database Systems (PODS)*, pages 280–290, Denver, CO, May 1991.

[BT91]   A. Brogi and F. Turini. Metalogic for Knowledge Representation. In *Proceedings KR 91*, 1991.

[BZ82]   J. Ben-Zvi. *The Time Relational Model*. PhD thesis, Computer Science Department of the University of California, Los Angeles, CA., 1982.

[CCT93]   J. Clifford, A. Croker, and A. Tuzhilin. Grouped and Ungrouped Historical Data Models: Expressive Power and Completeness. In *Proc. of the Int. Workshop on an Infrastructure for Temporal Databases*, Arlington, Texas, June 1993.

[Cha88]   S. Chaudhuri. Temporal Relationships in Databases. In *Proceedings of the 14th VLDB Conference*, Los Angeles, California, August 1988.

[CI88]   J. Chomicki and T. Imieliński. Temporal Deductive Databases and Infinite Objects. In *Proceedings of the 7th PODS*, pages 61–73, March 1988.

[CO93]   S. Cannan and G. Otten. *SQL: The Standard Handbook*. McGraw-Hill Book Company, 1993.

[CT85]   J. Clifford and A. Tansel. On an Algebra for Historical Relational Databases: Two Views. In *Proceedings of SIGMOD*, Austin, Texas, 1985.

[DS93]   C.E. Dyreson and R.T. Snodgrass. Valid-Time Indeterminacy. In *Proc. 9th Int. Conf. on Data Engineering IEEE*, pages 335–343, Vienna, Austria, 1993.

[Gad88]   S.K. Gadia. A Homogeneous Relational Model and Query Languages for Temporal Databases. *ACM Transactions on Database Systems*, 13(4):418–448, 1988.

[Gal90]   A. Galton. A Critical Examination of Allen's Theory of Action and Time. *Artificial Intelligence*, 42:159–188, 1990.

[GB93]   S.K. Gadia and G. Bhargava. SQL-like seamless query of temporal data. In *Proc. of the Int. Workshop on an Infrastructure for Temporal Databases*, Arlington, Texas, June 1993.

[GNP92]   S.K. Gadia, S.S. Nair, and Y. Poon. Incomplete information in relational temporal databases. In *Proceedings of the 18th VLDB Conference*, pages 395–406, Vancouver, Canada, August 1992.

[HJS93]   S. Hsu, Ch.S. Jensen, and R.T. Snodgrass. A Survey of Valid-Time Selection and Projection in Temporal Query Languages. In *The TSQL2 Language Specification*, 1993.

[JS93]   Ch.S. Jensen and R.T. Snodgrass. Three Proposals for a Third-Generation Temporal Data Model. In *Proc. of the Int. Workshop on an Infrastructure for Temporal Databases*, Arlington, Texas, June 1993.

[JSS93]   C. Jensen, M. Soo, and R.T. Snodgrass. Unification of Temporal Data Models. In *Proceedings of the Int. Conf. on Data Engineering*, Vienna, Austria, April 1993.

[KK92]   R.A Kowalski and J.S. Kim. A Metalogic programming approach to multi-agent knowledge and belief. In *Department of Computing, Imperial College, London*, 1992.

[KKR90]   P.C. Kanellakis, G.M. Kuper, and P.Z. Revesz. Constraint Query Languages. In *ACM SIGACT-SIGMOD-SIGART Symposium on Principles of Database Systems*, pages 299–313, Nashville, Tennessee, April 1990.

[Kli93]   H. Kline. An Update of the Temporal Database Bibliography. *Sigmod Record*, 22(4):66–80, 1993.

[Kou93]   M. Koubarakis. Representation and Querying in Temporal Databases: the Power of Temporal Constraints. In *Proc. 9th Int. Conf. on Data Engineering IEEE*, pages 327–334, Vienna, Austria, 1993.

[KS86]   R. Kowalski and M. Sergot. A Logic-Based Calculus of Events. *New Generation Computing*, 4(1), 1986.

[KSW90]   F. Kabanza, J-M Stevenne, and P. Wolper. Handling Infinite Temporal Data. In *Proceedings of PODS*, pages 392–403, Nashville, Tennessee, April 1990.

[Lor93]   N.A. Lorentzos. Axiomatic Generalization of the Relational Model to Support Valid Time Data. In *Proc. of the Int. Workshop on an Infrastructure for Temporal Databases*, Arlington, Texas, June 1993.

[McD82]   D. McDermott. A Temporal Logic for Reasoning about Processes and Plans. *Cognitive Sci.*, 6:101–155, 1982.

[Mei91]   I. Meiri. Combining Qualitative and Quantitative Constraints in Temporal Reasoning. *National Conference on Artificial Intelligence*, 1991.

[MPB92]   R. Maiocchi, B. Pernici, and F. Barbic. Automatic Deduction of Temporal Information. *ACM Transactions on Database Systems*, 17(4):647–688, 1992.

[MS93]   J. Melton and A.R. Simon. *Understanding the new SQL: A Complete Guide*. Morgan Kaufmann, 1993.

[NA89]   S.B. Navathe and R. Ahmed. A Temporal Relational Model and a Query Language. *Information Sciences*, 49:147–175, 1989.

[S+94a]   R.T. Snodgrass et al. A Consensus Glossary of Temporal Database Concepts. *Sigmod Record*, 23(1):52–64, 1994.

[S+94b]   R.T. Snodgrass et al. TSQL2 Language Specification. *Sigmod Record*, 23(1):65–85, 1994.

[Sar90]   N.L. Sarda. Extensions to SQL for Historical Databases. *IEEE Transactions on Knowledge and Data Engineering*, 2(2), 1990.

[SGM93]   R.T. Snodgrass, S. Gomez, and L.E. McKenzie. Aggregates in the Temporary Query Language TQuel. *IEEE Transactions on Knowledge and Data Engineering*, 5(5), 1993.

[Sho87]   Y. Shoham. Temporal Logics in AI: Semantical and Ontological Considerations. *Artificial Intelligence*, 33:89–104, 1987.

[SM93]   S.M. Sripada and P. Möller. SQL^C: The ChronoBase Temporal Query Language. Internal Report, ECRC, Munich, 1993.

[Sno84]   R.T. Snodgrass. The Temporal Query Language TQuel. In *Proceedings of the 3rd PODS*, pages 204–213, Waterloo, Canada, April 1984.

[Soo91]   M.D. Soo. Bibliography on Temporal Databases. *Sigmod*, 20(1):14–23, 1991.

[SRBK94]   S.M. Sripada, B.L. Rosser, J.M. Bedford, and R.A. Kowalski. Temporal Database Technology for Air Traffic Flow Management. In *Proc. of the 1st Int. Conf. Applications of Databases*, Sweden, June 1994.

[Sri88]   S.M. Sripada. A logical framework for temporal deductive databases. In *Proceedings of the 14th VLDB Conference*, pages 171–182, Los Angeles, California, August 1988.

[Sri91]   S.M. Sripada. *Temporal Reasoning in Deductive Databases*. PhD thesis, Department of Computing, Imperial College of Science and Technology, 1991.

[Sri92]   S.M. Sripada. On Snapshot, Rollback, Historical and Temporal Deductive Databases. In *Computer and Data Management COMAD'92*, pages 83–93, Bangalore, India, Dec 1992.

[Sri93a]   S.M. Sripada. A Metalogic Programming Approach to Reasoning about Time in Knowledge Bases. In *Proceedings of IJCAI'93*, France, 28 August - 3 September 1993.

[Sri93b]   S.M. Sripada. Design of the ChronoBase Temporal Deductive Database System. In *Proc. of the Int. Workshop on an Infrastructure for Temporal Databases*, Arlington, Texas, June 1993.

[SS87]   A. Segev and A. Shoshani. Logical Modeling of Temporal Data. In *Proceedings of ACM-SIGMOD*, San Francisco, December 1987.

[VK86]   M. Vilian and H. Kautz. Constraint Propagation Algorithms for Temporal Reasoning. In *Proceedings AAAI-86*, pages 377–382, 1986.

Contributors

Krzysztof R. Apt
CWI, P.O. Box 94079
1090 GB Amsterdam, The Netherlands
and
Dept. of Mathematics and
Computer Science
University of Amsterdam
Plantage Muidergracht 24
1018 TV Amsterdam, The Netherlands
apt@cwi.nl

Jonas Barklund
Computing Science Dept.
Box 311
S-751 05 Uppsala, Sweden
jonas@csd.uu.se

Katrin Boberg
Computing Science Dept.
Box 311
S-751 05 Uppsala, Sweden
katrin@csd.uu.se

Antony F. Bowers
Department of Computer Science
University of Bristol
Queen's Building, University Walk
Bristol BS8 1TR, U.K.
bowers@compsci.bristol.ac.uk

Antonio Brogi
Dipartimento di Informatica
Università di Pisa
Corso Italia 40
56125 Pisa, Italy
brogi@di.unipi.it

Luigia Carlucci Aiello
Dipartimento di Informatica e Sistemistica
Univ. di Roma "La Sapienza"
Via Salaria 113, I-00198 Roma, Italy
aiello@dis.uniroma1.it

Marta Cialdea
Dipartimento di Discipline Scientifiche:
Chimica ed Informatica
Terza Università di Roma
Via C. Segre 2, I-00146 Roma, Italy
cialdea@dis.uniroma1.it

Simone Contiero
Dipartimento di Informatica
Università di Pisa
Corso Italia 40
56125 Pisa, Italy
contiero@di.unipi.it

Carlos V. Damásio
Dept. Informática
Fac. Ciências e Tecnologia
Universidade Nova de Lisboa
Quinta da Torre
2825 Monte da Caparica, Portugal
cd@fct.unl.pt

Pierangelo Dell'Acqua
Computing Science Dept.
Box 311
S-751 05 Uppsala, Sweden
pier@csd.uu.se

Corin A. Gurr
Human Communication Research Centre
Edinburgh University
2 Buccleuch Place
Edinburgh EH8 9LW, U.K.
corin@cogsci.edinburgh.ac.uk

Yuejun Jiang
Department of Computing
Imperial College
180 Queen's Gate
SW7 2AZ London, United Kingdom
yj@doc.ic.ac.uk

Marianne Kalsbeek
Faculty of Mathematics and
Computer Science
University of Amsterdam
Plantage Muidergracht 24
1018TV Amsterdam, the Netherlands
marianne@fwi.uva.nl

Robert A. Kowalski
Department of Computing
Imperial College
180 Queen's Gate
London SW7 2BZ, UK
rak@doc.ic.ac.u

Bern Martens
Department of Computer Science
Katholieke Universiteit Leuven
Celestijnenlaan 200A
B-3001 Heverlee, Belgium
bern@cs.kuleuven.ac.be

Petra Möller
European Computer-Industry
Research Centre GmbH
Arabellastr. 17
81925 Munich, Germany
petra@ecrc.de

Daniele Nardi
Dipartimento di Informatica e Sistemistica
Univ. di Roma "La Sapienza"
Via Salaria 113, I-00198 Roma, Italy
nardi@dis.uniroma1.it

Wolfgang Nejdl
Lehrstuhl für Informatik V
RWTH Aachen
52056 Aachen, Germany
nejdl@informatik.rwth-aachen.de

Luís M. Pereira
Dept. Informática
Fac. Ciências e Tecnologia
Universidade Nova de Lisboa
Quinta da Torre
2825 Monte da Caparica, Portugal
lmp@fct.unl.pt

Marco Schaerf
Dipartimento di Informatica e Sistemistica
Univ. di Roma "La Sapienza"
Via Salaria 113, I-00198 Roma, Italy
schaerf@dis.uniroma1.it

Contributors

Danny De Schreye
Department of Computer Science
Katholieke Universiteit Leuven
Celestijnenlaan 200A
B-3001 Heverlee, Belgium
dannyd@cs.kuleuven.ac.be

Michael Schroeder
Lehrstuhl für Informatik V
RWTH Aachen
52056 Aachen, Germany
schroede@informatik.rwth-aachen.de

Suryanarayana M. Sripada
European Computer-Industry
Research Centre GmbH
Arabellastr. 17
81925 Munich, Germany
sripada@ecrc.de

Frank Teusink
CWI, P.O. Box 94079
1090 GB Amsterdam, The Netherlands
frankt@cwi.nl

Franco Turini
Dipartimento di Informatica
Università di Pisa
Corso Italia, 40
56125 Pisa, Italy
turini@di.unipi.it

Margus Veanes
Computing Science Dept.
Box 311
S-751 05 Uppsala, Sweden
margus@csd.uu.se